Document Image Processing

Document Image Processing

Special Issue Editors

Ergina Kavallieratou
Laurence Likforman-Sulem

MDPI • Basel • Beijing • Wuhan • Barcelona • Belgrade

MDPI

Special Issue Editors

Ergina Kavallieratou
University of the Aegean
Greece

Laurence Likforman-Sulem
Télécom ParisTech & Université Paris-Saclay
France

Editorial Office
MDPI
St. Alban-Anlage 66
Basel, Switzerland

This is a reprint of articles from the Special Issue published online in the open access journal *Journal of Imaging* (ISSN 2313-433X) from 2017 to 2018 (available at: http://www.mdpi.com/journal/jimaging/special_issues/document_image_processing)

For citation purposes, cite each article independently as indicated on the article page online and as indicated below:

LastName, A.A.; LastName, B.B.; LastName, C.C. Article Title. *Journal Name* **Year**, *Article Number*, Page Range.

ISBN 978-3-03897-105-4 (Pbk)
ISBN 978-3-03897-106-1 (PDF)

Contents

About the Special Issue Editors

Ergina Kavallieratou, Associate Professor. Ergina Kavallieratou was born in Kefalonia, Greece, in 1973. She received her Diploma in Electrical and Computer Engineering in 1996 from the Polytechnic School of the University of Patras and her PhD in Handwritten Optical Character Recognition and Document Image processing from the same department in 2000. She has worked as a guest researcher in the Dept. of Telecommunications Engineering of the Polytechnic School of Madrid; in the Institute of Communication Acoustics of Ruhr-Universitaet Bochum, Germany; in the Computer Science & Engineering, Lehigh University, USA; at CVC, Universidad de Barcelona; and at Institut Mines-Télécom/Télécom ParisTech. During the years 2002–2004, she was an Assistant Professor of Audio Processing in the Dept. of Audio and Musical Instruments Technology in the Technological Educational Institute of Ionian Islands, Greece. She taught in the Greek Open University, 2001–2013. Since September 2004, she has been a member of the teaching staff of the department of Information and Communication System Engineering, University of the Aegean, and has been an Associate Professor since 2017. Her research interests include Optical Character Recognition, Document Image Analysis, Computer Vision and Pattern Recognition, and Robotics.

Laurence Likforman-Sulem, Associate Professor, graduated in engineering from ENST-Bretagne (Ecole Nationale Superieure des Telecommunications) in 1984 and received her PhD from ENST-Paris in 1989 and her HDR (Habilitation a Diriger des Recherches) from Pierre & Marie Curie University in 2008. She has been an Associate Professor at Telecom ParisTech in the IDS (Image Data Signal) Department since 1991, where she serves as a senior instructor in Pattern Recognition and handwriting recognition. Laurence Likforman-Sulem is an IEEE Senior Member and a founding member of the francophone GRCE (Groupe de Recherche en Communication Ecrite). She chaired the program committee of CIFED held in Fribourg, Switzerland, in 2006; the program committees of two DRR Conferences (Document Recognition and Retrieval) held in 2009 and 2010 in San Jose, California; and the ASAR (Arabic and Derived Script Analysis and Recognition) workshop held in 2017 in Nancy, France.

Journal of **Imaging**

MDPI

Editorial

Document Image Processing

Laurence Likforman-Sulem [1,*] and Ergina Kavallieratou [2]

[1] Institut Mines-Télécom/Télécom ParisTech, Université Paris-Saclay, 75013 Paris, France
[2] Department Information and Communication Systems Engineering, University of the Aegean, Samos 83200, Greece; kavallieratou@aegean.gr
[*] Correspondence: laurence.likforman@telecom-paristech.fr or likforman@telecom-paristech.fr

Received: 15 June 2018; Accepted: 15 June 2018; Published: 22 June 2018

Keywords: document image processing; preprocessing; document restoration; binarization; slant removal; text-line segmentation; handwriting recognition; indic/arabic/asian scripts; OCR; Video OCR; word spotting; retrieval; document datasets; performance evaluation; document annotation tools

The Special Issue "Document Image Processing" in the *Journal of Imaging* aims at presenting approaches which contribute to access the content of document images. These approaches are related to low level tasks such as image preprocessing, skew/slant corrections, binarization and document segmentation, as well as high level tasks such as OCR, handwriting recognition, word spotting or script identification. This special issue brings together 12 papers that discuss such approaches. The first three articles deal with historical document preprocessing. The work by Hanif et al. [1] aims at removing bleed-through using a non-linear model, and at reconstructing the background by an inpainting approach based on non-local patch similarity. The paper by Almeida et al. [2] proposes a new binarization approach that includes a decision-based process for finding the best threshold for each RGB channel. In the paper by Kavallieratou et al. [3], a segmentation-free approach based on the Wigner-Ville distribution is used to detect the slant of a document and correct it.

Once a document image is preprocessed, a next step described in the paper by Ghosh et al. [4] consists in separating text components from non-text ones, using a classifier based on LBP features. Following steps may consist in recognizing text components or searching from word queries. In the paper by Nashwan et al. [5] a holistic-based approach for the recognition of printed Arabic words is proposed, coupled with an efficient dictionary reduction. In the work by Nagendar et al. [6] it is shown that using a query specific fast Dynamic Time Warping distance, improves the Direct Query Classifier (DQC) word spotting system.

Deep neural network-based approaches are now widely used in the domain of document image processing, especially for the recognition of textual elements. The following papers also follow this trend. In the work by Jangid and Srivastava [7], deep convolutional networks trained layer-wise, are applied to the recognition of Devanagari characters. The paper by Kesiman et al. [8] is dedicated to southeast Asian scripts written on palm leafs. Character and word images are recognized by CNNs (Convolutional Neural Networks) and RNNs (Recurrent Neural Networks), respectively. Several binarization and text-line segmentation approaches are also benchmarked on these specific documents. The work by Granell et al. [9] describes an efficient text-line recognition system, based on CNN and stacks of RNNs, that has been developed for the recognition of historical Spanish documents. These documents include out-of-vocabulary ancient words which are handled by a language model based on sub-lexical units.

Annotated datasets are necessary to train systems or to evaluate the various tasks related to document image processing. In several papers published in this special issue, new datasets are released as well as open-source tools that are able to generate synthetic images. A dataset of indic scripts is released in the paper by Mukhopadhyay et al. [10] and first results are provided with this dataset. The DocCreator software described in the paper by Journet et al. [11] creates additional document samples from input ones, using a degradation model. Such augmented data are used to train deep learning systems or to evaluate system performance. Document images can be extended to videos including text. The paper by Zayenne et al. [12] describes open-source tools for multiple document processing tasks: annotation of Arabic news videos, evaluation of text detection and text recognition. Authors also release the Activ2.0 database of Arabic videos and make it publicly available.

The guest editors would also like to thank all the authors that have submitted papers to this special issue, all the reviewers for their contribution, and the Journal of Imaging Editors.

Author Contributions: The two authors have equally contributed to the writing of this editorial.

Conflicts of Interest: The authors declare no conflict of interest.

References

1. Hanif, M.; Tonazzini, A.; Savino, P.; Salerno, E. Non-Local Sparse Image Inpainting for Document Bleed-Through Removal. *J. Imaging* **2018**, *4*, 68. [CrossRef]
2. Almeida, M.; Lins, R.D.; Bernardino, R.; Jesus, D.; Lima, B. A New Binarization Algorithm for Historical Documents. *J. Imaging* **2018**, *4*, 27. [CrossRef]
3. Kavallieratou, E.; Likforman-Sulem, L.; Vasilopoulos, N. Slant Removal Technique for Historical Document Images. *J. Imaging* **2018**, *4*, 80. [CrossRef]
4. Ghosh, S.; Lahiri, D.; Bhowmik, S.; Kavallieratou, E.; Sarkar, R. Text/Non-Text Separation from Handwritten Document Images Using LBP Based Features: An Empirical Study. *J. Imaging* **2018**, *4*, 57. [CrossRef]
5. Nashwan, F.M.A.; Rashwan, M.A.A.; Al-Barhamtoshy, H.M.; Abdou, S.M.; Moussa, A.M. A Holistic Technique for an Arabic OCR System. *J. Imaging* **2018**, *4*, 6. [CrossRef]
6. Nagendar, G.; Ranjan, V.; Harit, G.; Jawahar, C.V. Efficient Query Specific DTW Distance for Document Retrieval with Unlimited Vocabulary. *J. Imaging* **2018**, *4*, 37. [CrossRef]
7. Jangid, M.; Srivastava, S. Handwritten Devanagari Character Recognition Using Layer-Wise Training of Deep Convolutional Neural Networks and Adaptive Gradient Methods. *J. Imaging* **2018**, *4*, 41. [CrossRef]
8. Kesiman, M.W.A.; Valy, D.; Burie, J.-C.; Paulus, E.; Suryani, M.; Hadi, S.; Verleysen, M.; Chhun, S.; Ogier, J.-M. Benchmarking of Document Image Analysis Tasks for Palm Leaf Manuscripts from Southeast Asia. *J. Imaging* **2018**, *4*, 43. [CrossRef]
9. Granell, E.; Chammas, E.; Likforman-Sulem, L.; Martínez-Hinarejos, C.-D.; Mokbel, C.; Cîrstea, B.-I. Transcription of Spanish Historical Handwritten Documents with Deep Neural Networks. *J. Imaging* **2018**, *4*, 15. [CrossRef]
10. Mukhopadhyay, A.; Singh, P.K.; Sarkar, R.; Nasipuri, M. A Study of Different Classifier Combination Approaches for Handwritten Indic Script Recognition. *J. Imaging* **2018**, *4*, 39. [CrossRef]
11. Journet, N.; Visani, M.; Mansencal, B.; Van-Cuong, K.; Billy, A. DocCreator: A New Software for Creating Synthetic Ground-Truthed Document Images. *J. Imaging* **2017**, *3*, 62. [CrossRef]
12. Zayene, O.; Touj, S.M.; Hennebert, J.; Ingold, R.; Ben Amara, N.E. Open Datasets and Tools for Arabic Text Detection and Recognition in News Video Frames. *J. Imaging* **2018**, *4*, 32. [CrossRef]

Journal of
Imaging

MDPI

Article

Non-Local Sparse Image Inpainting for Document Bleed-Through Removal

Muhammad Hanif *, Anna Tonazzini, Pasquale Savino and Emanuele Salerno

Institute of Information Science and Technologies, Italian National Research Council, 56124 Pisa, Italy; anna.tonazzini@isti.cnr.it (A.T.); pasquale.savino@isti.cnr.it (P.S.); emanuele.salerno@isti.cnr.it (E.S.)
* Correspondence: muhammad.hanif@isti.cnr.it

Received: 14 January 2018; Accepted: 26 April 2018; Published: 9 May 2018

Abstract: Bleed-through is a frequent, pervasive degradation in ancient manuscripts, which is caused by ink seeped from the opposite side of the sheet. Bleed-through, appearing as an extra interfering text, hinders document readability and makes it difficult to decipher the information contents. Digital image restoration techniques have been successfully employed to remove or significantly reduce this distortion. This paper proposes a two-step restoration method for documents affected by bleed-through, exploiting information from the recto and verso images. First, the bleed-through pixels are identified, based on a non-stationary, linear model of the two texts overlapped in the recto-verso pair. In the second step, a dictionary learning-based sparse image inpainting technique, with non-local patch grouping, is used to reconstruct the bleed-through-contaminated image information. An overcomplete sparse dictionary is learned from the bleed-through-free image patches, which is then used to estimate a befitting fill-in for the identified bleed-through pixels. The non-local patch similarity is employed in the sparse reconstruction of each patch, to enforce the local similarity. Thanks to the intrinsic image sparsity and non-local patch similarity, the natural texture of the background is well reproduced in the bleed-through areas, and even a possible overestimation of the bleed through pixels is effectively corrected, so that the original appearance of the document is preserved. We evaluate the performance of the proposed method on the images of a popular database of ancient documents, and the results validate the performance of the proposed method compared to the state of the art.

Keywords: ancient document restoration; image inpainting; bleed-through removal; sparse representation

1. Introduction

Archival, ancient manuscripts constitute the primary carrier of most authentic information starting from the medieval era, serving as history's own closet, carrying stories of enigmatic, unknown places or incredible events that took place in the distant past, many of which are still to be revealed. These manuscripts are of great interest and importance for historians, and provide insight into culture, civilisation, events and lifestyles of our past. With the passage of time, these documents have been exposed to different types of progressive degradations, such as spots or ink fading, due to fragile nature of the writing media, and bad storage or environmental conditions. This degradation process limits the use of these ancient classics, and some of the deteriorated documents had a very narrow escape from total annihilation. Specifically, in the manuscripts written on both sides of the sheet, often the ink had seeped through and appears as an unpleasant degradation pattern on the reverse side. Ink penetration through the paper is mainly due to aging, humidity, ink chemical properties or paper porosity [1], and can range from faint to severe. In the literature, this kind of degradation is termed as bleed-through, and impairs the legibility and interpretation of the document contents [2]. Therefore, it is of great significance to remove the bleed-through contamination and restore the integrity of the original manuscripts. An example of bleed-through removal is shown in Figure 1. Earlier, physical restoration methods were applied to deal with bleed-through degradation, but unfortunately those methods were costly, invasive, and sometimes caused permanent, irreversible damage to the documents.

In recent years, digital preservation of the documental heritage has been the focus of intensive digitisation and archiving campaigns, aimed at its distribution, accessibility and analysis. With digitization prevailing, in addition to conservation, the computing technologies applied to the digital images of these documents have quickly become a powerful and versatile tool to simplify their study and retrieval, and to facilitate new insights into the document's contents. Digital image processing techniques can be applied to these electronic document versions, to perform any alteration to the document appearance, while preserving the original intact. Specifically, digital image processing techniques have been attempted for the virtual restoration of documents affected by bleed-through, with some impressive results. In addition, to improve the document readability, the removal of the bleed-though degradation is also a critical preprocessing step in many tasks such as feature extraction, optical character recognition, segmentation, and automatic transcription.

Figure 1. An example of bleed-through removal.

Bleed-through removal is a challenging task mainly due to the possible significant overlap between the original text and the bleed-through pattern, and the wide variation of its extent and intensity. In literature, bleed-through removal is addressed as a classification problem, where the document image is subdivided into three components: background (the paper support), foreground (the main text), and bleed-through [1]. Broadly speaking, the existing methods in this domain can be divided into two main categories: blind or single-sided, and non-blind or double-sided. In blind methods, the image of a single side is used, whereas the non-blind methods require the information of both the recto and verso sides of the document. Most of the earlier methods rely on the intensity information of the image and perform restoration based on the grayscale or color (red, green, blue) intensity distributions. The intensity based methods involve thresholding [3]; however, intensity information alone is insufficient as there is often a significant overlap between the foreground and bleed-through intensity profiles [4]. In addition, thresholding may also destroy other useful document features, such as stamps, annotations, or paper watermarks. Thus, intensity based thresholding is not suitable when the aim is to preserve the original appearance of the document. To overcome these drawbacks, some methods incorporate spatial information by exploiting the neighbouring structure.

Among the blind methods, in [5], an independent component analysis (ICA) method is proposed to separate the foreground, background, and bleed-through layers from an RGB image. A dual-layer Markov random field (MRF) is suggested in [6], whereas, in [7], a conditional random field (CRF) method is proposed. A multichannel based blind bleed-through removal is suggested in [8] using color decorrelation or color space transformations, whereas, in [9], a recursive unsupervised segmentation approach is applied to the data space first decorrelated by principal component analysis (PCA). In [10], bleed-through removal is addressed as a blind source separation problem, solved by using a Markov random field (MRF) based local smoothness model. Similarly, an expected maximization (EM)-based approach is suggested in [11].

As per the non-blind methods, a model based approach using differences in the intensities of recto and verso side is outlined in [12]. The same model is extended in [13] using variational models with spatial smoothness in the wavelet domain. A non-blind ICA method is outlined in [14]. Other methods of this category are proposed in [15–17]. The performance of the non-blind methods depends on the accurate registration of recto and verso sides of the document, which is a non-trivial pre-processing step.

For a plausible restoration of documents with bleed-through, in addition to bleed-through identification, finding a suitable replacement for the affected pixels is also essential. The restored image generated in most of the above methods is either binary, pseudo-binary (uniform background and varying foreground intensities), or textured (the bleed-through regions are replaced with an estimate of the local mean background intensity or with a random pattern). An estimate of the local mean background is used in [6,18], but such methods are good for manuscripts with a reasonably smooth background while producing visible artifacts for documents with a highly textured background. In [7], a random-fill inpainting method is suggested to replace the bleed-through pixels with background pixels randomly selected from the neighbourhood. However, the random pixel selection produces salt and pepper like artifacts in regions with large bleed-through. In [6,16], as a preliminary step, a "clean" background for the entire image is estimated, but this is usually a very laborious task. In bleed-through removal, the desired restored image is the one where the foreground and background texture is preserved as much as possible. Instead, most of the bleed-through removal methods usually concentrate on foreground text preservation, neglecting the background texture. In order to enhance the quality of the restored image, the identification of bleed-through pixels and the estimation of a tenable replacement for them should be addressed with equal attention.

Image inpainting, which refers to filling in missing or corrupted regions in an image, is a well studied and challenging topic in computer vision and image processing [19,20]. In image inpainting, the goal is to find an estimate for those regions in order to reconstruct a visually pleasant and consistent image [21]. Recently, sparse representation based image inpainting methods are reported with exquisite results [22,23]. These methods find a sparse linear combination for each image patch using an overcomplete dictionary, and then estimate the value of missing pixels in the patch. This linear sparse representation is computed adaptively, by using a earned dictionary and sparse coefficients, trained on the image at hand. A dictionary learning based method has been used for document image resolution enhancement [24], denoising [25], and restoration [26]. In addition to sparsity, non-local self-similarity is another significant property of natural images [27,28]. A number of non-local regularization terms, exploiting the non-local self-similarity, are employed in solving inverse problems [29,30]. Fusing image sparsity with non-local self-similarity produces better results in recently reported image restoration techniques [31,32]. The underlying assumption in such methods is that similar patches share the same dictionary atoms.

In this paper, we present a two-step method to restore documents affected by bleed-through using pre-registered recto and verso images. First, the bleed-through pattern is selectively identified on both sides; then, sparse image inpainting is used to find suitable fill in for the bleed-through pixels. In general, any off-the-shelf bleed-through identification methods can be used in the first step. Here, we adopt the algorithm described in [33], which is simple and very fast. Although efficient in locating the bleed-through pattern, the method in [33] lacks a proper strategy to replace the unwanted bleed-through pixels. The simple replacement with the predominant background gray level value causes unpleasant imprints of the bleed-through pattern, visible in the restored image. An interpolation based inpainting technique for such imprints is presented in [34], but the filled-in areas are mostly smooth. Here, we use a sparse image representation based inpainting, with non-local similar patches, to find a befitting fill-in for the bleed-through pixels. This sparse inpainting step, which constitutes the main contribution of the paper, enhances the quality of the restored image and preserves well the natural paper texture and the text stroke appearance. The optimization problem of sparse patch inpainting is formulated using the non-local similar patches, to account for the neighbourhood consistency, and orthogonal matching pursuit (OMP) is used to find the sparse approximation.

The rest of this paper is organized as follows. The next section briefly introduces sparse image representation and dictionary learning. Section 3 presents the non-blind bleed-through identification method. The proposed sparse image inpainting technique is described in Section 4. In Section 5, we comment on a set of experimental results, illustrating the performance of the proposed method and its comparison with state-of-the-art methods. The concluding remarks are given in Section 6.

2. Sparse Image Representation

Recently, sparse representation emerged as a powerful tool for efficient representation and processing of high-dimensional data. In particular, sparsity based regularization has achieved great success, offering solutions that outperform classical approaches in various image and signal processing applications. Among the others, we can mention inverse problems such as denoising [35,36], reconstruction [22,37], classification [38], recognition [39,40], and compression [41,42]. The underlying assumption of methods based on sparse representation is that signals such as audio and images are naturally generated by a multivariate linear model, driven by a small number of basis or regressors. The basis set, called dictionary, is either fixed and predefined, i.e., Fourier, Wavelet, Cosine, etc., or adaptively learned from a training set [43]. While the underlying key constraint of all these methods is that the observed signal is sparse, explicitly meaning that it can be adequately represented using a small set of dictionary atoms, the particularity of those based on adaptive dictionaries is that the dictionary is also learned to find the one that best describes the observed signal.

Given a data set $\mathbf{Y} = [\mathbf{y}_1, \mathbf{y}_2, ..., \mathbf{y}_N] \in \mathbb{R}^{n \times N}$, its sparse representation consists of learning an overcomplete dictionary, $\mathbf{D} \in \mathbb{R}^{n \times K}$, $N > K > n$, and a sparse coefficient matrix, $\mathbf{X} \in \mathbb{R}^{K \times N}$ with non-zero elements less than n, such that $\mathbf{y}_i \approx \mathbf{D}\mathbf{x}_i$, by solving the optimization problem given as

$$\min_{\mathbf{D},\mathbf{X}} ||\mathbf{Y} - \mathbf{D}\mathbf{X}||_F^2 \text{ s.t. } \| \mathbf{x}_i \|_p \leq m,$$

where the $\mathbf{x}_i$ are the column vectors of $\mathbf{X}$, m is the desired sparsity level, and $\| \cdot \|_p$ is the ℓ_p-norm, with $0 \leq p \leq 1$.

Most of these methods consist of a two stage optimization scheme: sparse coding and dictionary update [43]. In the first stage, the sparsity constraint is used to produce a sparse linear approximation of the observed data, with respect to the current dictionary $\mathbf{D}$. Finding the exact sparse approximation is an NP-hard (non-deterministic polynomial-time hard) problem [44], but using approximate solutions has proven to be a good compromise. Commonly used sparse approximation algorithms are Matching Pursuit (MP) [45], Basis Pursuits (BP) [46], Focal Underdetermined System Solver (FOCUSS) [47], and Orthogonal Matching Pursuit (OMP) [48]. In the second stage, based on the current sparse code, the dictionary is updated to minimize a cost function. Different cost functions have been used for the dictionary update, for example, the Frobenius norm with column normalization has been widely used. Sparse representation methods iterate between the sparse coding stage and the dictionary update stage until convergence. The performance of these methods strongly depends on the dictionary update stage, since most of them share a similar sparse coding [43].

As per the dictionary that leads to sparse decomposition, although working with pre-defined dictionaries may be simple and fast, their performance might be not good for every task, due to their global-adaptivity nature [49]. Instead, learned dictionaries are adaptive to both the signals and the processing task at hand, thus resulting in a far better performance [50].

For a given set of signals $\mathbf{Y}$, dictionary learning algorithms generate a representation of signal $\mathbf{y}_i$ as a sparse linear combination of the atoms $\mathbf{d}_k$ for $k = 1, ..., K$,

$$\hat{\mathbf{y}}_i = \mathbf{D}\mathbf{x}_i. \tag{1}$$

Dictionary learning algorithms distinguish themselves from traditional model-based methods by the fact that, in addition to $\mathbf{x}_i$, they also train the dictionary $\mathbf{D}$ to better fit the data set $\mathbf{Y}$. The solution is generated by iteratively alternating between the sparse coding stage,

$$\hat{\mathbf{x}}_i = \arg\min_{\mathbf{x}_i} \| \mathbf{y}_i - \mathbf{D}\mathbf{x}_i \|^2; \text{ subject to } \| \mathbf{x}_i \|_0 \leq m \tag{2}$$

for $i = 1, ..., N$, where $\|.\|_0$ is the ℓ_0-norm, which counts the non-zero elements in x, and the dictionary update stage for the $\mathbf{X}$ obtained from the sparse coding stage

$$\mathbf{D} = \arg\min_{\mathbf{D}} \| \mathbf{Y} - \mathbf{DX} \|_F^2 . \tag{3}$$

Dictionary learning algorithms are often sensitive to the choice of m. The update step can either be sequential (one atom at a time) [51,52], or parallel (all atoms at once) [53,54]. A dictionary with sequential update, although computationally a bit expensive, will generally provide better performance than the parallel update, due to the finer tuning of each dictionary atom. In sequential dictionary learning, the dictionary update minimization problem (3) is split into K sequential minimizations, by optimizing the cost function (3) for each individual atom while keeping fixed the remaining ones. Most of the proposed algorithms have kept the two stage optimization procedure, the difference appearing mainly in the dictionary update stage, with some exceptions having a difference in the sparse coding stage as well [43]. In the method proposed in [51], which has become a benchmark in dictionary learning, each column $\mathbf{d}_k$ of $\mathbf{D}$ and its corresponding row of coefficients $\mathbf{x}_k^{row}$ are updated based on a rank-1 matrix approximation of the error for all the signals when $\mathbf{d}_k$ is removed

$$\begin{aligned} \{\mathbf{d}_k, \mathbf{x}_k^{row}\} &= \arg\min_{\mathbf{d}_k, \mathbf{x}_k^{row}} \| \mathbf{Y} - \mathbf{DX} \|_F^2 \\ &= \arg\min_{\mathbf{d}_k, \mathbf{x}_k^{row}} \| \mathbf{E}_k - \mathbf{d}_k \mathbf{x}_k^{row} \|_F^2, \end{aligned} \tag{4}$$

where $\mathbf{E}_k = \mathbf{Y} - \sum_{i=1, i \neq k}^{K} \mathbf{d}_i \mathbf{x}_i^{row}$. The singular value decomposition (SVD) of $\mathbf{E}_k = \mathbf{U}\Delta\mathbf{V}^\top$ is used to find the closest rank-1 matrix approximation of $\mathbf{E}_k$. The $\mathbf{d}_k$ update is taken as the first column of $\mathbf{U}$, and the $\mathbf{x}_k^{row}$ update is taken as the first column of $\mathbf{V}$ multiplied by the first element of Δ. To avoid the loss of sparsity in $\mathbf{x}_k^{row}$ that would be created by the direct application of the SVD on $\mathbf{E}_k$, in [51], it was proposed to modify only the non-zero entries of $\mathbf{x}_k^{row}$ resulting from the sparse coding stage. This is achieved by taking into account only the signals $\mathbf{y}_i$ that use the atom $\mathbf{d}_k$ in Equation (4), or, by taking, instead of the SVD of $\mathbf{E}_k$, the SVD of $\mathbf{E}_k^R = \mathbf{E}_k \mathbf{I}_{w_k}$, where $w_k = \{i|1 \leq i \leq N; \mathbf{x}_k^{row}(i) \neq 0\}$, and $\mathbf{I}_{w_k}$ is the $N \times |w_k|$ submatrix of the $N \times N$ identity matrix obtained by retaining only those columns whose index numbers are in w_k.

3. Bleed-Through Identification

The algorithm used to recognise the pixels that belong to the bleed-through pattern makes use of both sides of the document, i.e., the recto and the verso images, and suitably compares their intensities in a pixel-by-pixel modality. Hence, it is essential that two corresponding, opposite pixels exactly refer to the same piece of information. In other words, at location (i, j), to the pixel in a side, let us say a bleed-through pixel, must correspond, in the opposite side, the foreground pixel that has generated it, and vice versa. In order to ensure this matching, one of the two images needs to be reflected horizontally, and then the two images must be perfectly aligned [55].

The way in which we perform the comparison between pairs of corresponding pixels is motivated by some considerations about the physical phenomenon. Indeed, through experience, we observed that, in the majority of the manuscripts examined, due to paper porosity, the seeped ink has also diffused through the paper fiber. Hence, in general, the bleed-through pattern is a smeared and lighter version of the opposite text that has generated it. Note that this assumption does not mean that, on the same side, bleed-through is lighter than the foreground text. In fact, on each side, the intensity of bleed-through is usually very variable, which is highly non-stationary, and sometimes can be as dark as the foreground text.

Other considerations can be made by reasoning in terms of "quantity of ink". Indeed, it is apparent that the quantity of ink should be zero in the background, i.e., the unwritten paper, no matter the color of the paper, maximum in the darker and sharper foreground text, and minimum in the lighter and smoother corresponding bleed-through text. As a measure for "quantity of ink" having such properties, we use the concept of optical density, which is related to the intensity as follows:

$$d(i,j) = D(s(i,j)) = -log\left(\frac{s(i,j)}{b}\right), \tag{5}$$

where $s(i,j)$ is the image intensity at pixel (i,j), and b represent the most frequent (or the average) intensity value of the background area in the image.

Thus, based on the physically-motivated assumptions above, we adopt a linear, non-stationary model in the optical densities, to describe the superposition between background, foreground and bleed-through in the two observed recto and verso images:

$$d_r^{obs}(i,j) = d_r(i,j) + q_v(i,j)D(h_v(i,j) \otimes s_v(i,j)),$$
$$d_v^{obs}(i,j) = d_v(i,j) + q_r(i,j)D(h_r(i,j) \otimes s_r(i,j)), \tag{6}$$

for each pixel (i,j). In Equation (6), d^{obs} is the observed optical density, and d is the ideal optical density of the free-of-interferences image, with the subscripts r and v indicating the recto and verso side, respectively. D is the operator that, when applied to the intensity, returns the optical density according to Equation (5), and $\otimes$ indicates convolution between the ideal intensity s and a unit volume Point Spread Function (PSF), h, describing the smearing of ink penetrating the paper. At present, we assume stationary PSFs, empirically chosen as Gaussian functions, although a more reliable model for the phenomenon of the ink spreading should consider non-stationary operators. Finally, the space-variant quantities q_r and q_v have the physical meaning of attenuation levels of the density (or ink penetration percentage), from one side to the other.

The proposed algorithm locates the bleed-through pixels in each side as those whose optical density is lower than that of the corresponding pixels in the opposite side, i.e., of the foreground that has generated the bleed-through. Thus, on the basis of Equation (6), at each pixel, we first compute the following ratios:

$$q_r(i,j) = \frac{d_v^{obs}(i,j)}{D(h_r(i,j) \otimes s_r^{obs}(i,j)) + \epsilon},$$
$$q_v(i,j) = \frac{d_r^{obs}(i,j)}{D(h_v(i,j) \otimes s_v^{obs}(i,j)) + \epsilon}. \tag{7}$$

Since the equations above are intended to identify bleed-through pixels, they are derived from the model in Equation (6) assuming that the ideal optical density $d(i,j)$ is zero on the side at hand. As a consequence of this assumption, the opposite, ideal density, should correspond to that of the foreground text, and then coincide with the density of the blurred observed intensity s^{obs}. Then, for all pixels, we maintain the smallest between the two computed attenuation levels, and set to zero the other. This allows for correctly discriminating the two instances of foreground on one side and bleed-through in the other, so that, all pixels where $q_r > 0$ are classified as bleed-through in the verso side, whereas those where $q_v > 0$ are classified as bleed-through in the recto side.

However, it is apparent that, with the criterion above, we can obtain wrong positive attenuation levels, on one of the two sides, in correspondence of some background pixels and some occlusion pixels, i.e., where the two foreground texts superimpose on each other. This happens because, in the cases background–background and foreground–foreground, the two densities are almost the same, around zero in the first case and around the maximum density in the other, with small oscillations that make unpredictable the value of the ratios.

To correct this possible overestimation of the bleed-through pixels, we set to zero the attenuation level when the densities d_r^{obs} and d_v^{obs} are both low (or high, respectively) and close to each other. We experimentally verified that this procedure works well in most cases. On the other hand, even if some pixels remain misclassified as bleed-through, the sparse inpainting algorithm that we propose here is able to properly replace them with the original, correct values. As detailed in the next section, the inpainting algorithm also incorporates information of similar neighbouring patches, thus making possible the distinction between false bleed-through pixels in the background and false bleed-through pixel in the foreground.

4. Sparse Bleed-Through Inpainting

After successful identification of the bleed-through pixels, the next task is to find a suitable replacement for them. In this paper, we treat the bleed-through pixels as missing or corrupt image regions, and use sparse image inpainting to estimate proper fill-in values, which are consistent with the known uncorrupted surrounding pixels.

In recent years, image inpainting techniques have been widely used in image restoration, target removal, and compression. Generally, image inpainting techniques can be divided into two groups: diffusion based methods and exemplar based methods [21]. The diffusion based methods use a parametric model or partial differential equations, which extend the local structure from the surrounding to the internal of the region to be repaired [56]. In [57], a weighted average of known neighbourhood pixels is used to replace the missing pixels, using a fast matching method. A diffusion based method, with total variational approach, is presented in [58]. In [59], a multi-color image inpainting is outlined using anisotropic smoothing. The methods in this category are suitable for non-textured images with small missing regions.

In the exemplar based methods, an image block is selected as a unit, and the information is replicated from the known part of the image to the unknown region. In [20], a patch priority based inpainting is suggested that extends known image patches to the missing parts of the image. A non-local exemplar based method is suggested in [60], where the missing patches are estimated as means of selected non-local patches. Comparatively, the exemplar based methods are faster and exhibit better performance, but use only a single best matching block to estimate the unknown pixels. However, pure texture synthesis fails to preserve the structure information of the image, which constitutes its basic outline. A combination of diffusion and exemplar based inpainting is suggested in [61] to repair the structure and texture layers separately. This greedy kind of approach often introduces artifacts and also consumes more time in finding the best match for each image patch [21].

Recently, sparse representation based image inpainting algorithms have been reported with impressive results [62]. As sparse representation works on image patches, the main idea is to find the optimal sparse representation for each image patch and then estimate the missing pixels in a patch using the sparse coefficients of the known pixels. A sparse image inpainting method, using samples from the known image part, is presented in [23]. A fusion of an exemplar-based technique and sparse representation is presented in [22] to better preserve the image structure and the consistency of the repaired patch with its surroundings. In [62], a sparse representation method based on structure similarity (SSIM) of image patches is presented, where the dictionary training and the sparse coefficient estimation are based on the SSIM index.

Mathematically, the image inpainting problem is formulated as the reconstruction of the underlying complete image (in a column vector form) $\mathbf{C} \in \mathbb{R}^W$ from its observed incomplete version $\mathbf{I} \in \mathbb{R}^L$, where $L < W$. We assume a sparse representation of $\mathbf{C}$ over a dictionary $\mathbf{D} \in \mathbb{R}^{n \times K}$: $\mathbf{C} \approx \mathbf{DX}$. The incomplete image $\mathbf{I}$ and the complete image $\mathbf{C}$ are related through $\mathbf{I} = \mathbf{MC}$, where $\mathbf{M} \in \mathbb{R}^{L \times W}$ represents the layout of the missing pixels. In formulas it is:

$$\begin{aligned} \mathbf{I} &= \mathbf{MC} \\ &\approx \mathbf{M(DX)}. \end{aligned} \tag{8}$$

Assuming that a well trained dictionary $\mathbf{D}$ is available, the problem boils down to the estimation of sparse coefficients $\hat{\mathbf{X}}$ such that the underlying complete image $\hat{\mathbf{C}}$ is given by $\hat{\mathbf{C}} = \mathbf{D\hat{X}}$. To learn the dictionary $\mathbf{D}$, a training set $\mathbf{Y}$ is created by extracting overlapping patches of size $\sqrt{p_s} \times \sqrt{p_s}$ from the image at location $j = 1, 2, \ldots, P$, where p_s is the patch size and P is the total number of patches.

Then, we have $\mathbf{y}_j = \mathbf{R}_j(\mathbf{I})$, where $\mathbf{R}_j(.)$ is an operator that extracts the patch $\mathbf{y}_j$ from the image $\mathbf{I}$, and its transpose, denoted by $\mathbf{R}_j^T(.)$, is able to put back a patch into the j-th position in the reconstructed image. Considering that patches are overlapped, the recovery of $\mathbf{C}$ from $\{\hat{\mathbf{y}}_j\}$ can be obtained by averaging all the overlapping patches, as follows:

$$\mathbf{C} = \sum_{j=1}^{P} \mathbf{R}_j^T(\hat{\mathbf{y}}_j)./ \sum_{j=1}^{P} \mathbf{R}_j^T(\mathbf{1}_{p_s}). \tag{9}$$

4.1. Group Based Bleed-Through Patch Inpainting

Traditional sparse patch inpainting, where the missing pixel values are estimated using the known pixels from the corresponding patch only, ignores the relationship between neighbouring patches when estimating the missing pixels [31]. Incorporating information of similar neighbouring patches assists in the estimation of missing pixels and guarantees smooth transition by exploiting the local similarity typical of natural images. Following this line, we used a non-local group based patch inpainting approach here. For each patch to be inpainted, we search for similar patches within a limited neighborhood using Euclidean distance as similarity criterion, calculated as given below:

$$dist_{patch} = \sqrt{(Px_{ref} - Px_{new})^2 + (Py_{ref} - Py_{new})^2},$$

where Px_{ref}, Py_{ref} and Px_{cur}, Py_{cur} represents the horizontal and vertical position of central pixel in the reference and current patch, respectively.

For each patch $\mathbf{y}$ with bleed-through pixels, we select L non-local similar patches within an $N_s \times N_s$ neighbouring window. The similar patches are grouped together in a matrix, $\mathbf{y}_G \in \mathbb{R}^{p_s \times L}$. In each patch, we have known pixels and missing or bleed-through pixels. Let Ω be an operator that extracts the known pixels in a patch and $\tilde{\Omega}$ an operator that extracts the missing pixels, so that $\Omega(\mathbf{y})$ represents the known pixels and $\tilde{\Omega}(\mathbf{y})$ represents the missing pixels in a patch $\mathbf{y}$. An illustration of such pixels' extraction is given in Figure 2.

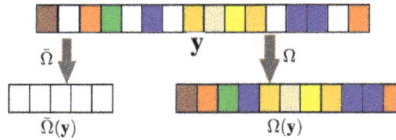

Figure 2. Extracting known and bleed-through pixels in a patch.

Similarly, for a group of patches, $\Omega(\mathbf{y}_G)$ extracts the known pixels of all patches, averaging multiple entries at the same pixel location, and $\tilde{\Omega}(\mathbf{y}_G)$ represents the missing pixels. Given a well-trained dictionary $\mathbf{D}$, the sparse reconstruction of patches with bleed-through pixels can be formulated as

$$\hat{x} = \arg\min_x \| \Omega(\mathbf{y}_G) - \Omega(\mathbf{D}x) \|^2 + \alpha \| x \|_0, \tag{10}$$

where α is a small constant. The first term of Equation (10) represents the data-fidelity and the second term is the sparse regularization. After obtaining the sparse coefficients $\hat{x}$ using Equation (10), an estimate for the bleed-through pixels can be obtained using

$$\tilde{\Omega}(\mathbf{y}) = \tilde{\Omega}(\mathbf{D}\hat{x}). \tag{11}$$

Using the reconstructed patches, an estimated, bleed-through free image is obtained by means of patch averaging, according to Equation (9).

In this paper, we learned a dictionary **D** from the training set **Y** created from the overlapping patches of an image with bleed-through, using the method described in Section 2. For optimization, we used only complete patches from **Y**, i.e., the patches with no bleed-through pixels, selected from both background areas and foreground text. This choice of '*clean patches*' speeds up the training process and excludes the '*non-informative*' bleed-through pixels. After dictionary training, the sparse coefficients in Equation (10) are estimated using the OMP algorithm presented in [48]. The order in which the bleed-through patches are inpainted has a significant impact on the final restored image. Thus, similarly to [20], high priority is given to patches with structure information in the known part. This patch priority scheme enables a smooth transition of structure information from the known part to the unknown (bleed-through) part of the patch.

5. Experimental Results

In this section, we discuss the performance of our method in order to validate its effectiveness. We compared the proposed method with other state-of-the-art methods including [7,16]. For evaluation, we used images from the well known database of ancient documents presented in [63,64]. This database contains 25 pairs of recto-verso images of ancient manuscripts affected by bleed-through, along with ground truth images. In the ground truth images, the foreground text is manually labeled. For the proposed method, the input images are first processed for bleed-through detection as discussed in Section 3.

The dictionary training data set **Y** is constructed by selecting the overlapping patches of size 8×8 with no bleed-through pixels from the input image. We learned an overcomplete dictionary **D** of size 64×256 from **Y**, with sparsity level m $= 5$ and $\alpha = 0.26$. We used discrete cosine transform (DCT) matrix as an initial dictionary. For each patch to be inpainted, the sparse coefficients are estimated using the learned dictionary and OMP. The sparse coefficients of each patch, denoted by x_j, where j indicates the number of the patch, are then used to estimate the fill-in values for the bleed-through areas. In terms of computational complexity, the dictionary training step comparatively consumes more time. The K-SVD algorithm requires K-times singular valve decomposition (SVD), with computational cost of $O(K^4)$, where K represents the number of atoms. The proposed method is implemented in the MATLAB2016a platform (The MathWorks, Inc., Natick, MA, USA) on a personal computer with core i5-6500 CPU at 3.20 Ghz and 8 GB of RAM. It took about 2 min for dictionary learning, and 57 s for inpainting an image of 720×940 pixels.

In bleed-through restoration, the efficacy is generally evaluated qualitatively, as in real cases the original clean image is not available. A visual comparison of the proposed method with other state-of-the-art methods is presented in Figure 3. The reported results for [16] are obtained from the online available ancient manuscripts database (https://www.isos.dias.ie/). In the ground truth images, obtained from [7], foreground text and bleed through are manually labeled. As can be seen, the proposed method (Figure 3e) produces comparatively better results considering the given ground truth image. It efficiently removes the bleed-through degradation, leaves intact the foreground text, and preserves the original look of the document. The non-parametric method of [16] (Figure 3c), although retaining foreground text and background texture, leaves clearly visible bleed-through imprints in some cases. The recent method presented in [7] (Figure 3d) produces better results, but some strokes of the foreground text are missing.

A bleed-through free colour image, obtained by using the proposed method, is shown in Figure 4. In the case of color images, we applied the proposed inpainting method only in the luminance (luma) band, and a simple nearest-neighbour based pixel interpolation is used in the other two chrominance bands. The proposed method copes very well with bleed-through removal and the dictionary based inpainting preserves the original appearance of the document.

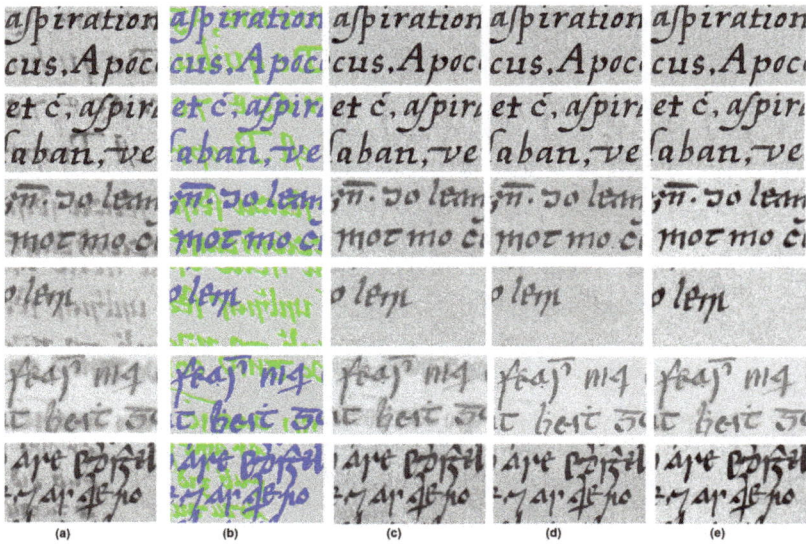

Figure 3. Visual comparison of bleed-through restoration. (**a**) input degraded image; (**b**) hand labeled ground truth image; (**c**) restored image by [16]; (**d**) restored image by [7]; (**e**) restored image by the proposed method.

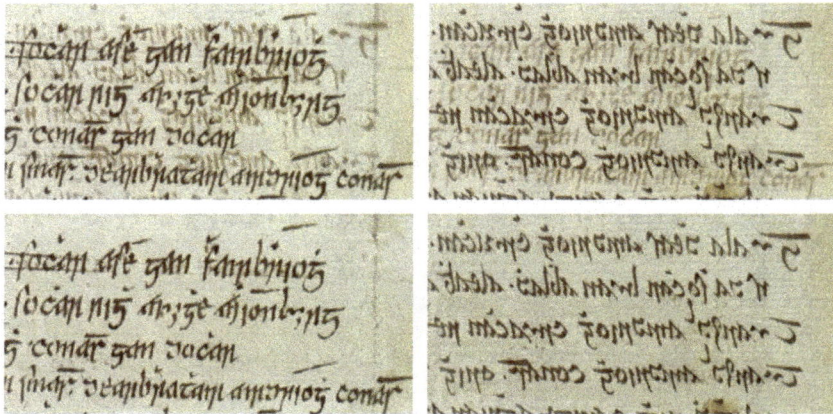

Figure 4. Ink bleed-through removal in a colour image: input image (**top**) and the restored image using the proposed method (**bottom**).

6. Conclusions

This paper presents a novel and general framework for high quality image restoration of documents affected by bleed-through. We use the bleed-through identification method presented in [33] in conjunction with group based sparse image inpainting, in order to obtain a non-blind document bleed-through the removal method. The non-stationary linear model in [33] efficiently locates the bleed-through pattern in recto-verso image pairs, but lacks a proper method to replace the unwanted bleed-through pixels. Finding a befitting fill-in for the degraded pixels is a crucial

J. Imaging **2018**, *4*, 68

task because the imprints due to assigned values that are not in accordance with the neighborhood have unpleasant visual effects, and destroy the original look of the restored document. The simple replacement with the predominant gray level value of the local background does not solve the problem. To remedy this issue, a non-local group based adaptive sparse image inpainting is suggested to estimate plausible fill-in values to replace the identified bleed-through pixels. The inclusion of non-local similar patches encourages the consistency in local fine textures, without blocking or smoothing artefacts. The proposed image inpainting method efficiently employs the intrinsic local sparsity and the non-local patch similarity. The performance of the proposed method is compared with other state-of-the-art methods on a database of recto-verso documents with bleed-through degradation.

Author Contributions: M.H. conceived and implemented the sparse representation inpainting method, run the experiments and wrote the paper. A.T. suggested the problem, devised the restoration algorithm in its whole, and contributed to write the paper. P.S. provided the bleed-through maps for the entire dataset and, with E.S., improved the bleed-through identification method and optimized the related algorithm. All authors participated in the evaluation of the results, and read and approved the final manuscript.

Acknowledgments: This work has been partially supported by the European Research Consortium for Informatics and Mathematics (ERCIM), within the "Alain Bensoussan" Fellowship Programme.

Conflicts of Interest: The authors declare no conflict of interest.

References

1. Fadoua, D.; Bourgeois, F.L.; Emptoz, H. Restoring Ink Bleed-Through Degraded Document Images Using a Recursive Unsupervised Classification Technique. In *Document Analysis Systems VII*; Lecture Notes in Computer Science; Springer: Berlin, Germany, 2006; Volume 3872, pp. 38–49.
2. Tan, C.L.; Cao, R.; Shen, P. Restoration of Archival Documents Using a Wavelet Technique. *IEEE Trans. Pattern Anal. Mach. Intell.* **2002**, *24*, 1399–1404.
3. Estrada, R.; Tomasi, C. Manuscript bleed-through removal via hysteresis thresholding. In Proceedings of the 10th International Conference on Document Analysis and Recognition, Barcelona, Spain, 26–29 July 2009; pp. 753–757.
4. Shi, Z.; Govindaraju, V. Historical document image enhancement using background light intensity normalization. In Proceedings of the 17th International Conference on Pattern Recognition, Cambridge, UK, 26–26 August 2004; pp. 473–476.
5. Tonazzini, A.; Bedini, L.; Salerno, E. Independent component analysis for document restoration. *Int. J. Doc. Anal. Recognit.* **2004**, *7*, 17–27. [CrossRef]
6. Wolf, C. Document ink bleed-through removal with two hidden Markov random fields and a single observation field. *IEEE Trans. Pattern Anal. Mach. Intell.* **2010**, *32*, 431–447. [CrossRef] [PubMed]
7. Sun, B.; Li, S.; Zhang, X.P.; Sun, J. Blind Bleed-Through Removal for Scanned Historical Document Image with Conditional Random Fields. *IEEE Trans. Image Process.* **2016**, *25*, 5702–5712. [CrossRef] [PubMed]
8. Tonazzini, A. Color space transformations for analysis and enhancement of ancient degraded manuscripts. *J. Pattern Recognit. Image Anal.* **2010**, *20*, 404–417. [CrossRef]
9. Drira, F.; Bourgeois, F.L.; Emptoz, H. *Restoring Ink Bleed-Through Degraded Document Images Using a Recursive Unsupervised Classification Technique*; Bunke, H., Spitz, A., Eds.; Springer: Berlin, Germany, 2006; Volume 3872, pp. 38–49.
10. Tonazzini, A.; Gerace, I.; Martinelli, F. Multichannel blind separation and deconvolution of images for document analysis. *IEEE Trans. Image Process.* **2010**, *19*, 912–925. [CrossRef] [PubMed]
11. Tonazzini, A.; Bedini, L.; Salerno, E. A Markov model for blind image separation by a mean-field EM algorithm. *IEEE Trans. Image Process.* **2006**, *15*, 473–482. [CrossRef] [PubMed]
12. Moghaddam, R.F.; Cheriet, M. Low quality document image modeling and enhancement. *Int. J. Doc. Anal. Recognit.* **2009**, *11*, 183–201. [CrossRef]
13. Moghaddam, R.F.; Cheriet, M. A variational approach to degraded document enhancement. *IEEE Trans. Pattern Anal. Mach. Intell.* **2010**, *38*, 1347–1361. [CrossRef] [PubMed]
14. Tonazzini, A.; Salerno, E.; Bedini, L. Fast correction of bleed-through distortion in grayscale documents by a blind source separation technique. *Int. J. Doc. Anal. Recognit.* **2007**, *10*, 17–27. [CrossRef]

15. Yi, H.; Brown, M.S.; Dong, X. User-assisted ink-bleed reduction. *IEEE Trans. Image Process.* **2010**, *19*, 2646–2658.
16. Rowley-Brooke, R.; Pitié, F.; Kokaram, A.C. A Non-parametric Framework for Document Bleed-through Removal. In Proceedings of the 2013 IEEE Conference on Computer Vision and Pattern Recognition (CVPR), Portland, OR, USA, 23–28 June 2013; pp. 2954–2960.
17. Merrikh-Bayat, F.; Babaie-Zadeh, M.; Jutten, C. Linear-quadratic blind source separating structure for removing show-through in scanned documents. *Int. J. Doc. Anal. Recognit.* **2011**, *14*, 319–333. [CrossRef]
18. Dubois, E.; Dano, P. Joint compression and restoration of documents with bleed-through. In Proceedings of the 2nd IS&T Archiving Conference, Washington, DC, USA, 26–29 April 2005; pp. 170–174.
19. Bertalmio, M.; Sapiro, G.; Caselles, V.; Ballester, C. Image Inpainting. In Proceedings of the 2000 SIGRAPH Conference, New Orleans, LA, USA, 23–28 July 2000; pp. 417–424.
20. Criminisi, A.; Perez, P.; Toyama, K. Region filling and object removal by exemplar-based image inpainting. *IEEE Trans. Image Process.* **2004**, *13*, 1200–1212. [CrossRef] [PubMed]
21. Guillemot, C.; Meur, O.L. Image inpainting: Overview and recent advances. *IEEE Signal Process. Mag.* **2014**, *31*, 127–144. [CrossRef]
22. Xu, Z.; Sun, J. Image inpainting by patch propagation using patch sparsity. *IEEE Trans. Image Process.* **2010**, *19*, 1153–1165. [PubMed]
23. Shen, B.; Hu, W.; Zhang, Y.; Zhang, Y. Image inpainting via sparse representation. In Proceedings of the 2009 IEEE International Conference on Acoustics, Speech, and Signal Processing, Taipei, Taiwan, 19–24 April 2009; pp. 697–700.
24. Walha, R.; Drira, F.; Lebourgeois, F.; Garcia, C.; Alimi, A.M. Joint denoising and magnification of noisy Low-Resolution textual images. In Proceedings of the International Conference on Document Analysis and Recognition, Tunis, Tunisia, 23–26 August 2015.
25. Hoang, T.; Barney Smith, E.; Tabbone, S. Sparsity-based edge noise removal from bilevel graphical document images. *Int. J. Doc. Anal. Recognit.* **2014**, *17*, 161–179. [CrossRef]
26. Kumar, V.; Bansal, A.; Tulsiyan, G.H.; Mishra, A.; Namboodiri, A.; Jawahar, C.V. Sparse Document Image Coding for Restoration. In Proceedings of the International Conference on Document Analysis and Recognition, Washington, DC, USA, 28 August 2013.
27. Buades, A.; Coll, B.; Morel, J. A non-local algorithm for image denoising. In Proceedings of the 2005 IEEE Computer Society Conference on Computer Vision and Pattern Recognition (CVPR'05), San Diego, CA, USA, 20–25 June 2005; pp. 60–65.
28. Smith, S.; Brady, J. Susan—A new approach to low level image processing. *Int. J. Comput. Vis.* **1997**, *23*, 45–78. [CrossRef]
29. Jung, M.; Bresson, X.; Chan, T.F.; Vese, L.A. Nonlocal Mumford–Shah regularizers for color image restoration. *IEEE Trans. Image Process.* **2011**, *20*, 1583–1598. [CrossRef] [PubMed]
30. Zhang, X.; Burger, M.; Bresson, X.; Osher, S. Bregmanized nonlocal regularization for deconvolution and sparse reconstruction. *SIAM J. Image Sci.* **2010**, *3*, 253–276. [CrossRef]
31. Zhang, J.; Zhao, D.; Gao, W. Group-based sparse representation for image restoration. *IEEE Trans. Image Process.* **2014**, *8*, 3336–3351. [CrossRef] [PubMed]
32. Dong, W.; Zhang, L.; Shi, G.; Li, X. Nonlocally centralized sparse representation for image restoration. *IEEE Trans. Image Process.* **2013**, *22*, 1620–1630. [CrossRef] [PubMed]
33. Tonazzini, A.; Savino, P.; Salerno, E. A non-stationary density model to separate overlapped texts in degraded documents. *Signal Image Video Process.* **2015**, *9*, 155–164. [CrossRef]
34. Gerace, I.; Palomba, C.; Tonazzini, A. An inpainting technique based on regularization to remove bleed-through from ancient documents. In Proceedings of the 2016 International Workshop on Computational Intelligence for Multimedia Understanding (IWCIM), Reggio Calabria, Italy, 27–28 October 2016; pp. 1–5.
35. Elad, M.; Aharon, M. Image denoising via sparse and redundant representations over leanred dictionaries. *IEEE Trans. Image Process.* **2006**, *15*, 3736–3745. [CrossRef] [PubMed]
36. Mairal, J.; Elad, M.; Sapiro, G. Sparse representation for color image restoration. *IEEE Trans. Image Process.* **2008**, *17*, 53–69. [CrossRef] [PubMed]
37. Ravishankar, S.; Bresler, Y. MR image reconstruction from highly undersampled k-space data by dictionary learning. *IEEE Trans. Med. Imag.* **2011**, *30*, 1028–1041. [CrossRef] [PubMed]

38. Soltani-Farani, A.; Rabiee, H.R.; Hosseini, S.A. Spatial aware dictionary learning for hyperspectral image classification. *IEEE Trans. Geosci. Remote Sens.* **2015**, *53*, 527–541. [CrossRef]
39. Wright, J.; Yang, A.; Ganesh, A.; Sastry, S.; Ma, Y. Robust face recognition via sparse representation. *IEEE Trans. Pattern Anal. Mach. Intell.* **2008**, *31*, 210–227. [CrossRef] [PubMed]
40. Jiang, Z.; Lin, Z.; Davis, L. Label Consistent K-SVD: Learning a Discriminative Dictionary for Recognition. *IEEE Trans. Pattern Anal. Mach. Intell.* **2013**, *35*, 2651–2664. [CrossRef] [PubMed]
41. Zhan, X.; Zhang, R.; Yin, D.; Huo, C. SAR image compression using multiscale dictionary learning and sparse representation. *IEEE Geosci. Remote Sens. Lett.* **2013**, *10*, 1090–1094. [CrossRef]
42. Bryt, O.; Elad, M. Compression of facial images using the K-SVD algorithm. *J. Vis. Commun. Image Represent.* **2008**, *19*, 270–283. [CrossRef]
43. Tosic, I.; Frossard, P. Dictionary learning. *IEEE Signal Process. Mag.* **2011**, *28*, 27–38. [CrossRef]
44. Tropp, J.A.; Wright, S.J. Computational methods for sparse solution of linear inverse problems. *Proc. IEEE* **2010**, *98*, 948–958. [CrossRef]
45. Mallat, S.G.; Zhang, Z. Matching Pursuits with Time-Frequency Dictionaries. *IEEE Trans. Signal Process.* **1993**, *41*, 3397–3415. [CrossRef]
46. Chen, S.; Donoho, D.; Saunders, M. Atomic Decomposition by basis pursuit. *SIAM J. Sci. Comput.* **1999**, *20*, 33–61. [CrossRef]
47. Gorodnitsky, I.; Rao, B. Sparse Signal reconstruction from limited data using FOCUSS: A re-weighted minimum norm algorithm. *IEEE Trans. Signal Process.* **1997**, *45*, 600–616. [CrossRef]
48. Tropp, J. Greed is Good: Algorithmic Results for Sparse Approximation. *IEEE Trans. Inf. Theory* **2004**, *50*, 2231–2242. [CrossRef]
49. Kreutz-Delgado, K.; Murray, J.; Rao, B.; Engan, K.; Lee, T.; Sejnowski, T. Dictionary Learning Algorithms for Sparse Representation. *Neural Comput.* **2003**, *15*, 349–396. [CrossRef] [PubMed]
50. Olshausen, B.; Field, D. Sparse coding with an overcomplete basis set: A strategy employed by V1? *J. Vis. Res.* **1997**, *37*, 3311–3325. [CrossRef]
51. Aharon, M.; Elad, M.; Bruckstein, A. K-SVD: An algorithm for designing overcomplete dictionaries for sparse representation. *IEEE Trans. Signal Process.* **2006**, *54*, 4311–4322. [CrossRef]
52. Rubinstein, R.; Zibulevsky, M.; Elad, M. Double sparsity: Learning sparse dictionaries for sparse signal approximation. *IEEE Trans. Signal Process.* **2010**, *58*, 1553–1564. [CrossRef]
53. Engan, K.; Aase, S.O.; Hakon-Husoy, J. Method of Optimal directions for frame design. In Proceedings of the 1999 IEEE International Conference on Acoustics, Speech, and Signal Processing, Phoenix, AZ, USA, 15–19 March 1999; pp. 2443–2446.
54. Hanif, M.; Seghouane, A.K. Maximum likelihood orthogonal dictionary learning. In Proceedings of the 2014 IEEE Workshop on Statistical Signal Processing (SSP), Gold Coast, VIC, Australia, 29 June–2 July 2014; pp. 141–144.
55. Savino, P.; Tonazzini, A. Digital restoration of ancient color manuscripts from geometrically misaligned recto-verso pairs. *J. Cult. Herit.* **2016**, *19*, 511–521. [CrossRef]
56. Bertalmio, M.; Bertozzi, A.; Sapiro, G. Navier-stokes, fluid dynamics, and image and video inpainting. In Proceedings of the 2001 IEEE Computer Society Conference on Computer Vision and Pattern Recognition, Kauai, HI, USA, 8–14 December 2001; pp. 355–362.
57. Telea, A. An image inpainting technique based on the fast marching method. *J. Graph. Tool* **2004**, *9*, 23–24. [CrossRef]
58. Chan, T.; Shen, J. Local inpainting models and TV inpainting. *SIAM J. Appl. Math.* **2001**, *61*, 1019–1043.
59. Tschumperl, D. Fast anisotropic smoothing of multi-valued images using curvature-preserving PDE's. *Int. J. Comput. Vision* **2006**, *1*, 65–82. [CrossRef]
60. Wong, A.; Orchard, J. A nonlocal-means approach to exemplar-based inpainting. In Proceedings of the 15th IEEE International Conference on Image Processing, San Diego, CA, USA, 12–15 October 2008; pp. 2600–2603.
61. Bertalmio, G.S.M.; Vese, L.; Osher, S. Simultaneous structure and texture image inpainting. *IEEE Trans. Image Process.* **2003**, *12*, 882–889. [CrossRef] [PubMed]
62. Ogawa, T.and Haseyama, M. Image inpainting based on sparse representations with a perceptual metric. *EURASIP J. Adv. Signal Process.* **2013**, *2013*, 179. [CrossRef]

63. Irish Script On Screen Project. 2012. Available online: http://www.isos.dias.ie (accessed on 5 January 2012).
64. Rowley-Brooke, R.; Pitié, F.; Kokaram, A.C. A ground truth bleed-through document image database. In Proceedings of the Theory and Practice of Digital Libraries, Paphos, Cyprus, 23–27 September; Zaphiris, P., Buchanan, G., Rasmussen, E., Loizides, F., Eds.; Springer: Berlin, Germany, 2012; Volume 7489, pp. 185–196.

Journal of
Imaging

MDPI

Article

A New Binarization Algorithm for Historical Documents

Marcos Almeida [1,*], Rafael Dueire Lins [2,3], Rodrigo Bernardino [4], Darlisson Jesus [4] and Bruno Lima [1]

[1] Departamento de Eletrônica e Sistemas, Centro de Tecnologia, Universidade Federal de Pernambuco, Recife-PE 50670-901, Brazil; brunocesar182@hotmail.com
[2] Centro de Informática, Universidade Federal de Pernambuco, Recife-PE 50740-560, Brazil; rdl.ufpe@gmail.com
[3] Departamento de Estatística e Informática, Universidade Federal Rural de Pernambuco, Recife-PE 52171-900, Brazil
[4] Programa de Pós-Graduação em Engenharia Elétrica, Universidade Federal de Pernambuco, Recife-PE 50670-901, Brazil; rbbernardino@gmail.com (R.B.); dmj.ufpe@gmail.com (D.J.);
*   Correspondence: mmar@ufpe.br; Tel.: +55-81-2126-7129

Received: 31 October 2017; Accepted: 16 January 2018; Published: 23 January 2018

Abstract: Monochromatic documents claim for much less computer bandwidth for network transmission and storage space than their color or even grayscale equivalent. The binarization of historical documents is far more complex than recent ones as paper aging, color, texture, translucidity, stains, back-to-front interference, kind and color of ink used in handwriting, printing process, digitalization process, etc. are some of the factors that affect binarization. This article presents a new binarization algorithm for historical documents. The new global filter proposed is performed in four steps: filtering the image using a bilateral filter, splitting image into the RGB components, decision-making for each RGB channel based on an adaptive binarization method inspired by Otsu's method with a choice of the threshold level, and classification of the binarized images to decide which of the RGB components best preserved the document information in the foreground. The quantitative and qualitative assessment made with 23 binarization algorithms in three sets of "real world" documents showed very good results.

Keywords: documents; binarization; back-to-front interference; bleeding

1. Introduction

Document image binarization plays an important role in the document image analysis, compression, transcription, and recognition pipeline [1]. Binary documents claim for far less storage space and computer bandwidth for network transmission than color or grayscale documents. Historical documents drastically increase the degree of difficulty for binarization algorithms. Physical noises [2] such as stains and paper aging affect the performance of binarization algorithms. Besides that, historical documents were often typed, printed or written on both sides of sheets of paper and the opacity of the paper is often such as to allow the back printing or writing to be visualized on the front side. This kind of "noise", first called back-to-front interference [3], was later known as bleeding or show-through [4]. Figure 1 presents three examples of documents with such a noise extracted from the three different datasets used in this paper in the assessment of the proposed algorithm. If the document is exhibited either in true-color or gray-scale, the human brain is able to filter out that sort of noise keeping its readability. The strength of the interference present varies with the opacity of the paper, its permeability, the kind and degree of fluidity of the ink used, its storage, age, etc. Thus, the difficulty for obtaining a good binarization performance capable of filtering-out such a noise increases enormously, as a new set of hues of paper and printing colors appear. The direct application of binarization algorithms may yield a completely unreadable document, as the interfering ink of the backside of the paper overlaps with the binary one in the foreground. Several document image compression schemes for color images are based on "adding color" to a binary image. Such compression strategy is unable to handle documents with back-to-front interference [5]. Optical Character Recognizers (OCRs) are also unable to work properly for such documents. Several algorithms were developed specifically to binarize documents with back-to-front interference [3,4,6–9]. There is no binarization technique to be an all case winner as many parameters may interfere in the quality of the resulting image [9]. The development of new binarization algorithms is still an important research topic. International competitions on binarization algorithms, such as DIBCO - Document Image Binarization Competition [10], are an evidence of the relevance of this area.

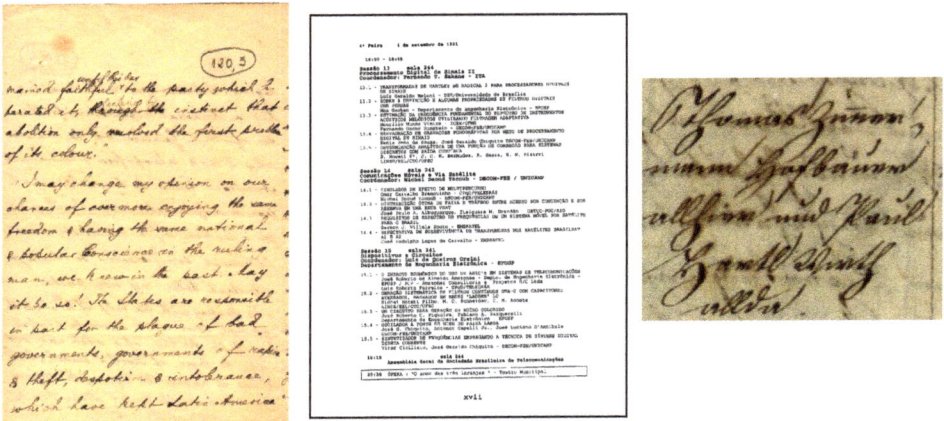

Figure 1. Images with back-to-front interference from the three test sets used in this paper: Nabuco bequest (**left**), LiveMemory (**center**) and DIBCO (**right**).

This paper presents a new global filter [1] to binarize documents, which is able to remove the back-to-front noise in a wide range of documents. Quantitative and qualitative assessments made in a wide variety of documents from three different "real-world" datasets (typed, printed and handwritten, using different kinds of paper, ink, etc.) allow to witness the efficiency of the proposed scheme.

2. The New Algorithm

The algorithm proposed here is performed in four steps: 1. decision-making for finding the vector of parameters of the image to be filtered, 2. filtering the image using a bilateral filter, 3. splitting the image into the RGB components, and performing their binarization using a method inspired by Otsu's algorithm for each RGB channel, and 4. choice of which of the RGB components best preserved the document information in the foreground, which is considered the final output of the algorithm. Figure 2 presents the block diagram of the proposed algorithm. The functionality of each block is detailed as follows.

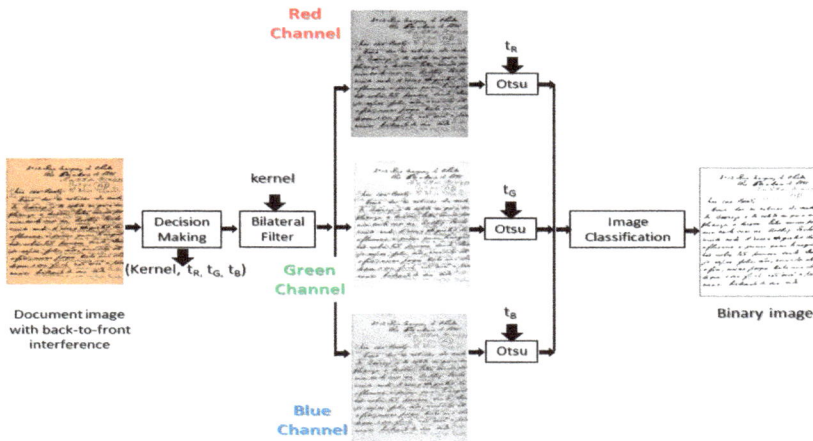

Figure 2. Block diagram of the proposed algorithm.

2.1. The Decision Making Block

The decision making block takes as input the image to be binarized and outputs a vector with four parameters: the value of the kernel (*kernel*) for the bilateral filter and three threshold values (t_R, t_G, t_B) that will be later used in the modified Otsu filtering.

The training of the binarization process proposed here is made with synthetic images which were generated as explained in Section 2.2. After filtering, the matrix of co-occurrence probabilities between the original image and of the binary image was calculated for each of the images in the document training set, whose generation is explained below.

The probabilistic structure applied in the analysis to each of the images in the training set is similar to the transmission of binary data in a Binary Asymmetric Channel, as shown in Figure 3. The probabilities $P(f/b)$ and $P(b/f)$ represent an additive noise in communication channels in information theory, here it represents the inability of the algorithm to correct the back-to-front interference of the image tested in the binarization process. The probabilities $P(b/b)$ and $P(f/f)$ are calculated from the pixel-to-pixel comparison of the binarized image generated by the proposed algorithm with the ground-truth image.

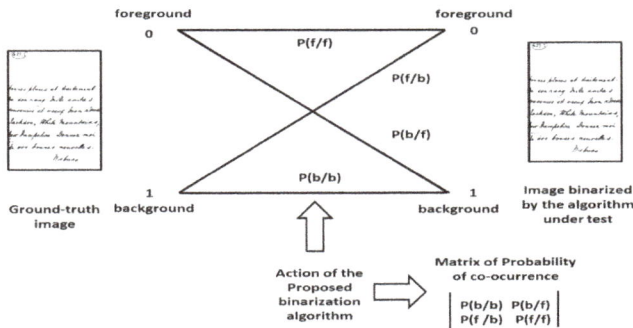

Figure 3. Generation of the co-occurrence matrix for each of the images in the training set.

The background-background probability is a function that needs to be optimized in the decision-making block, mapping background pixels (paper) from the original image onto white pixels of the binary image. It depends of all the parameters of the original image texture, strength of the back to front interference (simulated by the coefficient α), paper translucidity, etc. for each RGB channel. Thus, one can represent this dependence as:

$$P(b/b) = f(\alpha, R, G, B). \tag{1}$$

The optimal threshold t_c^* for each channel is calculated in the decision-making block, the index c can be R, G or B, maximizing $P(b/b)$:

$$t_c^* = MaxP(b/b), \tag{2}$$

subject to a given criterion $P(f/f) \geq M$. The criterion used here was M = 97%, that is at most 3% of the foreground pixels may be incorrectly mapped. During the training phase, the best t_c^* will be chosen from the three channels, which best maximizes the $P(b/b)$ for each of the images in the training set. The matrix of co-occurrence probability is calculated and the decision maker chooses the best binary image. The decision-making block was trained with 32,000 synthetic images in such a way to, given a real image to be binarized, it finds the optimal threshold parameters.

2.2. Generating Synthetic Images

The Decision-Making Block needs training to "learn" about the optimal threshold parameters and the value of the kernel to be used in the bilateral filter. Such training must be done using controlled images which are synthesized to mimic the different degrees of back-to-front interference, paper aging, paper translucidity, etc. Figure 4 presents the block diagram for the generation of synthetic images. Two binary images of documents of different nature (typed, handwritten with different pens, printed, etc.) are taken: F—front and V—verso (back). The front image is blurred with a weak Gaussian filter to simulate the digitalization noise [1], the hues that appear in after document scanning.

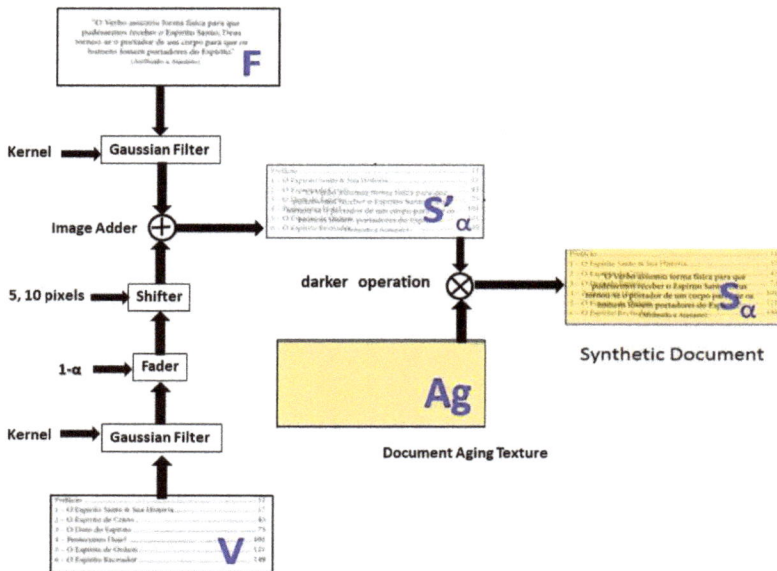

Figure 4. Block diagram of the scheme for the generation of synthetic images for the Decision-Making Block.

The verso image is "blurred" by passing through two different Gaussian filters that simulate the low-pass effect of the translucidity of the verso as seen in the front part of the paper. Two different parameters were used to simulate two different classes of paper translucidity. The "blurred" verso image is now faded with a coefficient α varying between 0 and 1 in steps of 0.01. Then, a circular shift of the lines of the document is made of either 5 or 10 pixels, to minimize the chances of the front and verso lines coincide entirely. Finally, the two images are overlapped by performing a "darker" operation pixel-by-pixel in the images. Paper texture is added to the image to simulate the effect of document aging. The texture pattern was extracted from document from late 19th century to the year 2000. The analysis of 3450 documents representative of a wide variety of documents of such a period was analyzed yielding 100 different clusters of textures. The synthetic texture to be applied to the image to simulate paper aging is generated using those 100 clusters by image quilting [11] and randomly, as explained in reference [9]. The training performed in the current version of the presented algorithm was made with 16 of those 200 synthetic textures. The total number of images used for training here was thus 16 (textures), times 10 (0 < α < 1 in steps of 0.10), times 2 blur parameters for the Gaussian filters, times 100 different binary images, totaling 32,000 images. Details of the full generation process of the synthetic image database are out of the scope of this paper and may be found in reference [9].

2.3. The Bilateral Filter

The bilateral filter was first introduced by Aurich and Weule [12] under the name "nonlinear Gaussian filter". It was later rediscovered by Tomasi and Manduchi [13] who called it the "bilateral filter" which is now the most commonly used name according to reference [14].

The bilateral filter is a technique to smoothen images while preserving their edges. The filter output at each pixel is a weighted average of its neighbors. The weight assigned to each neighbor decreases with both the distance values among pixels of the image plane (the spatial domain S) and the distance on the intensity axis (the range domain R). The filter applies spatial weighted averaging without smoothing the edges. It combines two Gaussian filters; one filter works in the spatial domain, while the other filter works in the intensity domain. Therefore, not only the spatial distance but also the intensity distance is important for the determination of weights. The bilateral filter combines two stages of filtering. These are the geometric closeness (i.e., filter domain) and the photometric similarity (i.e., filter range) among the pixels in a window of size N × N. Let $I(x,y)$ be a 2D discrete image of size N × N, such that $\{x,y\} \in \{0, 1, ..., N-1\} \times \{0, 1, ..., N-1\}$. Assume that $I(x,y)$ is corrupted by an additive white Gaussian noise of variance σ_n^2. For a pixel (x,y), the output of a bilateral filter can be as described by Equation (1):

$$I_{BF}(x,y) = \frac{1}{K} \sum_{i=x-d}^{x+d} \sum_{j=y-d}^{x+d} G_s(i;x,j;y) G_r[I(i,j), I(x,y)] I(i,j), \tag{3}$$

where $I(x,y)$ is the pixel intensity in the image before applying the bilateral filter, $I_{BF}(x,y)$ is the resulting pixel intensity after applying the bilateral filter and d is a non-negative integer such that $(2d + 1) \times (2d + 1)$ stands for the size of the neighborhood window. Let G_s and G_r be the domain and the range components, respectively, which are defined as:

$$G_s(i;x,\ j;y) = e^{-\frac{|(i-x)^2+(j-y)^2|}{2\sigma_s^2}} \tag{4}$$

and

$$G_r(I(i,j);I(x,y)) = e^{-\frac{|I(i,j)-I(x,y)|^2}{2\sigma_r^2}} \tag{5}$$

The normalization constant K is given as:

$$K = \frac{1}{\sum_{i=x-d}^{x+d} \sum_{j=y-d}^{x+d} G_s(i;x,j;y)G_r[I(i,j),I(x,y)]}$$

(6)

Equations (4) and (5) show that the bilateral filter has three parameters: σ_s^2 (the filter domain), σ_r^2 (the filter range), and the third parameter is the window size N × N [15].

The geometric spread of the bilateral filter is controlled by σ_s^2. If the value of σ_s^2 is increased, more neighbours are combined in the diffusion process yielding a "smoother" image, while σ_r^2 represents the photometric spreading. Only pixels with a percentage difference of less than σ_r^2 are processed [13].

2.4. Otsu Filtering

After passing through the bilateral filter, the image is split into its original (non-gamma corrected) Red, Green and Blue components, as shown in the block diagram in Figure 2. The kernel of the bilateral filter alters the balance of the colors in the original image in such a way to widen the differences between the color of the front and back-to-front interference. A modified version of Otsu [16] algorithm is applied to each RGB channel using the thresholds determined by the Decision Making Block, which may be considered as the "optimal" threshold for each RGB channel, and then three binary images are generated.

2.5. Image Classification

The image classification block was also trained with the synthetic images in such a way to analyze the three binary images generated in each of the channels and outputs the one that is considered the best one. This decision was also made by a naïve Bayes automatic classifier which was trained using the calculated co-occurrence matrix for each of the 32,000 synthetic images by comparing each of them with the original ground truth image, the Front image.

3. Experiments and Results

As already explained, the enormous variety of kinds of text documents makes extremely improbable that one single algorithm is able to satisfactorily binarize all kinds of documents. Depending on the nature (or degree of complexity) of the image several or no algorithm will be able to provide good results. This paper follows the assessment methodology proposed in reference [9], in which one compares the numbers of background and foreground pixels correctly matched with a ground-truth image. Twenty-three binarization algorithms were tested using the methodology described:

1. Mello-Lins [5]
2. DaSilva-Lins-Rocha [6]
3. Otsu [16]
4. Johannsen-Bille [17]
5. Kapur-Sahoo-Wong [18]
6. RenyEntropy (variation of [18])
7. Li-Tam [19]
8. Mean [20]
9. MinError [21]
10. Mixture-Modeling [22]
11. Moments [23]
12. IsoData [24]
13. Percentile [25]

14. Pun [26]
15. Shanbhag [27]
16. Triangle [28]
17. Wu-Lu [29]
18. Yean-Chang-Chang [30]
19. Intermodes [31]
20. Minimum (variation of [31])
21. Ergina-Local [32]
22. Sauvola [33]
23. Niblack [34]

A ground-truth image for each "real" world one is needed to allow a quantitative assessment of the quality of the final binary image. Only the DIBCO dataset [10] had ground-truth images available. This makes the assessment task of real-world images extremely difficult [35]. All care must be taken to guarantee the fairness of the process. The ground-truth images for the other datasets were generated by applying the 23 algorithms above and the bilateral algorithm to all the test images in the Nabuco [7] and LiveMemory [36] datasets. Visual inspection was made to choose the best binary image in a blind process, a process in which the people who selected the best image did not know which algorithm generated it. To increase the degree of fairness and the number of filtering possibilities, the three component images produced by the Decision Making block were all analyzed. The binary images chosen using the methodology above went through salt-and-pepper filtering and were used as ground-truth image for the assessment below. All the processing time figures presented in this paper are from Intel i7-4510U@ 2.00 GHzx2, 8 GB RAM, running Linux Mint 18.2 64-bit. All algorithms were coded in Java, possibly by their authors.

3.1. The Nabuco Dataset

The Nabuco bequest encompasses about 6500 letters and postcards written and typed by Joaquim Nabuco [7], totaling about 30,000 pages. Such documents are of great interest to whoever studies the history of the Americas, as Nabuco was one of the key figures in the freedom of black slaves, and was the first Brazilian Ambassador to the U.S.A. The documents of Nabuco were digitalized by the second author of this paper and the historians of the Joaquim Nabuco Foundation using a table scanner in 200 dpi resolution in true color (24 bits per pixel), back in 1992 to 1994. Due to serious storage limitations then, images were saved in the jpeg format with 1% loss. The historians in the project concluded that 150 dpi resolution would suffice to represent all the graphical elements in the documents, but choice of the 200-dpi resolution was made to be compatible with the FAX devices widely used then. About 200 of the documents in the Nabuco bequest exhibited back-to-front interference. The 15 document images used in this dataset were chosen for being representative of the diversity of documents in such a universe.

Table 1 presents the quantitative results obtained for all the documents in this dataset. $P(f/f)$ stands for the ratio between the number of foreground pixels in the original image mapped onto black pixels and the number of black pixels in the ground-truth image. Similarly, $P(b/b)$ is proportion between the number of background pixels in the original image mapped onto white pixels of the binary image and the number of white pixels in the ground-truth image. The figures for $P(b/b)$ and $P(f/f)$ are followed by "$\pm$" and the value of the standard deviation. The time corresponds to the mean processing time elapsed by the algorithm to process the images in this dataset. The results were ranked in $P(b/b)$ decreasing order.

The results presented in Table 1 shows the bilateral filter in third place for this dataset in terms of image quality, however the standard deviation is much lower than the two first. That implies that its quality is more stable for the various document images in this dataset. Figure 5 presents the document for which the bilateral filter presented the best and the worst results in terms of image quality with two zoomed areas from the original and the binarized document.

Table 1. Binarization results for images from Nabuco bequest.

AlgName	P(f/f)	P(b/b)	Time (s)
IsoData	98.08 ± 3.39	99.38 ± 0.60	0.0171
Otsu	98.08 ± 3.39	99.36 ± 0.63	0.0159
Bilateral	99.57 ± 1.23	99.29 ± 0.93	1.0790
Huang	99.40 ± 2.14	98.69 ± 0.88	0.0200
Moments	99.39 ± 1.34	98.40 ± 1.70	0.0160
Ergina-Local	99.99 ± 0.03	98.13 ± 0.64	0.3412
RenyEntropy	100.00	97.56 ± 1.17	0.0188
Kapoo-Sahoo-Wong	100.00	97.51 ± 1.07	0.0172
Yean-Chang-Chang	100.00	97.38 ± 1.26	0.0161
Triangle	100.00	95.94 ± 1.46	0.0160
Mello-Lins	98.61 ± 5.14	89.63 ± 24.43	0.0160
Mean	100.00	81.77 ± 5.99	0.0168
Johannsen-Bille	98.87 ± 2.97	59.77 ± 48.80	0.0164
Pun	100.00	55.44 ± 2.57	0.0185
Percentile	100.00	53.21 ± 1.33	0.0185
Sauvola	85.51 ± 12.93	99.95 ± 0.11	1.2977
Niblack	99.75 ± 0.34	77.06 ± 5.63	0.2135

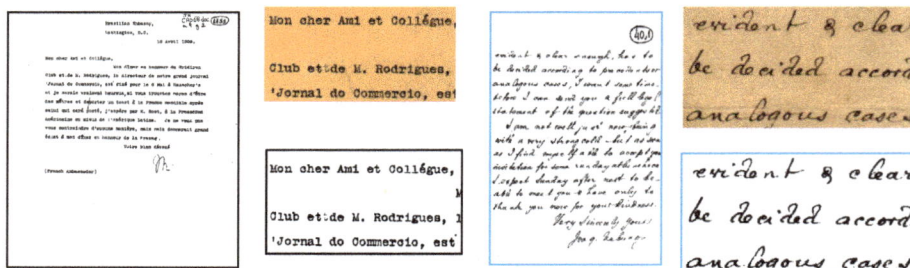

Figure 5. Historical documents from Nabuco bequest with the best ((**left**)—P(f/f) = 100, P(b/b) = 99.99) and the worst ((**right**)—P(f/f) = 89.76, P(b/b) = 99.98) binarization results for the bilateral filter with zooms from the original (**top**) and binary (**bottom**) parts.

3.2. The LiveMemory Dataset

This dataset encompasses 15 documents with 200 dpi resolution selected from the over 8000 documents from the LiveMemory project that created a digital library with all the proceedings of technical events from the Brazilian Telecommunications Society. The original proceedings were offset printed from documents either typed or electronically produced. Table 2 presents the performance results for the 12 best ranked algorithms. The bilateral filter obtained the best results in terms of image filtering. It is worth observing that in the case of the worst quality image (Figure 6, right) the performance degraded for all the algorithms. This behavior is due to the shaded area in the hard-bound spine of the volumes of the proceedings.

Table 2. Binarization results for images from the LiveMemory project.

AlgName	P(f/f)	P(b/b)	Time (s)
Bilateral	100.00	98.90 ± 1.07	3.3325
IsoData-ORIG	99.56 ± 0.69	98.61 ± 1.99	0.0734
Otsu	99.60 ± 0.68	98.57 ± 2.08	0.0735
Moments	99.99 ± 0.03	97.91 ± 1.87	0.0716
Ergina-Local	98.98 ± 2.82	97.62 ± 1.04	0.9917
Huang	99.93 ± 0.27	96.42 ± 4.20	0.0865
Triangle	100.00	94.24 ± 2.15	0.0728
Mean	100.00	83.58 ± 5.59	0.0747
Niblack	99.76 ± 0.76	78.31 ± 2.97	0.6710
Pun	100.00	55.28 ± 3.60	0.0800
Percentile	100.00	53.91 ± 1.96	0.0795
Kapur-Sahoo-Wong	98.62 ± 4.92	97.15 ± 1.44	0.0729

Figure 6. Images from LiveMemory with the best ((**left**)—P(f/f) = 100.00, P(b/b) = 99.99) and the worst ((**right**)—P(f/f) = 100.00, P(b/b) = 95.97) binarization results for the bilateral filter with zooms from the original (**top**) and binary (**bottom**) parts.

3.3. The DIBCO Dataset

This dataset has all the 86 images from the Digital Image Binarization Contest from 2009 to 2016. Table 3 presents the results obtained. The performance of the bilateral filter in this set may be considered good, in general. The overall performance of the bilateral filter was strongly degraded by the single image shown in Figure 7 (right) in which the P(f/f) of 25.93 drastically dropped the average result of the algorithm in this test set. It is important to remark that such an image is almost unreadable even for humans and that it degraded the performance of all the best algorithms.

Table 3. Binarization results for images from Document Image Binarization Competition (DIBCO).

AlgName	P(f/f)	P(b/b)	Time (s)
Ergina-local	91.37 ± 6.25	99.88 ± 1.89	0.1844
RenyEntropy	90.13 ± 14.19	96.77 ± 3.50	0.0125
Yean-Chang-Chang	90.61 ± 14.44	96.16 ± 4.35	0.0112
Moments	90.75 ± 9.91	95.80 ± 5.19	0.0112
Bilateral	92.99 ± 9.06	90.78 ± 16.01	0.6099
Huang	95.62 ± 6.37	84.22 ± 18.36	0.0147
Triangle	96.40 ± 5.72	80.80 ± 23.32	0.0113
Mean	99.35 ± 1.14	78.99 ± 9.35	0.0115
MinError	92.79 ± 23.46	74.29 ± 19.36	0.0115
Pun	99.68 ± 0.82	56.20 ± 6.18	0.0122
Percentile	99.71 ± 0.72	55.06 ± 3.58	0.0121
Sauvola	59.75 ± 30.06	99.58 ± 079	0.6933
Niblack	95.91 ± 2.31	78.61 ± 5.69	0.1241

4. Conclusions

Historical documents are far more difficult to binarize as several factors such as paper texture, aging, thickness, translucidity, permeability, the kind of ink, its fluidity, color, aging, etc. all may influence the performance of the algorithms. Besides all that, many historical documents were written or printed on both sides of translucent paper, giving rise to the back-to-front interference.

This paper presents a new binarization scheme based on the bilateral filter. Experiments performed in three datasets of "real world" historical documents with twenty-three other binarization algorithms. Image quality and processing time figures were provided, at least for the top 10 algorithms assessed. The results obtained showed that the proposed algorithm yields good quality monochromatic images that may compensate its high computational cost. This paper provides evidence that no binarization algorithm is an "all-kind-of-document" winner, as the performance of the algorithms varied depending of the specific features of each document. A much larger test set of synthetic about 250,000 images is currently under development, such a test set will allow much better training of the Decision Making and Image Classifier blocks of the bilateral algorithm presented. The authors are currently attempting to integrate the Decision Making and Image Classifier blocks in such a way to anticipate the choice of the best component image. This would highly improve the time performance of the proposed algorithm.

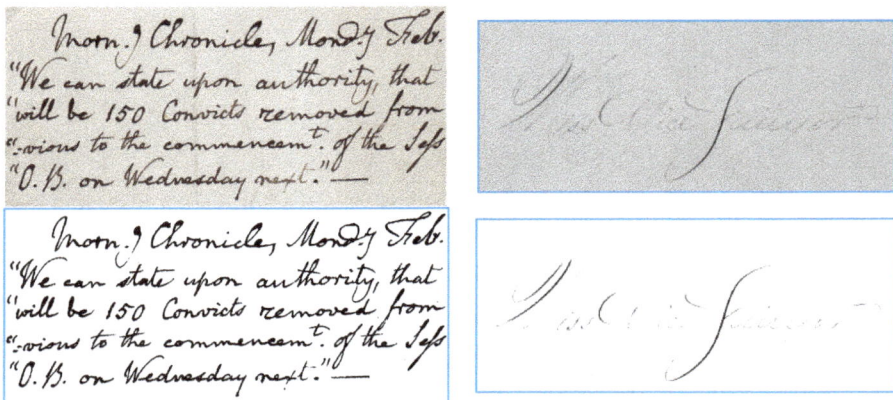

Figure 7. Two documents from DIBCO dataset: (**left-top**) original image (**left-bottom**) binary image obtained using the bilateral filter best result (P(f/f) = 97.05, P(b/b) = 99.88); (**right-top**) original image. (**right-bottom**) the worst binarization results for the bilateral filter (P(f/f) = 25.93, P(b/b) = 99.99).

The authors of this paper are promoting a paramount research effort to assess the largest possible number of binarization algorithms for scanned documents using over 5.4 million synthetic images in the DIB-Document Image Binarization platform. An image matcher, a more general and complex version of the Decision Making block, is also being developed and trained with that large set of images, in order to whenever fed with a real world image, to be able to match with the most similar synthetic one. Once that match is made, the most suitable binarization algorithms are immediately known. If this paper were accepted, all the test images and algorithms will be included in the DIB platform. The preliminary version of the DIB-Document Image Binarization platform and website is publicly available at https://dib.cin.ufpe.br/.

Acknowledgments: The authors of this paper are grateful for the referees whose comments much helped in improving the current version of this paper and to those researchers who made the code of their algorithms publicly available for testing and performance analysis and to the DIBCO team from making their images publicly available. The authors also acknowledge the partial financial support of to CNPq and CAPES—Brazilian Government.

Author Contributions: Marcos Almeida and Rafael Dueire Lins contributed in equal proportion to the development of the algorithm presented in this paper, which was written by the latter author. Bruno Lima was responsible for the first implementation of the algorithm proposed. Rodrigo Bernardino and Darlisson Jesus re-implemented the algorithm and were also responsible for all the quality and time assessment figures presented here.

Conflicts of Interest: The authors declare no conflict of interest.

References

1. Chaki, N.; Shaikh, S.H.; Saeed, K. *Exploring Image Binarization Techniques*; Springer: New Delhi, India, 2014.
2. Lins, R.D. A Taxonomy for Noise in Images of Paper Documents-The Physical Noises. In Proceedings of the International Conference Image Analysis and Recognition, Halifax, NS, Canada, 6–8 July 2009; Volume 5627, pp. 844–854.
3. Lins, R.D. An Environment for Processing Images of Historical Documents. *Microprocess. Microprogr.* **1995**, *40*, 939–942. [CrossRef]
4. Sharma, G. Show-through cancellation in scans of duplex printed documents. *IEEE Trans. Image Process.* **2001**, *10*, 736–754. [CrossRef] [PubMed]
5. Mello, C.A.B.; Lins, R.D. Generation of Images of Historical Documents by Composition. In Proceedings of the 2002 ACM Symposium on Document Engineering, New York, NY, USA, 8–9 November 2002; pp. 127–133.
6. Silva, M.M.; Lins, R.D.; Rocha, V.C. Binarizing and Filtering Historical Documents with Back-to-Front Interference. In Proceedings of the 2006 ACM Symposium on Applied Computing, New York, NY, USA, 23–27 April 2006; pp. 853–858.
7. Lins, R.D. Nabuco—Two Decades of Processing Historical Documents in Latin America. *J. Univers. Comput. Sci.* **2011**, *17*, 151–161.
8. Roe, E.; Mello, C.A.B. Binarization of Color Historical Document Images Using Local Image Equalization and XDoG. In Proceedings of the 12th International Conference on Document Analysis and Recognition, Washington, DC, USA, 25–28 August 2013; pp. 205–209.
9. Lins, R.D.; Almeida, M.A.M.; Bernardino, R.B.; Jesus, D.; Oliveira, J.M. Assessing Binarization Techniques for Document Images. In Proceedings of the ACM Symposium on Document Engineering, Valletta, Malta, 4–7 September 2017.
10. Pratikakis, I.; Zagoris, K.; Barlas, G.; Gatos, B. ICDAR 2017 Competition on Document Image Binarization (DIBCO 2017). In Proceedings of the 14th IAPR International Conference on Document Analysis and Recognition, Kyoto, Japan, 13–15 November 2017; pp. 2140–2379.
11. Efros, A.A.; Freeman, W.T. Image quilting for texture synthesis and transfer. In Proceedings of the 28th Annual Conference on Computer Graphics and Interactive Techniques (SIGGRAPH '01), New York, NY, USA, 12–17 August 2001; pp. 341–346.
12. Aurich, V.; Weule, J.B. Non-Linear Gaussian Filters Performing Edge Preserving Diffusion. In Proceedings of the DAGM Symposium, London, UK, 13–15 September 1995; pp. 538–545.
13. Tomasi, C.; Manduchi, R. Bilateral Filtering for Gray and Color Images. In Proceedings of the 6th International Conference on Computer Vision, Washington, DC, USA, 4–7 January 1998; pp. 836–846.
14. Paris, P.; Kornprobst, P.; Tumblim, J.; Durand, F. Bilateral Filtering: Theory and Applications. *Found. Trends Comput. Graph. Vis.* **2008**, *4*, 1–73. [CrossRef]
15. Shyam Anand, C.; Sahambi, J.S. Pixel Dependent Automatic Parameter Selection for Image Denoising with Bilateral Filter. *Int. J. Comput. Appl.* **2012**, *45*, 41–46.
16. Otsu, N. A Threshold Selection Method from Gray-Level Histograms. *IEEE Trans. Syst. Man Cybern.* **1979**, *9*, 62–66. [CrossRef]
17. Johannsen, G.; Bille, J.A. A Threshold Selection Method Using Information Measure. In Proceedings of the 6th International Conference on Pattern Recognition (ICPR'82), Munich, Germany, 19–22 October 1982; pp. 140–143.
18. Kapur, N.; Sahoo, P.K.; Wong, A.K.C. A New Method for Gray-Level Picture Thresholding Using the Entropy of the Histogram. *Comput. Vis. Graph. Image Process.* **1985**, *29*, 273–285. [CrossRef]
19. Li, C.H.; Tam, P.K.S. An iterative algorithm for minimum cross entropy thresholding. *Pattern Recognit. Lett.* **1998**, *19*, 771–776. [CrossRef]

20. Glasbey, C.A. An analysis of histogram-based thresholding algorithms. *Graph. Models Image Process.* **1993**, *55*, 532–537. [CrossRef]
21. Kittler, J.; Illingworth, J. Minimum error thresholding. *Pattern Recognit.* **1986**, *19*, 41–47. [CrossRef]
22. Mixture Modeling. ImageJ. Available online: http://imagej.nih.gov/ij/plugins/mixture-modeling.html (accessed on 20 January 2018).
23. Tsai, W.H. Moment-preserving thresholding: A new approach. *Comput. Vis. Graph. Image Process.* **1985**, *29*, 377–393. [CrossRef]
24. Doyle, W. Operation useful for similarity-invariant pattern recognition. *J. Assoc. Comput. Mach.* **1962**, *9*, 259–267. [CrossRef]
25. Pun, T. Entropic Thresholding, A New Approach. *Comput. Vis. Graph. Image Process.* **1981**, *16*, 210–239. [CrossRef]
26. Shanbhag, A.G.G. Utilization of Information Measure as a Means of Image Thresholding. *Comput. Vis. Graph. Image Process.* **1994**, *56*, 414–419. [CrossRef]
27. Zack, G.W.; Rogers, W.E.; Latt, S.A. Automatic measurement of sister chromatid exchange frequency. *J. Histochem. Cytochem.* **1977**, *25*, 741–753. [CrossRef] [PubMed]
28. Wu, U.L.; Songde, A.; Haqing, L.U.A. An Effective Entropic Thresholding for Ultrasonic Imaging. In Proceedings of the International Conference Pattern Recognition, Brisbane, Australia, 16–20 August 1998; pp. 1522–1524.
29. Yen, J.C.; Chang, F.J.; Chang, S. A New Criterion for Automatic Multilevel Thresholding. *IEEE Trans. Image Process.* **1995**, *4*, 370–378. [PubMed]
30. Ridler, T.W.; Calvard, S. Picture Thresholding Using an Iterative Selection Method. *IEEE Trans. Syst. Man Cybern.* **1978**, *8*, 630–632.
31. Prewitt, M.S.; Mendelsohn, M.L. The Analysis of Cell Images. *Ann. N. Y. Acad. Sci.* **1996**, *128*, 836–846. [CrossRef]
32. Kavallieratou, E.; Stamatatos, S. Adaptive binarization of historical document images. In Proceedings of the 18th International Conference on Pattern ICPR 2006, Hong Kong, China, 20–24 August 2006; Volume 3.
33. Sauvola, J.; Pietikainen, M. Adaptive document image binarization. *Pattern Recognit.* **2000**, *33*, 225–236. [CrossRef]
34. Niblack, W. *An introduction to Digital Image Processing*; Prentice-Hall: Upper Saddle River, NJ, USA, 1986.
35. Ntirogiannis, K.; Gatos, B.; Pratikakis, I. Performance Evaluation Methodology for Historical Document Image Binarization. *IEEE Trans. Image Process.* **2013**, *22*, 595–609. [CrossRef] [PubMed]
36. Lins, R.D.; Silva, G.F.P.; Torreão, G.; Alves, N.F. Efficiently Generating Digital Libraries of Proceedings with the LiveMemory Platform. In *IEEE International Telecommunications Symposium*; IEEE Press: Rio de Janeiro, Brazil, 2010; pp. 119–125.

Journal of
Imaging

MDPI

Article

Slant Removal Technique for Historical Document Images

Ergina Kavallieratou [1,*], Laurence Likforman-Sulem [2] and Nikos Vasilopoulos [1]

[1] Department Information and Communication Systems Engineering, University of the Aegean, Samos 83200, Greece; nvasilopoulos@aegean.gr
[2] Institut Mines-Télécom/Télécom ParisTech, Université Paris-Saclay, 75013 Paris, France; laurence.likforman@telecom-paristech.fr
* Correspondence: kavallieratou@aegean.gr

Received: 14 May 2018; Accepted: 5 June 2018; Published: 12 June 2018

Abstract: Slanted text has been demonstrated to be a salient feature of handwriting. Its estimation is a necessary preprocessing task in many document image processing systems in order to improve the required training. This paper describes and evaluates a new technique for removing the slant from historical document pages that avoids the segmentation procedure into text lines and words. The proposed technique first relies on slant angle detection from an accurate selection of fragments. Then, a slant removal technique is applied. However, the presented slant removal technique may be combined with any other slant detection algorithm. Experimental results are provided for four document image databases: two historical document databases, the TrigraphSlant database (the only database dedicated to slant removal), and a printed database in order to check the precision of the proposed technique.

Keywords: slant removal; document image processing; document image page

1. Introduction

In handwriting, slant removal is a necessary component of the text normalization procedure in systems that perform recognition (e.g., optical character recognition (OCR) [1] or word-spotting [2]), in order to improve the training procedure (less samples, lower computational cost). Moreover, writer identification/verification systems also use slant estimation and/or detection [3]. After ideal slant removal processing, the text should appear with the vertical stokes parallel to the perpendicular axis of the page. Due to its importance, many researchers have already developed techniques for slant removal [4–17].

The available techniques may be divided into three categories:

1. Techniques that estimate the slant by averaging the angles of the near-vertical strokes [4–7].
2. Techniques that analyze projection histograms [8,9] and detect the slant based on a pre-defined criterion (e.g., a parameter maximization or minimization).
3. Techniques that are based on the statistics of chain-coded contours [10–12].

Considering the application these techniques can handle, they can be further classified into:

1. Uniform slant estimation and removal techniques [4–12]: they deal with uniform slant all over the text.
2. Non-uniform slant correction techniques [13–16]: they handle the characters apart and deal with the existence of several slants, simultaneously.

Recently, Brink et al. [3] categorized the proposed techniques by angle-frequency and repeated-shearing approaches that are described as follows:

1. Angle-frequency approach: Down-strokes are first located based on such criteria as the minimum vertical extent or velocity. Next, the angle of the local ink direction is measured at these locations and the resulting angles are agglomerated in a histogram. From this histogram, the slant angle is determined. This is a one-step procedure.
2. Repeated-shearing approach: This method is based on the assumption that the projection of dark pixels is maximized along an axis parallel to the slant angle. The basic principle is to repeatedly shear images of individual text lines, varying the shear angle, and optimizing the vertical projection of dark pixels. This approach is clearly more time consuming, but proves more accurate, as indicated by its popularity.

The first category will be referred to here as 'slant estimation' (one-step procedure), and the second category is referred to as slant detection, since this method searches among many, for the most common angle. Slant estimation techniques are presented in [4–7], whereas a slant detection technique is presented in [9]. According to Brink et al. [3], the slant detection techniques are the most popular with the most precise results. The technique described in [9] is also used in that paper where extensive experiments over slant are performed. Last but not least, in the specific experiments, the pages were sheared entirely, since the alternative line or word segmentation is characterized as "less reliable and breaks ink traces at region boundaries" [3]. The proposed techniques up to now require line or word segmentation in order to be applied. In Figure 1, an example of the slant removal algorithm described in [9], is presented. The image is from the IAM Handwriting Database (IAM-DB), and the application of the algorithm requires image segmentation into text lines (Figure 1, horizontal stripes). For this example, text line segmentation could succeed since text lines are spaced enough. It is not the case for the document image shown in Figure 2 (17th century) which includes touching ascenders and descenders and noise in the inter-line space. Since all existing algorithms perform slant removal on word or text line level, a segmentation-free approach is desirable for difficult to segment documents. Moreover, avoiding the text-line segmentation processing is computationally less expensive.

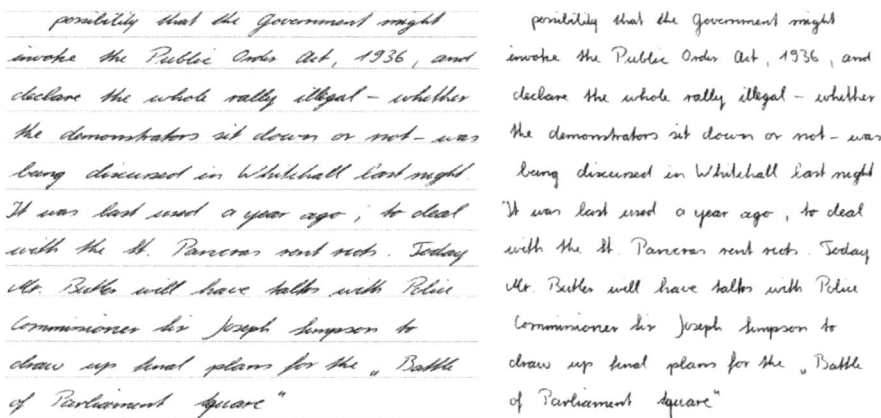

Figure 1. An example of a slant removal application resulting from the detection algorithm described in [9]. Text-line segmentation is required prior to slant estimation.

A preliminary approach has been described in [17], while in this paper the parameter set up is considered and described in detail. Moreover, the approach is extensively evaluated on new databases. The proposed technique is appropriate for slant detection and removal from document images with homogenous slant. It does not require page segmentation into text lines or words. This makes the proposed technique appropriate for historical documents, especially formal ones, since they ensure a uniform slant over the entire page. Moreover, the segmentation into text lines would create more noise. Usually, formal historical documents are written by well-educated people with a standard writing style and fixed slant. On the other hand, the proposed methodology is inappropriate for document images containing unconstrained writing and several slant angles. Methodologies, such as the one described in [6], are more appropriate in such cases. Thus, experiments are performed on several databases:

- the TrigraphSlant database [18] (the only available database for slant estimation),
- two databases of historical documents (George Washington [19] and Barcelona historical, handwritten marriages database BH2M [20])
- a synthetic printed database where slants are fully determined.

The contribution of the current work consists of:

1. To the best of our knowledge, this is the first time that a slant removal technique is proposed, able to be applied to the entire page, without requiring text line or word segmentation.
2. It does not generate extra noise, due to line and/or word segmentation that would remain in the page after slant removal, which is accomplished by shifting the entire page uniformly and ensuring text homogeneity. Most of the existed techniques apply to ideal databases, like IAM-DB (Figure 1) that is appropriately made for line and word segmentation. In the case of historical documents (Figure 2), the final result would be full of dots and strokes because of the segmentation.
3. Instructions are given over the best application to document page, after detailed results.

In Section 2.1, a short description of the elaborated slant detection algorithm [9] is presented. The proposed technique is described in detail in Section 2.2, where the parameters that are examined in detail are analyzed. The experimental results are presented and analyzed in Section 3 while the conclusions are discussed in Section 4.

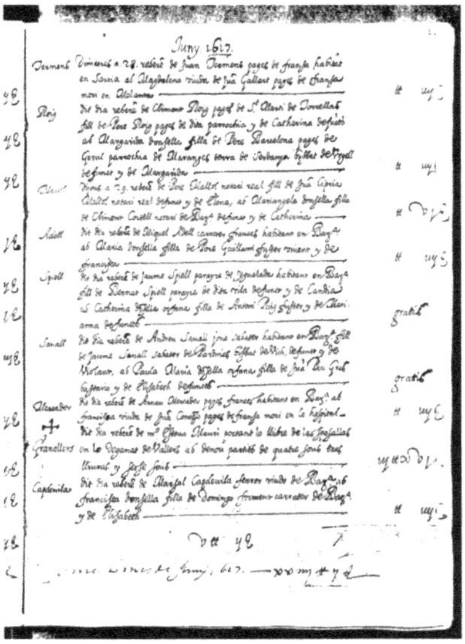

Figure 2. A document from the Barcelona historical, handwritten marriages database (BH2M) [20].

2. Materials and Methods

2.1. Slant Detection Algorithm

As already mentioned, in the proposed technique, the slant detection algorithm that is presented in detail in [9] is used. The reason is that an algorithm for the detection is needed and this one has proved to be popular and successful [3].

This specific algorithm makes use of the difference between the ascenders and descenders. In case the text is vertical, the difference is larger (Figure 3). To detect that, it uses the Wigner Ville distribution (WVD) [21], a space-frequency distribution of Cohen's class, which is given by the formula

$$W(s,f) = \int\limits_{-\infty}^{+\infty} z(s + \tau/2)z * (s - \tau/2)e^{-2i\pi f\tau}d\tau, \tag{1}$$

where s the signal, f the frequency, $z(s)$ represents the analytical signal associated with the discrete signal h(s), that in the paper [9] is the vertical projection profile of the word.

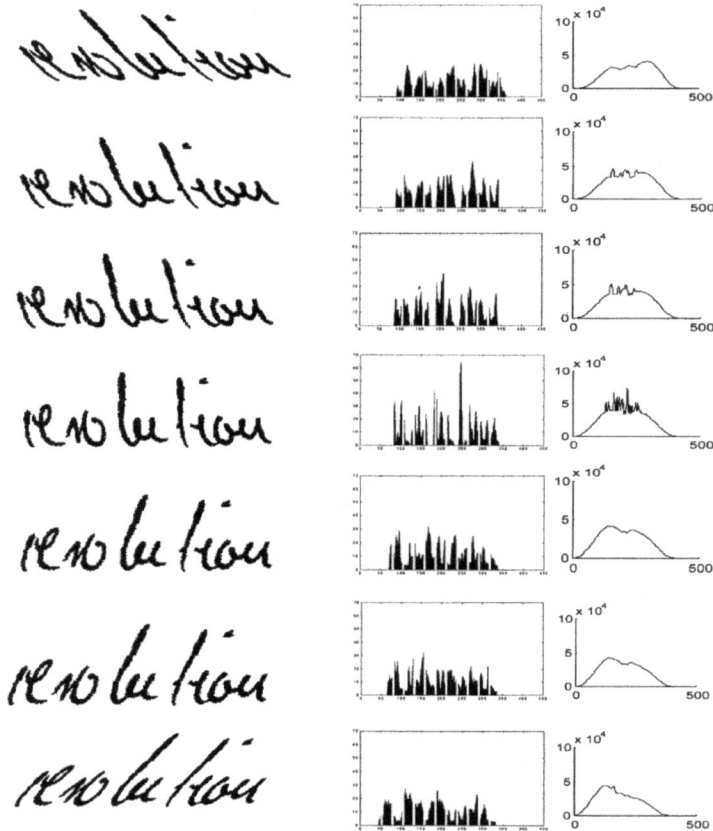

Figure 3. From left to right: slanted words, corresponding histograms and maximum intensity curves of WVD (Wigner–Ville distribution).

In more detail, the detection algorithm [9] consists of the steps:

1. The word image is artificially slanted to both, left and right, under different slant detection angles. The maximum slant angle is approximately 45 degrees and the slant angle step depends on the height of the text image.
2. For each of the extracted word images, the vertical projection profile is calculated.
3. The WVD is calculated for all the above projected profiles.
4. The curves of maximum intensity of the WVDs are extracted, just by keeping the maximum value of each curve of the space-frequency distribution, for the specific slant.
5. The curve of maximum intensity with the greatest peak, corresponding to the projected profile with the most intense alternations is selected.
6. The corresponding word image is selected as the most non-slanted word.

The above procedure is repeated twice, once for a big step size of 10 degrees (*BigStep*) where the area around an ANGLE1 is selected closer to the slant and the second time for a smaller step size of 1 degree, where a more detailed detection is performed and a more exact area ANGLE2 is detected.

This way, the computational cost is reduced, since the first detection is performed between fewer possible angles in order to localize roughly the area ANGLE1, before a more accurate detection (ANGLE2) is performed in this specific area for a step size of one degree. A slant of less than one degree is not considered important enough to be examined. Finally, the detected angle (*Detected_Slant*) is given by

$$Detected\_Slant = ANGLE1 \times BigStep + ANGLE2 \times 1 \qquad (2)$$

2.2. Proposed Slant Removal Technique

The proposed technique is based on the slant detection algorithm presented in Section 2.1, but in our case, it is applied to text fragments instead of words (Figure 4). It is based on the fact that in historical documents there is a uniform slant that extends throughout the entire document image. Since no segmentation is performed, fragments of text are used instead of words.

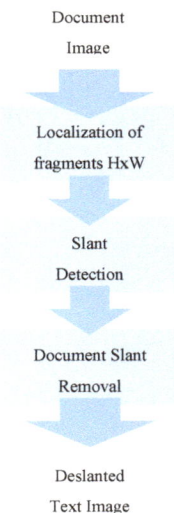

Figure 4. The proposed slant removal technique applied to fragments of text corrects the entire page without segmentation.

As previously mentioned, in the past, educated persons took special care when writing, resulting in a high degree of stability in the slant of their writing style. Thus, in order to detect the slant of the text in a historical document page, a few fragments of text are considered. Although one sample could theoretically be enough, several ones are in practice necessary to ensure coverage of pages with sparse text or special formatting, such as columns, arrays, etc. To localize appropriate fragments the following way is followed: a page is scanned from left to right, top to bottom using a window of size *HxW* (heightXwidth), starting from the pixel position (skip, skip) in order to skip scanning or other noise. Skip can be general e.g., 1/5 of document width (here), or be determined depending on the collection. All black pixels (*black_pixels*) inside the window are counted. The area inside the window is retained if Condition (3) is true and the scanning stops when the required number M of fragments is localized.

$$\frac{black\_pixels}{HxW} > R \tag{3}$$

The Condition (3) requires the text in the window to take up more than $R = 0.10$ of the area. The size of the window in these experiments, *HxW*, was selected as $H = 2$ mb and $W = 7$ mb, where mb is the main character body size in the page (height of the character body excluding ascenders and descenders). In the current paper, the following metrics and parameters are set up:

1. The text ratio R in the window;
2. The amount M of the fragments in use;
3. The height H of the window;
4. The width W of the window.

However, the same techniques are considered for:

1. The main body height detection [22], since it does not require line or word segmentation;
2. The slant detection procedure. Once the M fragments have been selected (Figure 5), the slant detection algorithm [9], described in Section 2, is applied and the slant angles are detected, one per fragment. The maximum and minimum slant angles are ignored as possible outliers, while *slant* is defined as the detected slant of the page. The entire document page is then corrected according to the slant angle by shifting each pixel so that

$$x_f = x_0 + round\left[y_0 \tan\left(\pi\frac{slant}{180}\right)\right] \tag{4}$$

$$y_f = y_0 \tag{5}$$

where (x_0, y_0) defines the initial position of the pixel and (x_f, y_f) is the final pixel position.

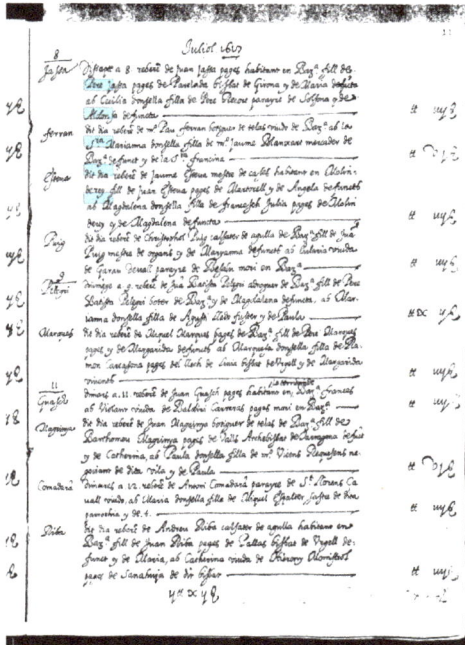

Figure 5. Possible localization of appropriate fragments on the page. The required number of fragments is M = 5.

3. Results

Since there is no previous similar work to compare with, a trial is made to perform an absolute evaluation in various ways. Thus, in order to perform our experiments and evaluate the parameters, four databases were used:

1. The TrigraphSlant database (DB) [18], in order to perform tests on a renowned DB for slant. However, in this DB each writer was asked to write two pages of his natural slant and two of force slants. Only the natural slant documents were used here (see Experimental Results).
2. The George Washington DB [19], in order to perform tests on a renowned DB of historical documents.
3. The BH2M: the Barcelona Historical Handwritten Marriages database [20], in order to perform tests on a second DB of historical documents.
4. The Print DB: printed documents with artificial slant, in order to check the accuracy of the technique. Moreover, since all the rest do not guarantee the existence of all the possible slants, special care was taken to include all possible slants, including 0 (no slant).

A validation set and a test set were created. The validation set consists of 20 document images from the TrigraphSlant DB, 4 from the Washington DB, 4 from the BH2M DB, and 80 from the Print DB. The test set consists of 60 document images from the TrigraphSlant DB, 16 from the Washington DB, 16 from the BH2M DB, and 298 from the Print DB.

The measure used to evaluate the technique is the root-mean-square error (RMSE)

$$RMSE = \sqrt{\frac{\sum_{d=1}^{N} \left(Slant_{gr}^{d} - Slant_{es}^{d} \right)^2}{N}} \qquad (6)$$

where $Slant_{gr}$ is the ground truth slant of document d and $Slant_{es}$ the slant estimated using the technique. This measure gives a comparative result that is independent of the right or left slant direction. N refers to the amount of documents. Next, a short description of the databases is given, while the setup on the parameters follows. As initial parameter values, the parameters used in [17] are used in our experiments, and as soon as the best parameter value is estimated, it is used further on. Finally, experimental results for the four databases are presented (see Section 3.9).

3.1. TrigraphSlant DB

The TrigraphSlant [18] database contains images of handwriting produced under normal and forced slant conditions. It includes 190 handwritten document images, written by 47 people. For each image, the slant has been estimated by two researchers (Axel and Rolland) from the average slant computed from 10 measurements on each document image. In Figure 6, an example of the TrigraphSlant database after the application of the proposed technique is shown.

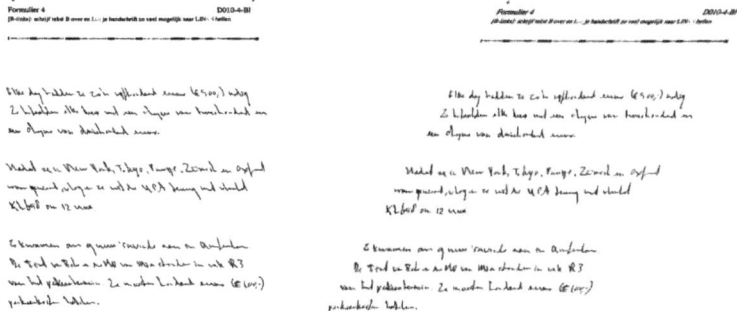

Figure 6. Application of the technique on a sample from the TrigraphSlant database (**left**); corrected by +28 degrees (**right**).

3.2. George Washington DB

This archive contains a set of 20 page images from the George Washington collection [19] at the Library of Congress in the United States. A process similar to that used for the TrigraphSlant database was followed. Ten slants were measured by two humans on each page and the mean of these measurements was considered to be the page slant. Figure 7 shows an example of the George Washington DB after the application of the technique.

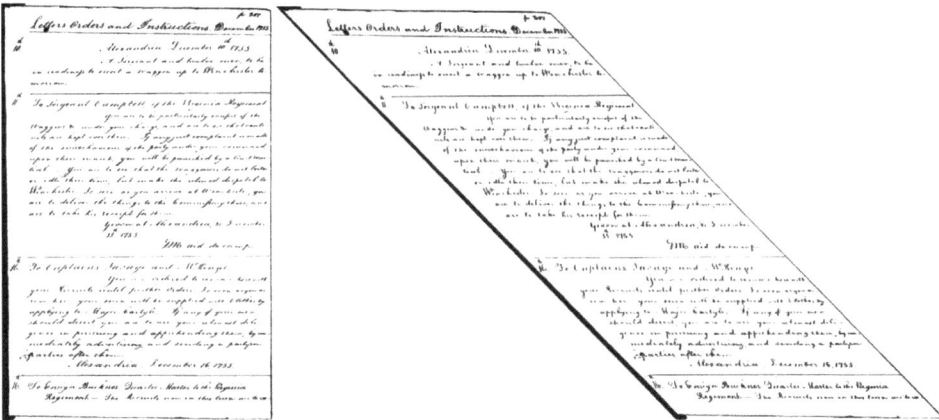

Figure 7. Application of our slant removal technique on a sample from the George Washington DB (**left**); corrected by $-42.5°$ (**right**).

3.3. BH2M DB

The BH2M database [20] has been created under the EU ERC project Five Centuries of Marriages (5CofM). It includes the Archives of the Barcelona Cathedral: 244 books with information on approximately 550,000 marriages held between 1451 and 1905 in over 250 parishes. Each book was written by a different writer and contains information of the marriages during two years. Here pages for the 6th book are used. In Figure 8, an example of the BH2M DB is presented.

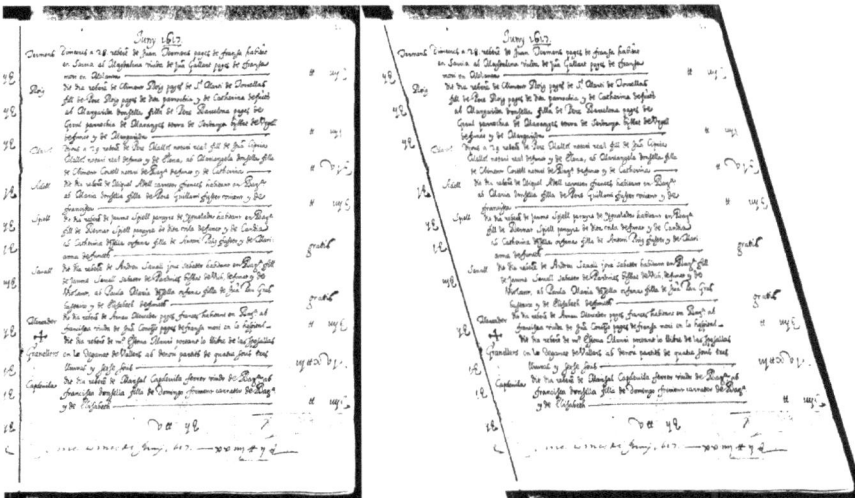

Figure 8. Application of our slant removal technique on sample from BH2M DB (**left**); corrected by $-17°$ (**right**).

3.4. PrintDB

The PrintDB consists of five printed document images that are artificially slanted over a range from $-45°$ to $+45°$, yielding a total of 455 slanted, printed document images. The exact slant is predetermined, making evaluation of the technique easier and more precise. The documents were made from parts of .pdf files to ensure precise slant values. The pages were selected to include different type of text types (including spare writing and single/double columns). All the text was slanted, keeping the original dimensions ((1/5) * A4_height * (1/2) * A4_width). Figure 9 shows an example of the PrintDB after the application of the technique.

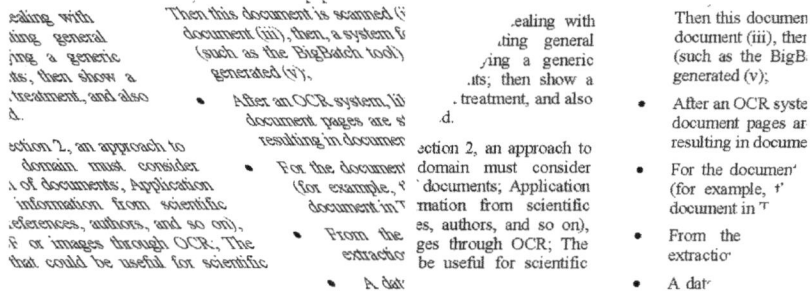

Figure 9. Application of our slant removal technique on sample of PrintDB (**left**); corrected by 35° (**right**).

3.5. Set-Up of the Text Ratio R Parameter

The amount of text in the window is very important in order to detect the slant angle. Little text includes very little information, while too much text would increase the computational cost. In [17], the rate of 10% was used as approved amount of text in the selected windows just by test and trial. Here, more detailed experiments are performed. The text ratio of each window is counted and compared to the slant detection error. The experiment was performed for every window on several images of the validation set: 5 document images from the TrigraphSlant DB, 1 from the Washington DB, 1 from the BH2M DB, and 20 from the Print DB. Since the handwritten images were of high resolution, the procedure was very time consuming and just few of them were used. On the other hand, the images of printed text were all used. In Figure 10, the curve of sum of slant square errors (SSE) with reference to the text ratio is presented for printed and handwritten text.

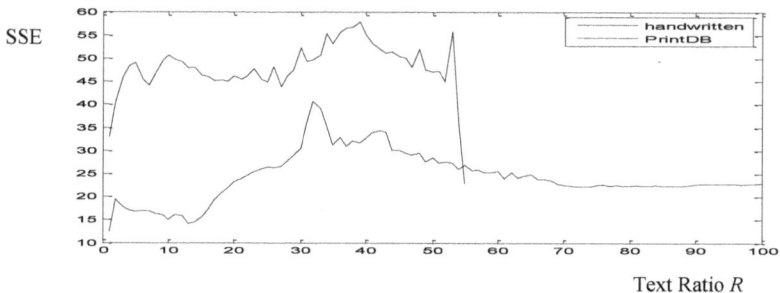

Text Ratio *R*

Figure 10. Sum of square errors according to text ratio *R* for the printed db (dotted line) and the three handwritten databases (broken line).

The main volume of windows, as it is shown in Figure 11, includes text up to 20%. Since the printed images with slant are an artificial database, the result derived from handwritten images were taken into account as a common R, that is 14% text in the image is the case of windows with minimum detection error. The majority of windows include text up to 20% (see Figure 11). Since the slanted printed images are an artificial database, the result derived from handwritten images were taken into account as a common $R = 14\%$ corresponding to the minimum detection error (see Figure 10).

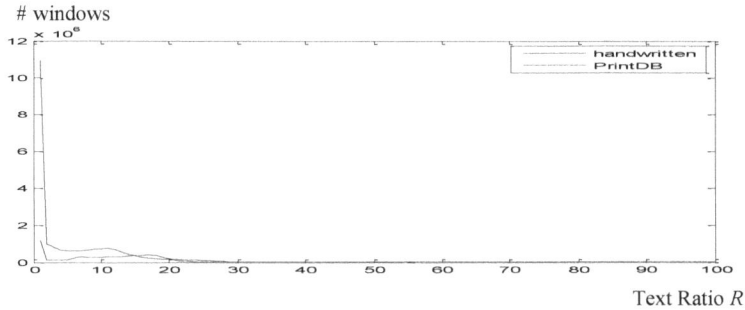

Figure 11. Number of windows according to text ratio in the window. The large amount close to zero corresponds to almost empty windows (background).

3.6. Set-Up of the Height H of the Window

In order to estimate the best number for the window height, the whole validation set was used in our program with the ratio equal to 0.14, as derived by the previous experiment, for values of the height from 1 to 10 mb (main-bodies). The sum of square errors (SSE) was considered as evaluation measure. The results are shown in Figure 12. The SSE is higher than in Figure 10, as here, all the document images of the validation set were used, contrary to Figure 10 where just few of them were considered due to the computational cost of the corresponding experiment.

As it is obvious in Figure 12, the value of two main bodies seems to give the best results (smaller SSE) as (the 1: one main body) was considered extremely small in case of low resolution.

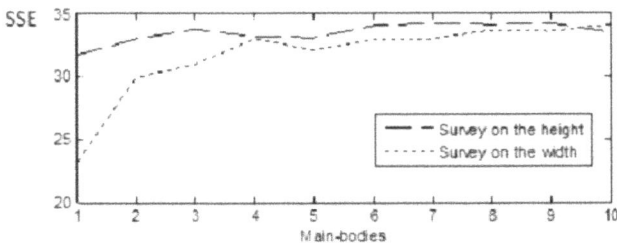

Figure 12. Sum of square errors for various widths and heights of the detected windows. The best choice is five.

3.7. Set-Up of the Width W of the Window

In order to estimate the best number for the window width, the whole validation set was also used in our program with the ratio $R = 0.14$ and $H = 2$ mbs, for values of the width from 1 to 10 main-bodies. Sum of square errors (SSE) was considered as evaluation measure. The results are shown in Figure 12. The SSE is much higher than in Figure 10, as here, all the document images of the validation set were used, contrary to Figure 10 where just few of them were considered due to the computational cost of the corresponding experiment.

As it is obvious in Figure 12, the value of just five main-bodies seems to give significantly better results (smaller SSE) which is far smaller than seven, initially used in [17]. The values of 1 and 2 that give very low error proved very low in the case of low resolution, while more characters could confuse the system.

3.8. Set-Up of the Number M of Fragments to Use

In order to estimate the best number for the window width, the whole validation set was also used in our program with the ratio $R = 0.14$, $H = 2$ mbs, and $W = 5$ mbs, for values of the amount of the fragments M of 4 and 5. More fragments could be used, however it is a problem in the case of small images or images of low resolution. The sum of square errors (SSE) was considered as evaluation measure for the scenarios:

- Four fragments, mean of the fragments: SSE on the evaluation set 563
- Five fragments, mean of the fragments: SSE on the evaluation set 513
- Five fragments, median of the fragments: SSE on the evaluation set 509

Obviously, the median was finally selected as the best case.

3.9. Experimental Results on the Databases

Having specified the parameters above, several experiments were conducted for the four databases, on each test set. In order to demonstrate the flexibility of the proposed technique, two more slant detection techniques were also used for the slant detection part, using the parameters that were specified by the use of our technique [9]. The technique [23] detects the slant by using the main part of the text having removed the horizontal parts, ascenders, and descenders. The technique [24] estimates the slant by using the peaks of the slanted words.

The experimental results for all databases are given in Table 1 through RMSE. For the TrigraphSlantDB: the RMSE is given only for the normal slants (between −45 and +45 degrees) but for the both estimators (Axel & Rolland). In the TrigraphSlant, each writer was asked to force different slants in two of the four documents, these are not presented since the results were strange due to unnatural slant >45 degrees.

Table 1. Experimental results

Database	Proposed Slant Detection [9]	Slant Detection [23]	Slant Detection [24]
TrigraphSlant (Axel estimat.)	7.08	8.30	6.99
TrigraphSlant (Rolland est.)	7.43	7.97	7.44
George Washington DB	3.44	6.53	3.40
BH2M DB	4.68	6.04	5.03
PrintDB	2.99	4.32	2.97

There were several significant differences in the RMSE values between the various DBs. Several reasons for this are:

- In the TrigraphSlant, the writing is modern and not as uniform as in the historical documents. When examined by human estimators, a standard deviation of 2.45 was observed.
- George Washington DB of historical documents is more uniform.
- BH2M presents more density which made our character main body size algorithm fail more times.
- PrintDB includes printed text that is artificially slanted, and therefore is uniform.
- We present in the following experiments in order to evaluate the improvement brought by our slant detection and removal technique on document analysis and recognition tasks. We thus conduct recognition experiments on printed documents with an OCR, and word spotting

experiments on handwritten documents, before and after slant removal. The recognition results for the handwriting of our databases were a failure, due to having historical documents or/and languages other than English. For the PrintDB database, in Figure 13, the character error rate vs. the artificial slant are shown, as obtained by a commercial OCR system (Adobe Acrobat).

For a page slant of less than −19 or greater than 37 degrees, it is very difficult to find a correspondence between the characters of the image and the OCR result. The corrected version is very well handled (0 degrees, English). Moreover, the OCR software handles right-slanted characters better than left-slanted characters. This is likely due to extra training for italics in the commercial software.

The computational cost is less than 5 s for a historical document image of size a little bigger than A4 and resolution 600 dpi in a computer with processor Intel(R) Core(TM) i5-4210U CPU @ 1.70GHz 2.40 GHz.

Since the proposed system was built in order to help our free-segmentation word-spotting system [25], we provide an example of word spotting task on handwritten documents. It is worth mentioning that an improvement of the recall of at least 20% is observed (Figure 14) for the 20 document images of George Washington DB and 100 queries. The improvement appears extremely high, which may be a result of the query being a part of the same page or collection. However, for slanted characters, the slant degree is not always fixed and in many cases, the slanted characters overlap with other characters.

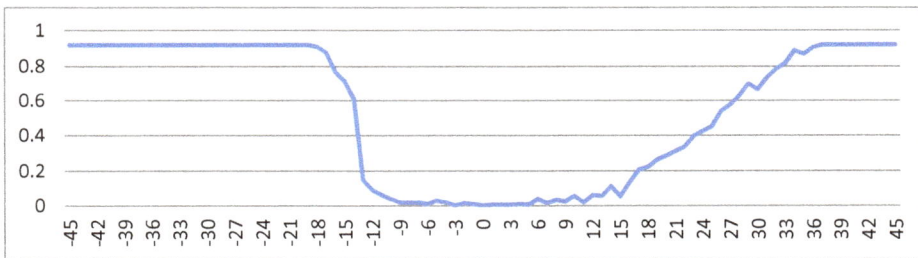

Figure 13. Character error rate vs. the degree of page slant as performed by a commercial OCR system.

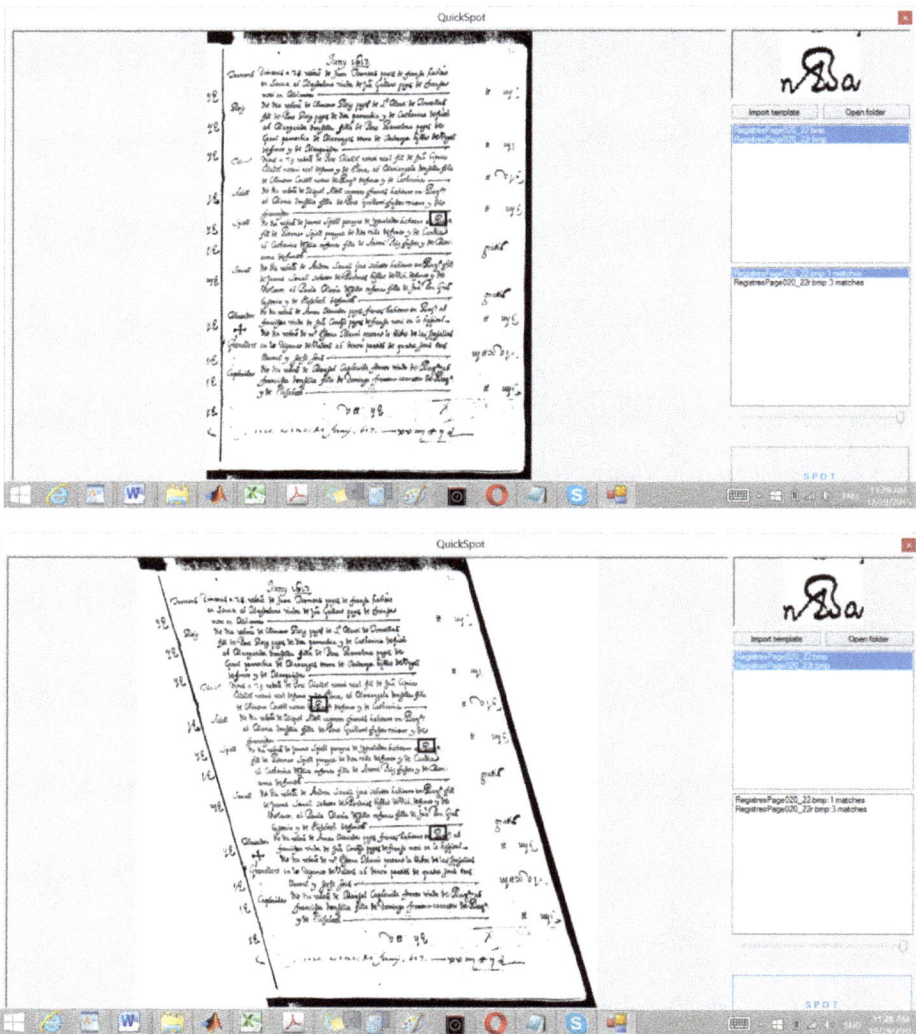

Figure 14. Application of the word-spotting system [25], before (**above**) and after (**below**) slant correction.

4. Conclusions

In this paper, a technique was presented for slant estimation and removal for the entire document page, without requiring line or word segmentation. The proposed technique is recommended for historical document images that include homogenous slant throughout the page. This segmentation-free technique guarantees the minimization of extra noise that could be introduced since the segmentation procedure, in historical document images, where the text is very dense, would leave behind small strokes and other kind of noise (points, lines).

Experimental results were provided for four databases: PrintDB, which contains artificially slanted printed text to prove the accuracy of the technique; TrigraphSlant, a known DB of slant text;

J. Imaging **2018**, *4*, 80

and the BH2M DB and the Washington DB to prove the application's utility on historical documents. The results proved quite satisfactory with a RMSE of less than three for the PrintDB, less than four for the Washington DB, less than five for BH2M, and less than eight for the normal slanted documents of the TrigraphSlant. The improvement in our word-spotting system for difficult historical documents was impressive.

The technique fails if the character main body size detection is not correct. Thus, a good main body size detection algorithm is required. Moreover, the proposed technique is appropriate only if the slant is homogenous throughout the entire document image. However, more slant removal algorithms could be used in combination with the proposed technique. This is among our future plans.

Author Contributions: The mentioned authors, E.K., L.L.-S. and N.V. have contributed to all stages of the work.

Funding: This research received no external funding.

Conflicts of Interest: The authors declare no conflict of interest.

References

1. Parvez, T.M.; Sabri, A.M. Arabic handwriting recognition using structural and syntactic pattern attributes. *Pattern Recognit.* **2013**, *46*, 141–154. [CrossRef]
2. José, A.R.-S.; Perronnin, F. Handwritten word-spotting using hidden Markov models and universal vocabularies. *Pattern Recognit.* **2009**, *42*, 2106–2116.
3. Brink, A.A.; Niels, R.M.J.; van Batenburg, R.A.; van den Heuvel, C.E.; Schomaker, L.R.B. Towards robust writer verification by correcting unnatural slant. *Pattern Recognit. Lett.* **2011**, *32*, 449–457. [CrossRef]
4. Bozinovic, R.; Srihari, S. Off-line cursive script word recognition. *IEEE Trans. Pattern Anal. Mach. Intell.* **1989**, *11*, 68–83. [CrossRef]
5. Kim, G.; Govindaraju, V. A lexicon driven approach to handwritten word recognition for real-time applications. *IEEE Trans. Pattern Anal. Mach. Intell.* **1997**, *19*, 366–379.
6. Shridar, M.; Kimura, F. Handwritten address interpretation using word recognition with and without lexicon. In Proceedings of the IEEE International Conference on Systems, Man and Cybernetics, Vancouver, BC, Canada, 22–25 October 1995; Volume 3, pp. 2341–2346.
7. Papandreou, A.; Gatos, B. Word slant estimation using non-horizontal character parts and core-region information. In Proceedings of the 10th IAPR International Workshop on Document Analysis Systems (DAS 2012), Gold Coast, QLD, Australia, 27–29 March 2012; pp. 307–311.
8. Alessandro, V.; Luettin, J. A new normalization technique for cursive handwritten words. *Pattern Recognit. Lett.* **2001**, *22*, 1043–1050.
9. Kavallieratou, E.; Fakotakis, N.; Kokkinakis, G. Slant estimation algorithm for OCR systems. *Pattern Recognit.* **2001**, *34*, 2515–2522. [CrossRef]
10. Britto, A., Jr.; Sabourin, R.; Lethelier, E.; Bortolozzi, F.; Suen, C. Improvement handwritten numeral string recognition by slant normalization and contextual information. In Proceedings of the 7th International Workshop on Frontiers in Handwriting Recognition, Amsterdam, The Netherlands, 11–13 September 2000; pp. 323–332.
11. Ding, Y.; Kimura, F.; Miyake, Y.; Shridhar, M. Accuracy improvement of slant estimation for handwritten words. In Proceedings of the International Conference on Pattern Recognition, Barcelona, Spain, 3–7 September 2000; Volume 4, pp. 527–530.
12. Ding, Y.; Ohyama, W.; Kimura, F.; Shridhar, M. Local slant estimation for handwritten English words. In Proceedings of the 9th International Workshop on Frontiers in Handwriting Recognition (IWFHR), Kokubunji, Tokyo, Japan, 26–29 October 2004; pp. 328–333.
13. Bertolami, R.; Uchida, S.; Zimmermann, M.; Bunke, H. Non-uniform slant correction for handwritten text line recognition. In Proceedings of the 9th International Conference on Document Analysis and Recognition, Parana, Brazil, 23–26 September 2007; pp. 18–22.
14. Taira, E.; Uchida, S.; Sakoe, H. Non-uniform slant correction for handwritten word recognition. *IEICE Trans. Inf. Syst.* **2004**, *E87-D*, 1247–1253.

15. Seiichi, U.; Eiji, T.; Hiroaki, S. Non uniform slant correction using dynamic programming. In Proceedings of the International Conference on Document Analysis and Recognition (ICDAR), Seattle, WA, USA, 10–13 September 2001.
16. Ziaratban, M.; Faez, K. Non-uniform slant estimation and correction for Farsi/Arabic handwritten words. *Int. J. Doc. Anal. Recognit. (IJDAR)* **2009**, *12*, 249–267. [CrossRef]
17. Kavallieratou, E. A slant removal technique for document page. In Proceedings of the IS&T/SPIE Electronic Imaging, San Francisco, CA, USA, 3–6 February 2013.
18. Available online:. Available online: http://www.unipen.org/trigraphslant.html (accessed on 14 May 2018).
19. Lavrenko, V.; Rath, T.M.; Manmatha, R. Holistic word recognition for handwritten historical documents. In Proceedings of the International Workshop on Document Image Analysis for Libraries (DIAL), Palo Alto, CA, USA, 23–24 January 2004; pp. 278–287.
20. Fernández-Mota, D.; Almazán, J.; Cirera, N.; Fornés, A.; Lladós, J. Bh2m: The barcelona historical, handwritten marriages database. In Proceedings of the 2014 22nd International Conference on Pattern Recognition (ICPR), Stockholm, Sweden, 24–28 August 2014; pp. 256–261.
21. Claasen, T.A.; Mecklenbrauker, W.F. The Wigner distribution: A tool for time-frequency signal analysis. *Phillips J. Res* **1980**, *35*(Pts 1, 2 and 3), 217–250, 276–300, 372–389.
22. Diamantatos, P.; Verras, V.; Kavallieratou, E. Detecting main body size in document images. In Proceedings of the 12th International Conference on Document Analysis and Recognition (ICDAR), Washington, DC, USA, 25–28 August 2013; pp. 1160–1164.
23. Zeeuw, F. Slant Correction Using Histograms. Ph.D. Thesis, Artifical Intelligence, University of Groningen, Groningen, The Netherlands, 2006.
24. Papandreou, A.; Gatos, B. Slant estimation and core-region detection for handwritten Latin words. *Pattern Recognit. Lett.* **2014**, *35*, 16–22.
25. Vasilopoulos;N.;Kavallieratou, E. A classification-free word-spotting system. In Proceedings of the IS&T/SPIE Electronic Imaging. International Society for Optics and Photonics, San Francisco, CA, USA, 3–6 February 2013.

Journal of
Imaging

MDPI

Article

Text/Non-Text Separation from Handwritten Document Images Using LBP Based Features: An Empirical Study

Sourav Ghosh [1,*], Dibyadwati Lahiri [1,*], Showmik Bhowmik [1], Ergina Kavallieratou [2] and Ram Sarkar [1]

[1] Department of Computer Science and Engineering, Jadavpur University, Kolkata, West Bengal 700032, India; showmik.cse@gmail.com (S.B.); raamsarkar@gmail.com (R.S.)
[2] Department of Information and Communication Systems Engineering, University of Aegean, Lesbos 811 00, Greece; kavallieratou@aegean.gr
* Correspondence: souravghosh2197@gmail.com (S.G.); dibyadwati.lahiri@gmail.com (D.L.)

Received: 15 December 2017; Accepted: 6 April 2018; Published: 12 April 2018

Abstract: Isolating non-text components from the text components present in handwritten document images is an important but less explored research area. Addressing this issue, in this paper, we have presented an empirical study on the applicability of various Local Binary Pattern (LBP) based texture features for this problem. This paper also proposes a minor modification in one of the variants of the LBP operator to achieve better performance in the text/non-text classification problem. The feature descriptors are then evaluated on a database, made up of images from 104 handwritten laboratory copies and class notes of various engineering and science branches, using five well-known classifiers. Classification results reflect the effectiveness of LBP-based feature descriptors in text/non-text separation.

Keywords: text/non-text separation; local binary pattern; handwritten document; document image processing; texture-based features

---

1. Introduction

Documents, in the modern day, are required to be stored in digitized form to increase their longevity, portability and security. In order to achieve this purpose, the development of a complete Document Image Processing System (DIPS) has become an utmost need. Along with the other steps, any DIPS needs to identify the texts present in a document image separately from the non-text components like tables, diagrams, graphic designs before processing the text through an Optical Character Recognition (OCR) engine [1–3]. The reason for this is very obvious: OCR engines do not process non-text components. Researchers, to date, have reported many solutions to this problem for printed documents [4–6]. However, the same is not true for regular handwritten documents; a rather limited amount of work is available in this area, to the best of our knowledge, among which two significant ones are [7,8]. In document image processing, researchers mostly use OCR technology in order to work on word and/or character level to provide a viable solution for information content exploitation [9].

In general, handwritten documents are unstructured i.e., in most cases, these documents do not follow any specific layout, unlike the printed documents. Thus, the appearance of text and non-text in handwritten documents is very chaotic. For example, text components often overlap with the non-text components. Furthermore, the building blocks (i.e., characters) of the text in handwritten documents do not follow the standard shape and size usually found in its printed counterpart. One of the key difficulties in the graphics recognition domain is also to work on complex and composite symbol recognition, retrieval and spotting [10]. Thus, the separation of text and non-text in handwritten documents is comparably complex than in printed documents.

Mostly, the reported solutions to the problem of text and non-text separation are done either at the region level [4] or at the connected component (CC) level [5,6]. Methods that implement text/non-text separation at the region level initially perform region segmentation and then classify each segmented region as either a text or graphics region. For classifying the segmented regions, researchers have mostly used texture based features like Gray Level Co-occurrence Matrix (GLCM) [4,11] Run-length based features [12,13] or white tiles based features [14]. However, region segmentation based methods are very sensitive to the segmentation results. Poor segmentation can cause a significant degradation in the classification result. On the other hand, as CC based methods work at the component level, they do not suffer from such a problem. Methods that follow a CC based approach use shape-based features [5,6]. In general, methods reported in this literature for text/non-text separation in handwritten documents have mostly followed the CC based approach [7,8]. It is worth mentioning here that, as historical handwritten manuscripts suffer from various quality degradation issues, techniques like binarization and CC extraction become very error prone. Thus, in some recent articles [15–18], researchers have followed a pixel based approach, which avoids the binarization and CC extraction steps.

From the available research work on this topic, it can be observed that texture features like GLCM (Gray Level Co-occurrence matrix) [4,11], Run-length encoding based features [12,13], Black-and-white transitional matrix based feature [19] have been commonly used by researchers to solve the text/non-text separation problem for printed documents, as well as to separate handwritten and printed text sections in documents [20]. In a recent work [8], a Rotation Invariant Uniform Local Binary Pattern (RIULBP) operator has also been used successfully to separate the text and non-text components in handwritten class-notes. Texture features have proven to be very useful in the field of text/non-text separation due to the fact that text regions and graphics regions in most cases have very different patterns, which can be exploited to differentiate between them. Motivated by this fact, in the present work, we have attempted to evaluate the performance of different Local Binary Pattern (LBP) based texture features to classify the components present in handwritten documents as text or non-text.

The key contributions of our paper are as follows:

1. We have given a detailed analysis of how accurately features extracted by different variants of the LBP operator from handwritten document images help in differentiating text components from non-text ones, which is one of the most challenging research areas in the domain of document image processing. For that purpose, we have considered five variants of LBP [21], namely, the basic LBP [22], improved LBP [23], rotation invariant LBP [22], uniform LBP [22], and rotation invariant and uniform LBP [22].
2. The contents of the dataset, used here for evaluation, have complex text and non-text components as well as variations in terms of scripts, as we have considered both Bangla and English texts. In addition to that, some of the documents have handwritten as well as printed texts.
3. We have also made a minor alteration to robust LBP [24] in order to develop robust and uniform LBP. A method to determine the appropriate threshold value used in this variant of LBP for handwritten documents has also been proposed.

2. Local Binary Patterns and Its Variants

LBP was first introduced by Ojala [25,26], as a computationally simple texture operator in a monochrome texture image.

The generalized definition of LBP, given in [22], used M sample points evenly placed on a circle of radius R with its center positioned at (x_{cen}, y_{cen}). The position (x_p, y_p) of the neighboring point p, where $p \in 0, 1, ..., M - 1$ is given by

$$(x_p, y_p) = (x_{cen} + R\cos(2\pi p/M), y_{cen} - R\sin(2\pi p/M)). \tag{1}$$

Let T be the feature vector representing the local texture:

$$T = func(\ I_{cen}, I_0, I_1, ..., I_M - 1),$$

where I_{cen} and I_p for $p \in \{0, 1, ..., M - 1\}$ represent gray values of the center pixel and the neighboring pixels, respectively. To achieve gray scale invariance, the texture operator is modified to consider the *difference* in intensities of the center pixel and its neighbors:

$$T = func(\ I_0 - I_{cen},\ I_1 - I_{cen},\ ...,\ I_{M-1} - I_{cen}).$$

Furthermore, to achieve a robustness against the scaling of grayscale, only the signs of difference in intensities are considered:

$$T = func(\ f(I_0 - I_{cen}),\ f(I_1 - I_{cen}),\ ...,\ f(I_{M-1} - I_{cen})).$$

Here,

$$f(x) = \begin{cases} 1, & \text{if } x \geq 0, \\ 0, & \text{if } x < 0. \end{cases} \tag{2}$$

Finally, the LBP operator, for the center pixel p^{cen} having intensity value I_{cen} with M neighbors $(X_1, X_2, ..., X_M)$ of intensities $(I_1, I_2, ..., I_M)$, respectively, can be defined below:

$$LBP_{(M,R)}(x_{cen}, y_{cen}) = \sum_{n=1}^{M} f(I_n - I_{cen}) \times 2^{n-1}. \tag{3}$$

LBP creates an M-bit string. Hence, for $M = 8$, the values of $LBP_{(M,R)}(x^{cen}, y^{cen})$ can vary from 0 to 255. The process is depicted in Figure 1.

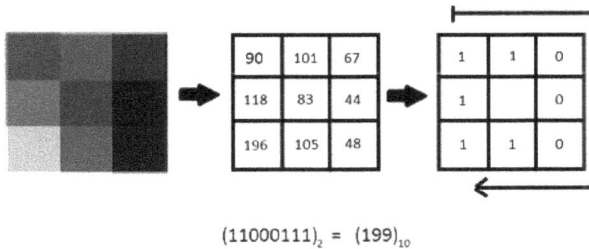

$$(11000111)_2 = (199)_{10}$$

Figure 1. Illustration of LBP value generation for a 3×3 gray image window, where $M = 8$, and radius = 1.

In order to efficiently extract texture features of various complexities, the original LBP operator has been modified to generate a number of variants.

2.1. Improved LBP (ILBP)

The main difference between ILBP [23] and simple LBP is that, instead of the intensity of the center pixel, the mean intensity value of all the pixels, including the center pixel, is used to find the intensity difference during binary pattern computation. In addition to that, while computing ILBP, the intensity of the center pixel is also compared with mean intensity. ILBP is formally defined as follows:

$$ILBP_{(M,R)}(x_{cen}, y_{cen}) = \sum_{n=0}^{M-1} f(I_n - I_{mean}) \times 2^n + f(I_{cen} - I_{mean}) \times 2^M, \tag{4}$$

$$I_{mean} = \frac{(\sum_{n=0}^{M-1} I_n) + I_{cen}}{M+1}.\tag{5}$$

The value of $f(x)$ is computed as given in Equation (2). As ILBP additionally considers the center pixel, thus the value of $ILBP_{(M,R)}(x_{cen}, y_{cen})$ can vary from 1 to 511 (see Figure 2).

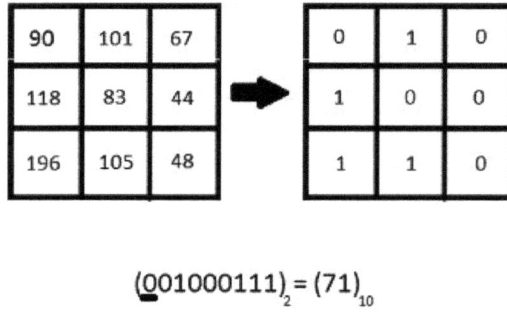

90	101	67
118	83	44
196	105	48

0	1	0
1	0	0
1	1	0

$$(\underline{0}01000111)_2 = (71)_{10}$$

Figure 2. Illustration of ILBP value generation for a 3 × 3 gray image window, where M = 8, Radius = 1 and I_{mean} = 94. The bit representing the center pixel has been underlined in the binary representation of the LBP value.

2.2. Rotation Invariant LBP (RILBP)

RILBP [22] is achieved by bit-wise rotation (circularly) of the binary patterns and then by selecting the minimum value. This is done to cancel out the effect of rotation on a texture, which changes the pattern, although the texture in consideration is essentially the same. RILBP can formally be defined as follows:

$$RILBP_{(M,R)}(x_{cen}, y_{cen}) = min\{Rot(LBP_{(M,R)}, i | 0 \leq i \leq M-1)\}.\tag{6}$$

Here, $Rot(A, i)$ is a function that takes an M-bit binary pattern 'A' and performs i time circular bit-wise right shift operation on 'A'. The entire process is shown in Figure 3.

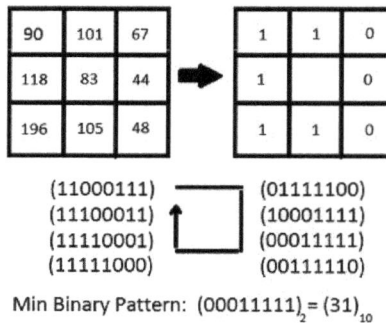

90	101	67
118	83	44
196	105	48

1	1	0
1		0
1	1	0

(11000111)   (01111100)
(11100011)   (10001111)
(11110001)   (00011111)
(11111000)   (00111110)

Min Binary Pattern: $(00011111)_2 = (31)_{10}$

Figure 3. Illustration of RILBP value generation for a 3 × 3 gray image window, where M = 8 and Radius = 1. The binary pattern is rotated clockwise here.

2.3. Uniform LBP (ULBP)

In ULBP [22], the binary patterns with less than or equal to two numbers of zero/one transitions are considered as uniform patterns and the rest are considered as non-uniform patterns. In this variant of LBP, all the non-uniform patterns are marked with the same label, whereas, for uniform patterns, different labels are used, one for each pattern. This is performed because it has been observed that certain patterns constitute a major portion of all texture features. ULBP uses, $M \times (M-1) + 3$ symbols to label the patterns.

2.4. Rotation Invariant and Uniform LBP (RIULBP)

In RIULBP [22], the patterns are chosen such that they are both rotation invariant and uniform. Similar to ULBP, here also all non-uniform rotation invariant patterns are placed in one separate bin. This variant of LBP can be formulated as

$$RIULBP_{(M,R)}(x_{cen}, y_{cen}) = \begin{cases} \sum_{n=1}^{M} f(I_n - I_{cen}), & \text{if } U(RILBP_{(M,R)}(x_{cen}, y_{cen})) \geq 2, \\ M + 1, & \text{otherwise.} \end{cases} \tag{7}$$

Here,

$$U(RILBP_{(M,R)}(x_{cen}, y_{cen})) = \left(\sum_{n=2}^{M} |f(I_n - I_{cen}) - f(I_{n-1} - I_{cen})| \right) + |f(I_M - I_{cen}) - f(I_1 - I_{cen})|. \tag{8}$$

2.5. Robust and Uniform LBP (RULBP)

In the present work, we have proposed a minor but significant modification to Robust LBP (RLBP) [24] to develop RULBP. In RLBP, the argument of the function $f(x)$ i.e., $(I_n - I_{cen})$ (see Equation (2)) is replaced with $(I_n - I_{cen} - th)$, where th acts as a threshold value. This essentially means that the value of I_n has to be greater than the center pixel's gray value I_{cen} by an amount th to produce a 1 (see Figure 4). This descriptor is devised with the idea of increasing the robustness to negligible changes in gray value. Therefore, the RLBP can be formally defined as follows:

$$RLBP_{(M,R)}(x_{cen}, y_{cen}) = \sum_{n=1}^{M} f(I_n - I_{cen} - th) \times 2^{n-1}. \tag{9}$$

In this work, we have given a notion of setting the value of th for text/non-text separation in handwritten documents and also incorporated the idea of 'uniform pattern' in RLBP to develop RULBP.

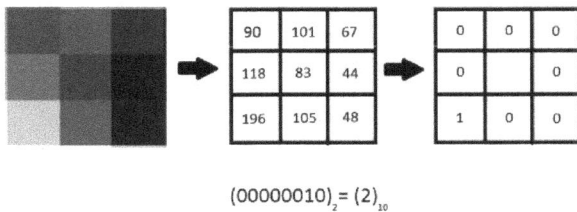

$$(00000010)_2 = (2)_{10}$$

Figure 4. Illustration of RLBP value generation for a 3×3 gray image window, where $M = 8$ and Radius = 1. Here, the value of $th = 90$.

2.5.1. Idea of 'Uniform Pattern'

To prove the effectiveness of LBP for texture classification [22], it has been shown that over 90 percent of the LBPs (generated using a segment of the image) present in a textured surface are 'uniform patterns'. Besides that, as 'uniform patterns' consider a very limited number of 0/1 transition, they can efficiently detect the common microfeatures like corner, edge and spots. Thus, in the present work, we have amalgamated the concept of 'uniform patterns' with RLBP to generate RULBP. The formal definition of RULBP is given below:

$$RULBP_{(M,R)}(x_{cen}, y_{cen}) = \begin{cases} \sum_{n=1}^{M} f(I_n - I_{cen} - th), & \text{if } U(RILBP_{(M,R)}(x_{cen}, y_{cen})) \geq 2, \\ M+1, & \text{otherwise.} \end{cases} \quad (10)$$

The value of $U(RLBP_{(M,R)}(x_{cen}, y_{cen}))$ is computed using Equation (8).

2.5.2. Selecting the Value of *th*

From Equation (9), it can be inferred that the threshold (*th*) in RLBP plays an important role and whose value might be application specific to some extent. Thus, in this work, we have attempted to rationalize it in the context of text/non-text separation in handwritten documents.

Most handwritten documents generally possess a large intensity variation at the stroke level due to the varied nature of writing instruments and non-uniformity in the amount of pressure applied while writing. This non-homogeneity over a single stroke can only be identified if we magnify the image (see the dark and bright patches within the stroke in Figure 5. For example, LBP for the 3×3 segment, marked in red, in Figure 5 is '00010001'. However, the visual perception of a human being considers this as a homogeneous region with all zeros '00000000'. This property of handwritten documents may generate erroneous LBP feature values, which, in turn, fail to distinguish the text components from the non-text ones. In order to solve such problems, a threshold '*th*' has been introduced in LBP to generate RLBP. This threshold ensures that two gray values that are not perceptibly different are not labeled differently. The problem with selecting a value of *th* is that, if the value is extremely large, then the entire region will behave like a homogeneous region with no intensity variation. This is because the binary pattern ,according to Equation (10), will be all zeros for every pixel. Therefore, we need to provide an upper limit, th_{max} , on the value of *th*.

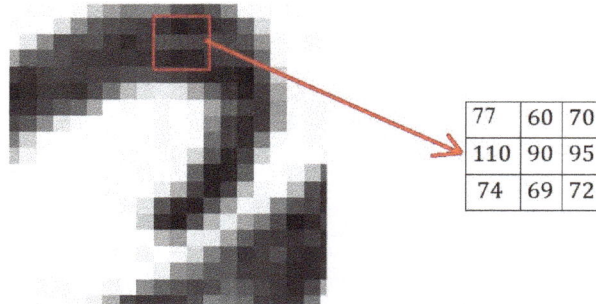

Figure 5. Magnified image of a stroke shows the variation in gray values. A 3×3 matrix shows the intensity values of the gray image segment marked in red.

To address this issue, we have set an upper limit, th_{max} , on the value of *th*. Generally, in a real-life handwritten document image, the intensity of the background pixels reside within a close proximity of the maximum intensity 255. Here, we assume that the intensity of the background pixels will be in a range of $[245, 255]$. Now, for each image, we find the highest gray-scale intensity ($I_{graymax}$) less than 245. We claim that the pixel P having this intensity value has to be a part of some writing stroke. th_{max} has to be such that, if we consider I_{cen} has a value $I_{graymax}$ and a neighboring pixel has a value 245, $f(x)$ as given in Equation (2) for $x = I_n - I_{cen} - th$ gives a value 1. Therefore, $th_{max} = 245 - I_{graymax}$. The value of *th* can be anything between th_{max} and 0. We have performed a weighted average of the threshold values in the range, with the weights increasing for higher values of *th* and found the ideal threshold th_{ideal} to be at around a value of 100. We have taken various threshold values from 5 to 115 and found experimentally that the accuracy of classification is maximum at about a threshold of 100. It is to be noted that we have set this hardcore threshold value after conducting a exhaustive experimentation on the images belonging to our dataset. A change in document images might change the threshold value a bit, but, we foretell that, this assumption would give the researchers a clear hint to set the threshold value for the document images they consider.

3. Method

The input color image is first converted to the grayscale image and then the connected components (CCs) are extracted for feature computation and classification. The entire process is depicted in Figure 6. For CC extraction, first the grayscale image is binarized and the bounding boxes (BBs) of all of the eight-connected components in the binarized image are calculated. Then, using these estimated bounding boxes, CCs from the corresponding grayscale image are extracted. As we are considering real-world handwritten documents, we need to be very careful about the noise present in these documents, which might affect the binarization and BB estimation process. Thus, for effective binarization, a background estimation and separation procedure is followed, prior to the actual binarization, using Otsu's method as given in [27]. During BB estimation from the binarized image, only the CCs having height and width greater than three pixels are considered to avoid noise. After extraction of the CCs from the grayscale image, six different LBP based features are computed. During feature computation, the radius R has been kept constant at 1 (i.e., the number of neighboring pixels $M = 8$). In order to compute a feature vector for each CC, we have generated a normalized histogram of those LBP values. The number of bins used depends on the particular LBP variant considered. Here, we should also point out that the LBP operators have been applied to each and every pixel of a CC, without any discrimination.

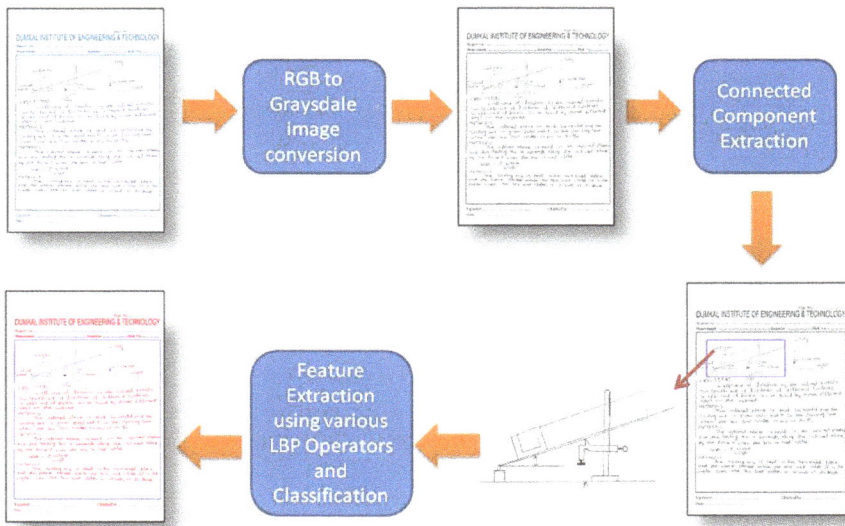

Figure 6. Flowchart of the entire text/non-text separation process.

4. Experimental Setup

Experimental setup for any pattern classification problem requires an annotated dataset, classifiers and a set of evaluation metrics. In this section, the data preparation procedure is described first, followed by details of the parameter values used by the classifiers. At the end, we present the evaluation metrics used in the experiment.

4.1. Database Preparation

It is found that the unavailability of a standard database may be one of the possible reasons for slow progress in some research areas, such as text/non-text separation from handwritten documents in spite of their importance. Keeping this fact in mind, in the present work, a database has been developed that consists of 104 handwritten engineering lab copies and class notes collected from an engineering college. These copies include textual contents along with a varying number of tables, graphic components and some printed texts. All of these lab copies are written by more than 20 students from different engineering and science streams. The age of the writers vary from 18 to 24. Please note that all these copies are written either in English or Bangla. The collected documents are scanned in 300 DPI (Dots per inch) using a flatbed scanner and then these scanned copies are stored as 24 bit 'BMP' files. A sample image from the current database is shown in Figure 7a and the corresponding ground truth image is shown in Figure 7b. In this work, from those 104 handwritten pages, a total of 66,058 CCs are extracted, out of which 25,011 are text components and 41,047 are non-text components.

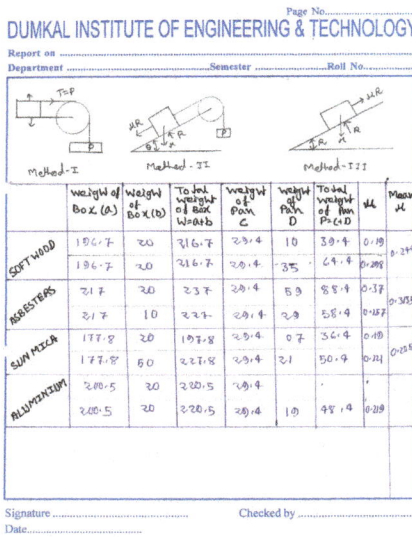

(a)

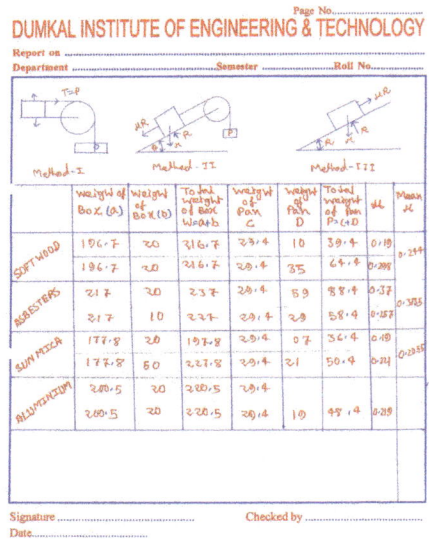

(b)

Figure 7. (a) sample image from our dataset; (b) ground truth of the given image (here, red represents text and blue represents non-text components).

4.2. Classifiers

For classification of the extracted CCs, five well-known classifiers are used in this work, namely, Naïve Bayes (NB), Multi-layer perceptron (MLP), K-nearest neighbor (K-NN), Random forest (RF) and Support Vector Machine (SVM). In the current experimental setup, performances of Simple LBP, ILBP, RILBP, ULBP and RIULBP descriptors with each of the considered classifiers for the present dataset are measured. Then, the classifier that performs better in all or most cases is used to justify the newly hypothesized 'uniform pattern' in RLBP i.e., RULBP. It is to be noted that one of the key parameters of RULBP is *th* whose value is subjective to the document image. Here, different trial runs are performed to choose the optimal value of *th*. In this work, Weka 3 [28], a data mining software (University of Waikato, Hamilton, New Zealand), has been used for classification and visualization purpose. The values of the classifiers' parameters used in the current experiment are given in Table 1.

Table 1. Detail values of the parameters used by the classifiers under consideration.

Classifier	Parameters with Values
NB	• Batch size: 100 • Normal distribution for numeric attributes
MLP	• Learning Rate for the back propagation algorithm: 0.3 • Momentum Rate: 0.2 • Number of epochs to train through: 500 • Learning Rate: 0.3
SMO	• Complexity constant C: 1 • Tolerance Parameter: 1.0×10^{-3} • Epsilon for round-off error: 1.0×10^{-12} • The random number seed: 1
K-NN	• K: 1 • Batch size: 100
RF	• Batch size: 100 • Minimum number of instances per leaf: 1 • Minimum numeric class variance proportion of train variance for split: 1.0×10^{-3} • The maximum depth of the tree: unlimited

4.3. Performance Metrics

The performances of the LBP variants are measured using the following conventional metrics:

$$Recall = \frac{TP}{TP + FN}, \tag{11}$$

$$Precision = \frac{TP}{TP + FP}, \tag{12}$$

$$FM = \frac{2 \times Recall \times Precision}{Recall + Precision}, \tag{13}$$

$$Accuracy = \frac{TP + TN}{Total\ number\ of\ samples} \times 100\%. \tag{14}$$

In Equations (11)–(14), TP, FP, TN and FN represent true positive, false positive, true negative and false negative, respectively. It is to be noted that all the experiments are done using 3-fold cross validation and the final results are computed after taking the average performance of the three folds.

5. Experimental Results

Detailed results for each LBP based feature descriptors except RULBP with each of the five classifiers for the current database are given in Table 2. From Table 2, it can be observed that the RF classifier outperforms others. Thus, classification results for RULBP with different threshold values are computed using RF classifier only. We also see that the RULBP operator gives the best accuracy in classification, among all the LBP variants considered. Detailed results depicting the performance of RULBP for different thresholds are given in Table 3. A pictorial comparison among the performances of different LBP operators using RF classifier is given in Figure 8. Figure 9 shows the image of a document containing text written in Bangla and classified using RULBP, which gives the best result among all LBP variants. In addition to this, a graphical comparison of the performance of various LBP variants are also presented in Figures 10 and 11. The data in Table 2 forms the basis for the points in Figure 10 while the data in Table 3 forms the basis for the points in Figure 11.

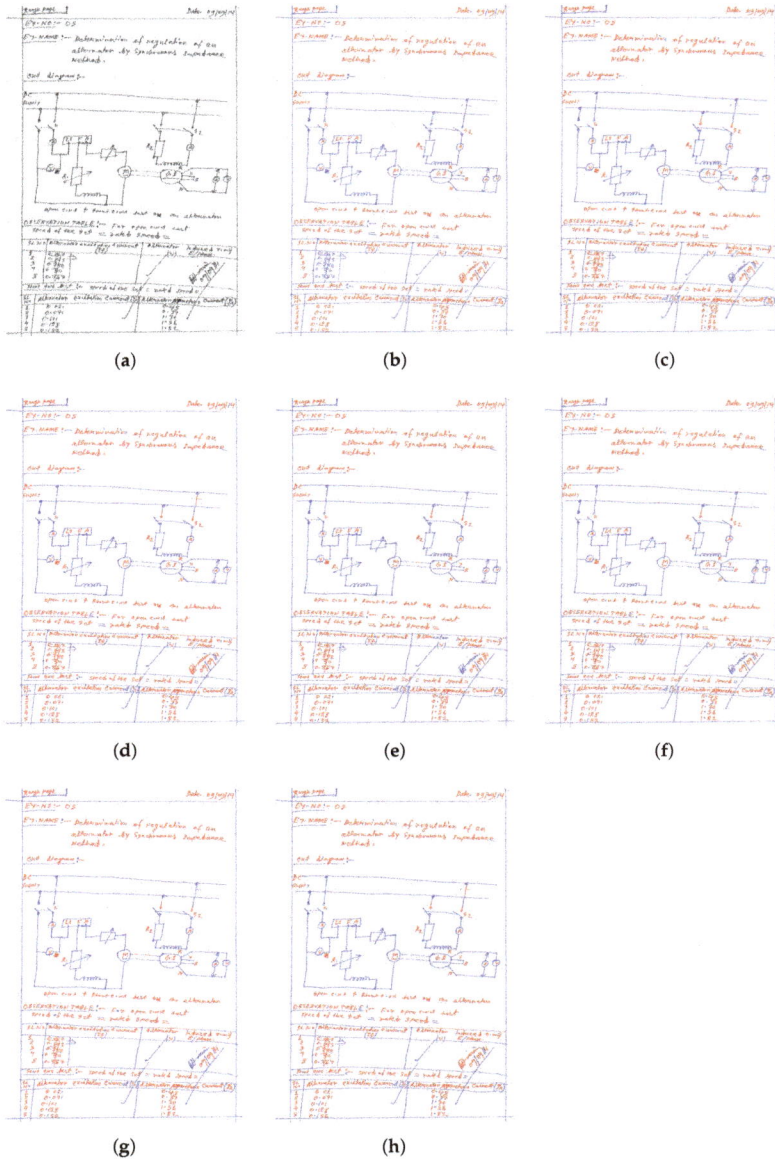

(a)

(b)

(c)

(d)

(e)

(f)

(g)

(h)

Figure 8. Pictorial comparison between the performances of different LBP based features with RF Classifier. Here, (**a**) grayscale image, (**b**) ground truth image, (**c**) result using LBP, (**d**) result using ILBP, (**e**) result using RILBP, (**f**) result using ULBP, (**g**) result using RIULBP, and (**h**) result using RULBP.

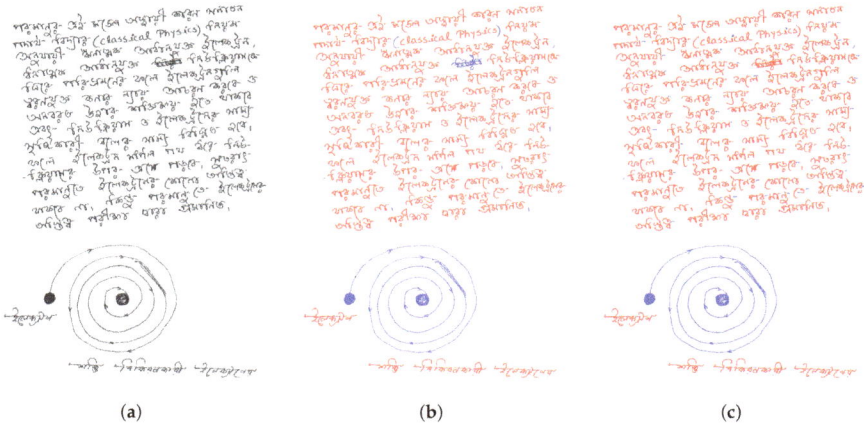

Figure 9. A Bangla handwritten document classified with RF classifier. Here, (**a**) grayscale image, (**b**) ground truth image, (**c**) result using RULBP.

Table 2. Performance measure for text/non-text separation, using various LBP features.

Feature	Feature Dimension	Classifier	Precision	Recall	F-Measure	Accuracy (in %)
LBP	256	NB	0.802	0.771	0.774	77.08
		MLP	0.529	0.54	0.534	54.04
		SMO	0.892	0.889	0.889	88.87
		K-NN	0.856	0.851	0.852	85.12
		RF	0.914	0.913	0.913	**91.33**
ILBP	511	NB	0.82	0.764	0.767	76.41
		MLP	0.386	0.621	0.476	62.13
		SMO	0.862	0.858	0.859	85.84
		K-NN	0.852	0.845	0.847	84.5
		RF	0.913	0.913	0.912	**91.31**
RILBP	36	NB	0.831	0.802	0.805	80.18
		MLP	0.908	0.907	0.905	90.66
		SMO	0.889	0.886	0.887	88.62
		K-NN	0.882	0.882	0.882	88.19
		RF	0.912	0.912	0.912	**91.23**
ULBP	59	NB	0.862	0.857	0.858	85.65
		MLP	0.912	0.912	0.912	91.22
		SMO	0.891	0.888	0.889	88.8
		KNN	0.901	0.901	0.901	90.13
		RF	0.918	0.918	0.917	**91.79**
RIULBP	10	NB	0.859	0.855	0.856	85.52
		MLP	0.907	0.907	0.906	90.71
		SMO	0.888	0.886	0.887	88.58
		KNN	0.886	0.886	0.886	88.62
		RF	0.909	0.909	0.908	**90.9**

Table 3. Classification results using various thresholds for RULBP. The classification accuracy gradually increases and attains a maximum at a *th* of 105 units.

Feature Dimension	Threshold (*th*)	Precision	Recall	F-Measure	Accuracy in %
	5	0.915	0.915	0.914	91.45
	25	0.915	0.915	0.915	91.52
	45	0.916	0.916	0.915	91.61
59	65	0.917	0.917	0.917	91.72
	85	0.919	0.918	0.918	91.84
	105	0.920	0.920	0.919	**91.96**
	115	0.919	0.918	0.918	91.84

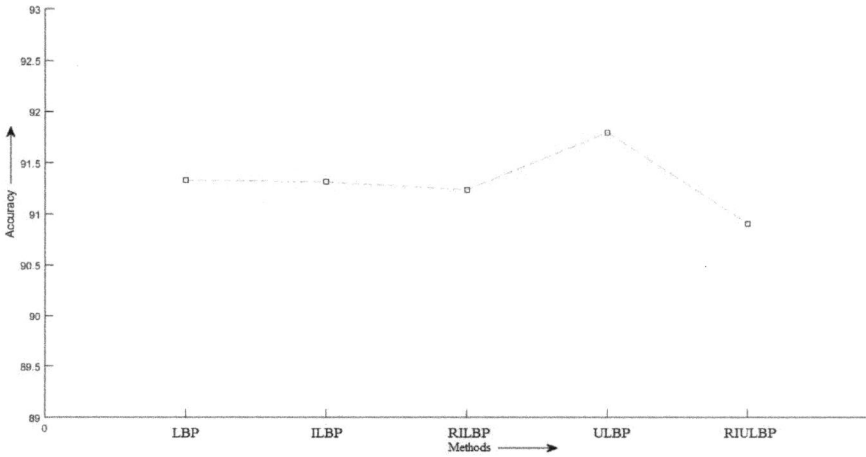

Figure 10. Graphical comparison of the performances of different LBP variants in classifying the texts and non-texts present in handwritten document images.

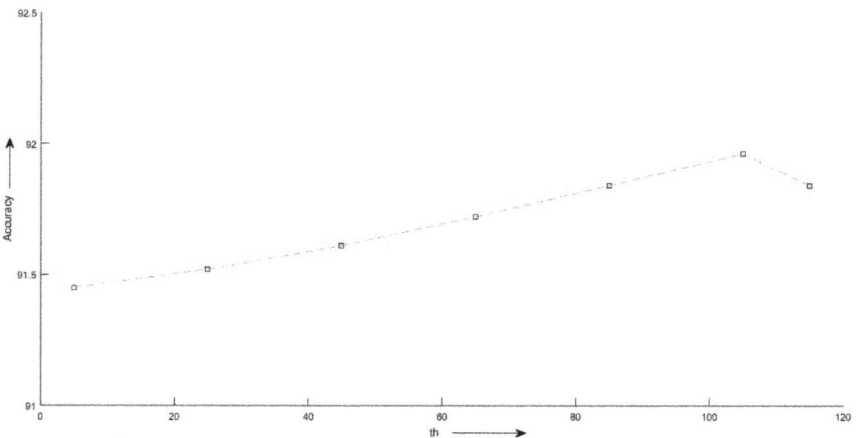

Figure 11. Graphical comparison of the performances of RULBP with different threshold in classifying the texts and non-texts present in handwritten document images.

In the literature, different texture feature descriptors have been used to separate text and non-text regions in printed documents. Here, we have considered two of the recent ones and compared their individual performances on our dataset, with the performance of the RULBP operator. One of the methods uses GLCM as feature descriptor [4] while the other uses Histogram of Oriented Gradients (HOG) [29]. Table 4 gives the accuracy of classification for each of the three feature descriptors using all five classifiers. It can be seen that the RULBP operator outperforms the other feature descriptors in most cases.

Table 4. Performance comparison in terms of recognition accuracy (in %) of GLCM, HOG and RULBP ($th = 105$) on the present dataset for five different classifiers.

Method	NB	MLP	SMO	KNN	RF
RULBP	50.38	**90.78**	**88.62**	**90.20**	**91.96**
GLCM	**77.92**	90.22	87.21	87.70	90.90
HOG	36.22	80.46	72.61	88.89	91.42

6. Conclusions

In the present work, our objective is to validate the utility of LBP based feature descriptors for the classification of text and non-text components present in handwritten documents, in a comprehensive way. We have experimentally shown that RLBP performs better than simple LBP, ILBP, RILBP, ULBP and RIULBP. However, a major issue in using RLBP is the selection of a suitable threshold, which might be domain specific. In the current research attempt, we have selected the optimal value of the threshold on the basis of a few observations, which is also validated through an experiment. We have provided a justification for this selection as well, which we believe would lead to deeper insight into the selection of the threshold used for LBP, especially in the case of handwritten documents. Excluding that, we have proposed a minor modification to RLBP by incorporating the concept of a 'uniform pattern' to develop RULBP, and it has been shown experimentally that RULBP performs better than RLBP. In the future, we would look for the other texture based features along with some other variants of LBP to see their utility in the current context. In the future, we plan to enlarge the database by incorporating various types of document images, which, in turn, would motivate more researchers to do some tangible work. It is worth mentioning here that, in order to analyze the texts written in different scripts, a script recognition module is required [30], since an OCR engine is script specific. Thus, our future plan is to incorporate the same in our model to make it more useful in a multi-script environment. Another area that we will look into is the generalization of the threshold value *th*, so that we may formulate a solid set of procedures that can be useful for any document, instead of using an empirical method to detect the same.

Acknowledgments: This work is partially supported by the Center for Microprocessor Application for Training Education and Research (CMATER) research laboratory of the Computer Science and Engineering Department, Jadavpur University, India, and PURSE-II and UPE-II Jadavpur University projects. Ram Sarkar is partially funded by a DST grant (EMR/2016/007213).

Author Contributions: The first three authors—Sourav Ghosh, Dibyadwati Lahiri and Showmik Bhowmik—have contributed equally to the paper. Ergina Kavallieratou and Ram Sarkar provided essential guidance and corrections at various stages of the work.

Conflicts of Interest: The authors have no conflict of interest.

Abbreviations

The following abbreviations are used in this manuscript:

LBP	Local Binary Pattern
GLCM	Gray-Level Co-Occurrence Matrix
CC	Connected Components
BB	Bounding Box

ILBP	Improved Local Binary Pattern
RILBP	Rotation Invariant Local Binary Pattern
ULBP	Uniform Local Binary Pattern
RIULBP	Rotation Invariant Uniform Local Binary Pattern
RULBP	Robust Uniform Local Binary Pattern
NB	Naive Bayes
MLP	Multilayer Perceptron
SMO	Sequential Minimal Optimization
k-NN	k-Nearest Neighbors
RF	Random Forest

References

1. Santosh, K.C.; Wendling, L. Character recognition based on non-linear multi-projection profiles measure. *Front. Comput. Sci.* **2015**, *9*, 678–690.
2. Santosh, K.C.; Iwata, E. Stroke-Based Cursive Character Recognition. In *Advances in Character Recognition*; InTechOpen: London, UK, 2012; Chapter 10.
3. Santosh, K.C.; Nattee, C.; Lamiroy, B. Relative Positioning Of Stroke-based Clustering: A New Approach To Online Handwritten Devnagari Character Recognition. *Int. J. Image Graph.* **2012**, *12*, 1250016, doi:10.1142/S0219467812500167.
4. Oyedotun, O.K.; Khashman, A. Document segmentation using textural features summarization and feedforward neural network. *Appl. Intell.* **2016**, *45*, 198–212.
5. Le, V.P.; Nayef, N.; Visani, M.; Ogier, J.-M.; De Tran, C. Text and non-text segmentation based on connected component features. In Proceedings of the 2015 13th International Conference on Document Analysis and Recognition (ICDAR), Tunis, Tunisia, 23–26 August 2015; Volume 45, pp. 1096–1100.
6. Tran, T.-A.; Na, I.-S.; Kim, S.-H. Separation of Text and Non-text in Document Layout Analysis using a Recursive Filter. *KSII Trans. Internet Inf. Syst.* **2015**, *9*, 4072–4091.
7. Sarkar, R.; Moulik, S.; Das, N.; Basu, S.; Nasipuri, M.; Kundu, M. Suppression of non-text components in handwritten document images. In Proceedings of the 2011 International Conference on Image Information Processing (ICIIP), Shimla, India, 3–5 November 2011; pp. 1–7.
8. Bhowmik, S.; Sarkar, R.; Nasipuri, M. Text and Non-text Separation in Handwritten Document Images Using Local Binary Pattern Operator. In *International Conference on Intelligent Computing and Communication*; Springer: Singapore, 2017; pp. 507–515.
9. Santosh, K.C. g-DICE: Graph mining-based document information content exploitation. *Int. J. Doc. Anal. Recognit. IJDAR* **2015**, *18*, 337–335, doi:10.1007/s10032-015-0253-z.
10. Santosh, K.C. Complex and Composite Graphical Symbol Recognition and Retrieval: A Quick Review. In *International Conference on Recent Trends in Image Processing and Pattern Recognition*; Springer: Singapore, 2016.
11. Vil'kin, A.M.; Safonov, I.V.; Egorova, M.A. Algorithm for segmentation of documents based on texture features. *Pattern Recognit. Image Anal.* **2013**, *23*, 153–159.
12. Park, H.C.; Ok, S.Y.; Cho, H. Word extraction in text/graphic mixed image using 3-dimensional graph model. In Proceedings of the ICCPOL, Tokushima, Japan, 24–26 March 1999; Volume 99, pp. 171–176.
13. Shih, F.Y.; Chen, S.-S. Adaptive document block segmentation and classification. *IEEE Trans. Syst. Man Cybern. Part B* **1996**, *26*, 797–802.
14. Antonacopoulos, A.; Ritchings, T.R.; De Tran, C. Representation and classification of complex-shaped printed regions using white tiles. In Proceedings of the Third International Conference on Document Analysis and Recognition, Montreal, QC, Canada, 14–16 August 1995; Volume 2, pp. 1132–1135.
15. Pintus, R.; Yang, Y.; Rushmeier, H. ATHENA: Automatic text height extraction for the analysis of text lines in old handwritten manuscripts. *J. Comput. Cult. Herit.* **2015**, *8*, 1.
16. Yang, Y.; Pintus, R.; Gobbetti, E.; Rushmeier, H. Automatic single page-based algorithms for medieval manuscript analysis. *J. Comput. Cult. Herit.* **2017**, *10*, 9.

17. Garz, A.; Sablatnig, R.; Diem, M. Layout analysis for historical manuscripts using sift features Document. In Proceedings of the 2011 International Conference on Document Analysis and Recognition (ICDAR), Beijing, China, 18–21 September 2011.

18. Garz, A.; Sablatnig, R.; Diem, M. Using Local Features for Efficient Layout Analysis of Ancient Manuscripts. In Proceedings of the European Signal Processing Conference, Barcelona, Spain, 29 Auguat–2 September 2011; pp. 1259–1263.

19. Wang, D.; Nihari, S.N. Classification of newspaper image blocks using texture analysis. *Comput. Vis. Graph. Image Process* **1989**, *47*, 327–352.

20. Belaïd, A.; Santosh, K.C.; d'Andecy, V.P. Handwritten and Printed Text Separation in Real Document. *arXiv* **2013**, arXiv:1303.4614.

21. Nanni, L.; Lumini, A.; Brahnam, S. Survey on LBP based texture descriptors for image classification. *Expert Syst. Appl.* **2012**, *39*, 3634–3641.

22. Ojala, T.; Pietikainen, M.; Maenpaa, T. Multiresolution gray-scale and rotation invariant texture classification with local binary patterns. *IEEE Trans. Pattern Anal. Mach. Intell.* **2002**, *24*, 971–987.

23. Jin, H.; Liu, Q.; Lu, H.; Tong, X. Face detection using improved LBP under Bayesian framework. In Proceedings of the Third International Conference on Image and Graphics (ICIG), Hong Kong, China, 18–20 December 2004; pp. 306–309.

24. Heikkila, M.; Pietikainen, M. A texture-based method for modeling the background and detecting moving objects. *IEEE Trans. Pattern Anal. Mach. Intell.* **2006**, *28*, 657–662.

25. Harwood, D.; Ojala, T.; Pietikäinen, M.; Kelman, S.; Davis, L. Texture classification by center-symmetric auto-correlation, using Kullback discrimination of distributions. *Pattern Recognit. Lett.* **1995**, *16*, 1–10.

26. Ojala, T.; Pietikäinen, M.; Harwood, D. A comparative study of texture measures with classification based on featured distributions. *Pattern Recognit.* **1996**, *29*, 51–59.

27. Das, B.; Bhowmik, S.; Saha, A.; Sarkar, R. An Adaptive Foreground-Background Separation Method for Effective Binarization of Document Images. In *Eighth International Conference on Soft Computing and Pattern Recognition*; Springer: Cham, Switzerland, 2016.

28. Witten, I.H.; Frank, E.; Hall, M.A.; Pal, C.J. The WEKA Workbench. Online Appendix for *Data Mining: Practical Machine Learning Tools and Techniques*, 4th ed.; Morgan Kaufmann: Burlington, MA, USA, 2016.

29. Sah, K.A.; Bhowmik, S.; Malakar, S.; Sarkar, R.; Kavallieratou, E.; Vasilopoulos, N. Text and non-text recognition using modified HOG descriptor. In Proceedings of the IEEE Calcutta Conference (CALCON), Kolkata, India, 2–3 December 2017; doi:10.1109/CALCON.2017.8280697

30. Obaidullah, S.M.; Santosh, K.C.; Halder, C.; Das, N.; Roy, K. Automatic Indic script identification from handwritten documents: Page, block, line and word-level approach. *Int. J. Mach. Learn. Cyber* **2017**, 1–20, doi:10.1007/s13042-017-0702-8

Journal of
Imaging

MDPI

Article

A Holistic Technique for an Arabic OCR System

Farhan M. A. Nashwan [1], Mohsen A. A. Rashwan [1], Hassanin M. Al-Barhamtoshy [2], Sherif M. Abdou [3,*] and Abdullah M. Moussa [1]

[1] Department of Electronics and Electrical Communications, Cairo University, Giza 12613, Egypt; far_nash@hotmail.com (F.M.A.N.); mrashwan@rdi-eg.com (M.A.A.R.); a.m.moussa@ieee.org (A.M.M.)
[2] Faculty of Computing and Information Technology, King Abdulaziz University, Jeddah 21589, Saudi Arabia; hassanin@kau.edu.sa
[3] Faculty of Computers & Information, Cairo University, Giza 12613, Egypt
* Correspondence: s.abdou@fci-cu.edu.eg; Tel.: +20-10-2661-4479

Received: 30 October 2017; Accepted: 22 December 2017; Published: 27 December 2017

Abstract: Analytical based approaches in Optical Character Recognition (OCR) systems can endure a significant amount of segmentation errors, especially when dealing with cursive languages such as the Arabic language with frequent overlapping between characters. Holistic based approaches that consider whole words as single units were introduced as an effective approach to avoid such segmentation errors. Still the main challenge for these approaches is their computation complexity, especially when dealing with large vocabulary applications. In this paper, we introduce a computationally efficient, holistic Arabic OCR system. A lexicon reduction approach based on clustering similar shaped words is used to reduce recognition time. Using global word level Discrete Cosine Transform (DCT) based features in combination with local block based features, our proposed approach managed to generalize for new font sizes that were not included in the training data. Evaluation results for the approach using different test sets from modern and historical Arabic books are promising compared with state of art Arabic OCR systems.

Keywords: Arabic OCR systems; holistic OCR approach; holistic OCR features; lexicon reduction

1. Introduction

Cursive scripts recognition has traditionally been handled by two major paradigms: a segmentation-based analytical approach and a word-based holistic approach. In the analytical approach, the input word is treated as a sequence of units (usually characters). Each unit is then individually recognized [1–4]. This approach has several disadvantages. The segmentation of cursive words is a challenging task and any errors in that process will increase the errors in the following recognition step. Also, many of the used fonts for cursive scripts extensively use ligatures where two or more letters are joined as a single glyph, which complicates the character level segmentation. Figure 1 shows some challenging samples of Arabic words.

Figure 1. Some examples of Arabic words that contain ligatures with manually segmented characters.

Cursively written word cannot be recognized without being segmented and cannot be segmented without being recognized [5]. This phenomenon, known as Sayre's paradox, pushes the community to search for more effective solutions to tackle the problem of classification. A more direct and efficient methodology can be provided using holistic recognition [6]. Holistic approach handles the whole word as a unified unit. A global feature vector is calculated for the indivisible input word sample which is then utilized to classify the word against a stored lexicon of words. Holistic recognition is inspired from what is known as the word superiority effect, which states that people have better recognition of letters presented within words as compared to isolated letters and to letters presented within non-words [7]. Holistic paradigms are not only effective, but also have the ability to maintain certain effects which are special to the class under operation such as coarticulation effects [8].

Several previous research efforts have investigated the holistic approach for Arabic cursive script recognition for both printed and handwritten types. Erlandson et al. [9] reported a word-level recognition system for machine-printed Arabic. They used an image-morphological based vector of features such as dots and hamzas, the direction of segments, the junctions and endpoints, direction of cavities, holes, descenders and intra-word gaps. All these features are computed for a query word image in the recognition phase and are matched against a pre-computed database of vectors from an Arabic words lexicon and that system achieved a word recognition rate of 65%. This accuracy was achieved with the integration of a lexicon pruning subsystem that is based on another recognition method that was developed under the same project for a training set of 8436 word images scanned at 300 dpi.

Al-Badr et al. [10] developed an Arabic holistic word recognition system based on a set of shape primitives that are detected with mathematical morphology operations. That system was trained using a single font with three types of documents: ideal (noise-free), synthetically degraded and scanned. The used feature extraction operators were very sensitive to the scanning noise and the degraded low resolution documents. That system achieved a recognition rate of 99.4% for noise-free documents. For synthetically degraded documents, the system accuracy decreased to 95.6% and to 73% for scanned documents. All these evaluations were performed using a limited lexicon that contained 4317 words [10].

Khorsheed and Clocksin [11] presented a technique for recognizing Arabic cursive words from scanned images of text by transforming each word in a certain lexicon into a normalized polar image, and then applied a two-dimensional Fourier transform to that polar image. Each word is represented by a template that includes a set of Fourier's coefficients, and for recognition, the system used a normalized Euclidean distance that measures the distance between the word under test and those templates. That system achieved a recognition rate of 90% for a lexicon size of 145 words and used 1700 word samples for training.

To get better performance, Khorsheed [12] presented a new system based on Hidden Markov Models (HMMs). In that system, each word was represented by a single HMM. The word models were trained using the word sample Fourier's spectrum. The experiments were conducted on four fonts, and the reported results are for Simplified Arabic and Arabic Traditional fonts only. The system achieved a higher recognition rate compared to the template-based recognizer. The highest achieved results for both fonts are: 90% as the first choice and 98% within the top-ten choices.

In a later work, Khorsheed [13] presented a cursive Arabic text recognition system based on HMM. This system was also segmentation-free with an easy-to-extract statistical features vector of length 60 elements, representing three different types of features. This system was trained with a data corpus which includes Arabic text of more than 600 A4-size sheets typewritten in six different computer-generated fonts: Tahoma, Simplified Arabic, Traditional Arabic, Andalus, Naskh and Thuluth. The highest achieved results were 88.7% and 92.4% for Andalus font in mono-model and tri-model, respectively. In another experiment, that system was trained with a multi-font data set that was selected randomly with same sample size from all fonts and tested with a data set consisting of 200 lines from each font, and achieved an accuracy of 95% using the tri-model.

In another effort, Krayem et al. [14] presented a word level recognition system using discrete hidden Markov classifier along with a block based discrete cosine transform. This system was trained by typewritten Arabic words in five fonts with size 14 points and lexicon size of 252 words. Vector quantization was used to map each feature vector to the closest symbol in the codebook. The multiple recognition hypotheses (N-best word lattice) of that system achieved a 97.65% accuracy. Also, the holistic approach was successfully used on the subword level. Nasrollahi and Ebrahimi [15] presented an approach to offline OCR for printed Persian subwords using wavelet packet transform. The proposed technique extracted font invariant and size invariant features from different subwords of four fonts and three sizes and compressed them using Principal Component Analysis (PCA). When tested on a subset of 2000 words of printed Persian text documents, that system achieved an accuracy of 97.9%.

In a later work [16], Slimane et al. organized the ICDAR2013 competition on multi-font and multi-size digitally represented Arabic text. The main characteristic of the winner system, Siemens system submitted by Marc-Peter Schambach et al., was the using of a three hidden layers neural network, that transforms a two-dimensional pixel plane into a sequence of class probabilities. the system have been applied on a subset of the APTI dataset [17] and managed to achieve an accuracy over 99%.

While the holistic approach avoids the challenging segmentation task of Arabic cursive scripts, it still has another challenge of dealing with large lexicon size of Arabic words. As the number of words in the lexicon grows, the recognition task becomes more computationally expensive. Most of the previously proposed holistic based Arabic OCR systems tested with small size vocabularies, but this is not practical for Arabic as a morphologically rich language with a huge vocabulary size.

In this paper, we propose a computationally efficient holistic Arabic OCR system for a large vocabulary size. For the sake of a practical approach, a lexicon reduction technique based on clustering the similar shape words is used to minimize the word recognition time. The proposed system utilizes a hybrid of several holistic features that combine global word level DCT-based features and local block based features. Using these types of features, the system manages to achieve Omni-font performance with font and size independence. Also, the presented system has a flexible architecture for integrating language modelling constraints by using a second rescoring pass for the top n-best word hypotheses. This rescoring operation provided a significant enhancement in the recognition accuracy of the system. The rest of the paper is organized as follows. Section 2 includes a description for the proposed holistic OCR system. The holistic DCT features used are described in Section 3. The developed lexicon reduction technique is illustrated in Section 4. Section 5 describes the language rescoring process used by the system. Section 6 presents system evaluation results and performance comparison with state of art commercial Arabic OCR systems. The final conclusions and prospects for future work are included in Section 7.

2. System Description

The developed holistic OCR system consists of two modules. The first one is the training module where the holistic features are extracted from the training set of the word images. The extracted features are used to build the set of clusters of similar word shapes. The generated words' clusters and their extracted features represent the knowledge base that is used in the recognition phase. The second module is the recognition module. In that module, after applying the preprocessing operations on the input image, the detected text blocks are segmented into lines and words. The features are extracted for each word image then the word cluster or best-n clusters, that have the minimum Euclidean distance with the test image vector, are assigned. The generated word list from the selected cluster is used to construct a word lattice for the possible recognition hypotheses of the whole line. This word lattice is rescored using n-gram language model to get the best recognition hypothesis. Figure 2 shows the block diagram of the proposed holistic OCR system.

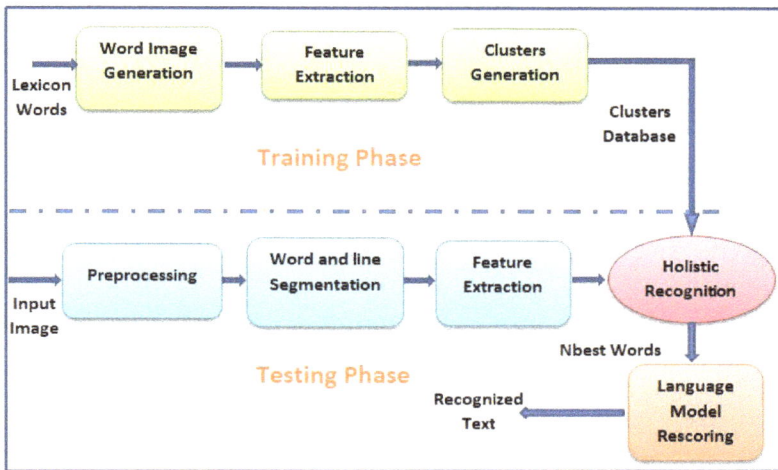

Figure 2. Block Diagram of the Holistic OCR System.

3. Feature Extraction

The main concept of the proposed algorithm is based on the property that the DCT transform compressed image is a decomposition vector which can uniquely represent the input image to be correctly reconstructed later at a decompression stage. In this work, the first 100–200 2D-DCT coefficients are used as word features that provide good approximation about the word image information. In our system, three features were experimented. Those features are: Discrete Cosine Transforms (DCT), Discrete Cosine Transforms 4-Blocks (DCT_4B), and a feature which is a combination of DCT and DCT_4B.

3.1. Discrete Cosine Transform (DCT)

The DCT features in our system are extracted via two dimensional DCT. The two dimensional DCT of an $M \times N$ image $f(x, y)$ is defined as follows:

$$T(u,v) = \frac{1}{\sqrt{MN}} C_u C_v \sum_{x=0}^{M-1} \sum_{y=0}^{N-1} f(x,y) cos(\frac{(2x+1)u\pi}{2M}) cos(\frac{(2y+1)v\pi}{2N}) \tag{1}$$

where $0 \leq x \leq M-1, 0 \leq y \leq N-1$

$$Cu = \begin{cases} \frac{1}{\sqrt{M}}, x = 0 \\ \frac{2}{\sqrt{M}}, 1 \leq x \leq M-1 \end{cases}, Cv = \begin{cases} \frac{1}{\sqrt{N}}, y = 0 \\ \frac{2}{\sqrt{N}}, 1 \leq y \leq N-1 \end{cases}.$$

After applying DCT to the whole word image, the features are extracted in a vector form by using the most significant DCT coefficients. The steps involved in DCT feature extraction as shown in Figure 3 are:

1. Apply the DCT to the whole word image.
2. Perform zigzag operation on the DCT coefficients I_{dct}.

The zigzag matrix I_z is a row vector matrix containing high frequency coefficients in its first N values that contain most word information. This forms features vector f_{dct} for each word.

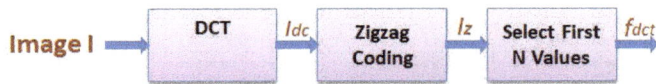

Figure 3. DCT based Feature Extraction.

3.2. Discrete Cosine Transform 4-Blocks (DCT_4B)

In this feature set, firstly we find the Centre of Gravity (COG) of image and make it as the starting point; in order to calculate the centre of gravity, the horizontal and vertical centre must be determined by the following equations:

$$C_x = \frac{M_{(1,0)}}{M_{(0,0)}} \tag{2}$$

$$C_y = \frac{M_{(0,1)}}{M_{(0,0)}} \tag{3}$$

where C_x is the horizontal centre and C_y the vertical centre of gravity and $M_{(p,q)}$ the geometrical moments of rank $p + q$:

$$M_{pq} = \sum_x \sum_y (\frac{x}{width})^p (\frac{y}{height})^q f(x,y). \tag{4}$$

The x and y determine the image word pixels. The division of x and y by the width and the height of the image, respectively, causes the geometrical moments to be normalized and be invariant to the size of the word [18]. This method uses features of COG and DCT at the same time, the first one as an auxiliary feature to divide the image into four parts and apply the second feature DCT on each part as a whole.

This feature set is extracted and implemented as follows:

1. Calculate the COG of the word image and make it as a starting point as explained in Equations (1)–(4).
2. Use the vertical and horizontal COG to divide the word image into four regions.
3. Apply the DCT to each part of the word image.
4. Perform zigzag operation on the DCT coefficients of each image part to get the first $N/4$ values that contain most word information on that word part.
5. Repeat Steps 3 and 4 sequentially for all the word parts, and then combine them together to form the feature vector of the word image.

3.3. Hybrid DCT and DCT_4B (DCT + DCT_4B)

This feature combines the two features DCT and DCT_4B.

4. Lexical Reduction and Clustering

To reduce the computation time for searching the whole lexicon in the recognition phase, the similar shape words are clustered together. The word search is performed in two steps. In the first one, the word cluster or the nearest n-clusters are determined then the best matching word inside that cluster are selected as the recognition output. For words clustering, we used the LBG algorithm [19] to cluster the words in each group depending on closeness of the word shapes from the point of view of the used features. For the clustering process, we used the same DCT and DCT_4B features that we use for the word recognition phase.

To measure the accuracy of the clustering step, and also lexical reduction, we used a clustering accuracy measure which counts the number of times the test word exists within the selected cluster/clusters per the tested words. For a vocabulary size of around 356,000 words of Simplified Arabic font (14 pt.), we tested the clustering accuracy using a test set of 3465 words and a codebook size of 1024. Table 1 shows the clustering accuracy rate of the tested words using the three implemented features when using varying number of clusters from one to 10.

Table 1. Clustering accuracy rate (percent) of Simplified Arabic font vs. number of clusters using three features (codebook size = 1024, lexicon $\simeq 356,000$).

Features	Number of Coefficients	Top1	Top2	Top3	Top4	Top5	Top6	Top7	Top8	Top9	Top10
DCT	160	84.7	96.0	98.4	98.9	99.1	99.4	99.5	99.6	99.7	99.7
DCT_4B	160	78.5	91.9	96.2	97.8	98.7	99.2	99.4	99.6	99.7	99.7
DCT+ DCT_4B	200	86.1	96.2	98.5	99.1	99.3	99.6	99.7	99.8	99.8	99.8

The results of Table 1 show that the DCT+DCT_4B feature is better than the other two. This hybrid feature benefited from the local and global feature of the DCT, so it achieved good results, especially in the noisy data. Figure 4 shows the relation between codebook size and clustering accuracy rate.

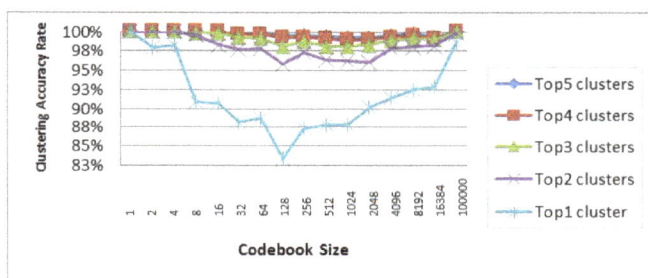

Figure 4. Clustering accuracy rate of Simplified Arabic font vs. codebook size number using DCT+ DCT_4B feature for different top clusters.

As shown in Figure 4, the clustering accuracy rate increases when using larger number of top-n clusters which is a logical consequence. When using a small number of clusters, each cluster contains large number of words which raises the possibility of finding the tested word within one of these clusters. When the number of clusters increase, the number of words in each cluster decrease, which reduces the clustering accuracy rate but at the same time the words within each cluster becomes more similar, which starts again to raise the clustering accuracy rate even up to the highest level when each cluster contains only one word.

5. Language Rescoring

To enhance the recognition accuracy, the top-hypotheses from the holistic recognition results are rescored using a language model. In our system, we used a 4-gram language model that was trained from a Giga-word Arabic training database [20]. The top n-hypotheses for each word are combined in a lattice format as shown in Figure 5, then we used the A* search technique to search for the best score path in that lattice using the 4-gram language model to select the best matching sentence according to the Arabic language constraints [21].

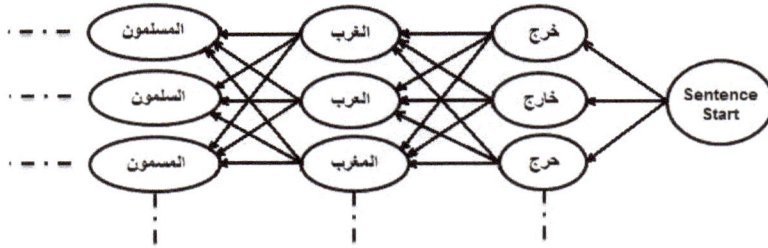

Figure 5. An example of a rescoring lattice.

6. Experiment Results

To train the proposed holistic Arabic OCR system, we used a lexicon of around 356,000 words selected from the news domain with high coverage for the Arabic Language. Using this lexicon, we generated a database of images for three fonts: Simplified Arabic, Traditional Arabic and Arabic Transparent, in 300 dpi with four different sizes.

To test the system, we used three different test datasets that represent different degrees of challenges:

1. Laser scanned text data set: This data set is composed of 1152 single words taken from newspaper articles and printed in three fonts and four different sizes in two types of qualities: clean and first copy.
2. Recent computerized books data set: A data set composed of 10 scanned pages from different recent computerized books that contain 2730 words.
3. Old un-computerized books: This data set consists of 10 scanned pages contain 2276 words from old books that are typewritten with not well known fonts.

Figure 6 illustrates some examples of the scanned images. In the first experiment, we evaluated our system using the laser scanned data set. Initially, we evaluated the system on a single font. The system was trained on a single font with single size but was tested on the same font with different sizes. We didn't use the language model with this dataset as it consists of single words. Table 2 illustrates the Word Recognition Rate (WRR) results for this experiment.

Figure 6. Some samples of the scanned images.

Table 2. Single font WRR (percent) for multi size fonts (12, 14, 16 and 20).

Training Font (Size)	Testing Font Name	Testing Font Size	Top 1	Top 2	Top 3	Top 4	Top 5
Simplified Arabic (14)	Simplified Arabic	12	98.26	99.48	99.91	99.96	99.96
		14	98.22	99.87	100	100	100
		16	98.39	99.44	99.78	99.83	99.83
		20	99	99.87	99.96	99.96	99.96
		Average	98.44	99.66	99.91	99.93	99.93
Arabic Transparent (14)	Arabic Transparent	12	98.13	99.61	99.96	100	100
		14	98.48	99.78	100	100	100
		16	98	99.74	99.96	100	100
		20	98.79	99.83	99.96	100	100
		Average	98.33	99.74	99.97	100	100
Traditional Arabic (16)	Traditional Arabic	12	97.57	99.65	99.83	99.96	99.96
		14	97.61	99.91	99.96	100	100
		16	97.39	99.43	99.78	99.83	99.83
		20	96.57	99.22	99.83	99.83	99.87
		Average	97.33	99.58	99.85	99.90	99.91

From the results in Table 2, we can see that the proposed system achieved very high accuracy and managed to generalize for new font sizes that were not included in the training data with best WRR of 98.44% for Simplified Arabic font and the lowest WRR of 97.33% for Traditional Arabic font. When considered the multiple recognition hypotheses, the top-5 WRR was almost 100%.

In the second experiment, the system was evaluated as omnifont by including several fonts and sizes from the laser scanned training data set. Table 3 includes the results for that evaluation.

Table 3. Multi-Fonts WRR (percent) for multi size fonts (12, 14, 16 and 20).

Training Font (Size)	Testing Font Name	Testing Font Size	Top 1	Top 2	Top 3	Top 4	Top 5
Simplified Arabic (14)	Simplified Arabic	12	98.26	99.61	100	100	100
Arabic Transparent (14)		14	98.13	99.87	100	100	100
Traditional Arabic (16)		16	98.35	99.44	99.78	99.83	99.83
		20	98.96	99.91	100	100	100
		Average	98.39	99.7	99.93	99.95	99.95
Simplified Arabic (14)	Arabic Transparent	12	98.35	99.65	99.96	100	100
Arabic Transparent (14)		14	98.96	99.87	100	100	100
Traditional Arabic (16)		16	98.74	99.83	100	100	100
		20	99.05	99.87	100	100	100
		Average	98.79	99.81	99.99	100	100
Simplified Arabic (14)	Traditional Arabic	12	97.57	99.65	99.83	99.96	99.96
Arabic Transparent (14)		14	97.61	99.91	99.96	100	100
Traditional Arabic (16)		16	97.39	99.43	99.78	99.83	99.83
		20	96.4	99.09	99.83	99.83	99.87
		Average	97.29	99.55	99.85	99.9	99.91

As we can see in Table 3, the proposed system managed to achieve for the multi-font and multi-size task almost the same WRR as the single font one. This result shows that the presented system can provide an omnifont performance.

In the third experiment, we evaluated our system using the recent and old books data sets. Table 4 shows the results of that evaluation.

Table 4. Multi-Fonts WRR (percent) for books.

Books Type	Top 1	Top 5	Top 10	Top 20
Recent Computerized	77.33	86.37	87.69	89.19
Old Uncomputerized	47.76	60.68	65.77	69.24

From the results in Table 4 we can see that our Arabic holistic OCR system achieved 77.3% WRR for recent books and 47.8% WRR for old books. Considering the top-10 hypotheses, the WRR for recent books increased to 87.7% and for old books increased to 65.7%. When considering top-20 hypotheses, the WRR increased to 89% and 69% for recent and old books, respectively. A data analysis for the recognition errors of the books data sets revealed several reasons that contributed to the reduction of the WRR. We found that this data sets included high Out Of Vocabulary (OOV) rate of around 6% for recent books and 7% for old books. It is known that the effect of the OOV is accumulative which means a single OOV word can result in recognition errors for more than one of its neighboring words. Another phenomenon that we noticed in these data set is the high rate of using the Kashida character, which was 4% for recent books and 6% for old books. The Kashida character resulted in altering the shapes of some characters which caused some word recognition errors. Also, we noticed that some fonts of the old books had large differences from the fonts used in training the system such as the Anglo-font which resulted in very low WRR for some pages.

When we applied a 4-gram language model rescoring for the books data sets using the top-10 hypothesis, we achieved 83% WRR for the recent books set and 53% WRR for the old books set. We got an absolute gain of 6% in WRR for both of the recent and old books data sets. This result show that a high percentage of the system recognition errors can be corrected using the top-n hypotheses and a language model.

In the fourth evaluation, we compared the performance of the proposed system with three commercial Arabic OCR systems, Sakhr, ABBYY and NovoDynamics, which represent the best performing Arabic OCR packages currently available. Table 5 shows these comparative results.

Table 5. Recognition rate (percent) of recent computerized and uncomputerized books. ED stands for Euclidean distance.

Books Type	NovoDynamics	Sakhr	ABBYY	Holistic (Using Top 15 with LM) Squared ED/Absolute ED
Computerized	88.45	82.17	54.33	82.97/84.76
Uncomputerized	78.15	54.94	29.22	53.21/58.04

The results in Table 5 show that, while using squared Euclidean distance as the distance measure, our system managed to achieve better performance than two systems, ABBYY and Sakhr, for the computerized books data set and achieved better performance than the ABBYY system for the uncomputerized books data set. When we used the absolute Euclidean distance, the recognition rate increased from 82.97% to 84.76% for the computerized books set and from 53.21% to 58.04% for the uncomputerized books set, and the proposed system outperformed Sakhr and ABBYY systems for both of the two datasets, although the NovoDynamics system outperfoms the proposed one. Our system is still much faster, as we will see in the next section.

As heavy computation is one of the main drawbacks for the holistic approach, we evaluated the runtime speed of the presented system. Table 6 shows the processing times of the proposed system before and after lexical reduction versus the number of selected word clusters. These experiments were run on Core i7 2.8 GHz machine with single thread execution.

Table 6. Processing time of word search and LM vs. words candidates.

Selected Words	Processing Time (s/word)
No Reduction	0.545
Lexicon Reduction (1 cluster)	0.0005
Lexicon Reduction (5 clusters)	0.0026
Lexicon Reduction (10 clusters)	0.0051

We can see from the displayed results in Table 6 that the computation cost of our developed holistic system is very practical. With lexical reduction, we managed to reduce the run time by a factor of 1000 and a one page with average number of 250 words can be computed in average time of 1.2 s compared to 1 s/page for Sakhr system, 2.3 s/page for NovoDynamics and 3.5 s/page for ABBY system.

7. Conclusions and Future Work

The holistic approaches provide effective solutions for the challenges of cursive scripts recognition such as Arabic OCR. The main drawback of such approaches is its complexity and heavy computation requirement especially for large vocabulary tasks. In this paper, we introduced a holistic Arabic OCR approach that is computationally efficient. A lexicon reduction technique based on clustering the similar shape words is utilized to reduce the word recognition time. The presented system makes use of a hybrid of several holistic features that combine global word level DCT based features and local block based features. Using this type of features, the system achieved Omni-font performance with size and font independence. Also, the suggested system has a flexible architecture to integrate language modelling constraints by using a second rescoring pass for the top n-best word hypotheses.

The proposed system has been tested using different sets of 1152 words with three different fonts and four font sizes and has achieved 99.3% WRR. It also has been tested using sets of 2730 words of recent computerized book's text and has attained more than about 84.8% WRR. Results of the holistic proposed system have been compared with known commercial Arabic OCR systems provided by the largest international and local companies, and the results were promising. In future work, we will investigate other holistic features like Wavelet Transform, Zernike Transform, Hough Transform and loci. Also, we will investigate other lexicon reduction techniques that benefit from linguistic information.

Acknowledgments: The teamwork of the "Arabic Printed OCR System" project was funded and supported by the NSTIP strategic technologies program in the Kingdom of Saudi Arabia- project no. (11-INF-1997-03). In addition, the authors acknowledge with thanks Science and Technology Unit, King Abdulaziz University for technical support.

Author Contributions: Farhan M. A. Nashwan and Mohsen A. A. Rashwan conceived and designed the experiments; Farhan M. A. Nashwan performed the experiments; Sherif M. Abdou and Mohsen A. A. Rashwan analyzed the data; Hassanin M. Al-Barhamtoshy contributed materials and analysis tools; Farhan M. A. Nashwan and Sherif M. Abdou wrote the paper; Abdullah M. Moussa substantively revised the paper.

Conflicts of Interest: The authors declare no conflict of interest. The funding sponsors had no role in the design of the study; in the collection, analyses, or interpretation of data; in the writing of the manuscript, and in the decision to publish the results.

References

1. Khorsheed, M.; Al-Omari, H. Recognizing cursive Arabic text: Using statistical features and interconnected mono-HMMs. In Proceedings of the 4th International Congress on Image and Signal Processing, Shanghai, China, 15–17 October 2011; Volume 5, pp. 1540–1543.
2. Abd, M.A.; Al Rubeaai, S.; Paschos, G. Hybrid features for an Arabic word recognition system. *Comput. Technol. Appl.* **2012** 3, 685–691.
3. Amara, M.; Zidi, K.; Ghedira, K. An efficient and flexible Knowledge-based Arabic text segmentation approach. *Int. J. Comput. Sci. Inf. Secur.* **2017**, *15*, 25–35.
4. Radwan, M.A.; Khalil, M.I.; Abbas, H.M. Neural networks pipeline for offline machine printed Arabic OCR. *Neural Process. Lett.* **2017**, 1–19, doi:10.1007/s11063-017-9727-y.
5. El rube', I.A.; El Sonni, M.T.; Saleh, S.S. Printed Arabic sub-word recognition using moments. *World Acad. Sci. Eng. Technol.* **2010**, *4*, 610–613.
6. Madhvanath S.; Govindaraju, V. The role of holistic paradigms in handwritten word recognition. *IEEE Trans. Pattern Anal. Mach. Intell.* **2001**, *23*, 149–164.

7. Falikman, M.V. Word superiority effects across the varieties of attention. *J. Rus. East Eur. Psychol* **2011**, *49*, 45–61.
8. Srimany, A.; Chowdhuri, S.D.; Bhattacharya, U.; Parui, S.K. Holistic recognition of online handwritten words based on an ensemble of svm classifiers. In Proceedings of the 11th IAPR International Workshop on Document Analysis Systems (DAS), Tours, France, 7–10 April 2014; pp. 86–90.
9. Erlandson, E.J.; Trenkle, J.M.; Vogt, R.C. Word-level recognition of multifont Arabic text using a feature vector matching approach. In Proceedings of the International Society for Optics and Photonics , San Jose, CA, USA, 7 March 1996; Volume 2660, pp. 147–166.
10. Al-Badr, B.; Haralick, R.M. A segmentation-free approach to text recognition with application to Arabic text. *Int. J. Doc. Anal. Recognit.* **1998**, *1*, 147–166.
11. Khorsheed, M.S.; Clocksin, W.F. Multi-font Arabic word recognition using spectral features. In Proceedings of the 15th International Conference on Pattern Recognition (ICPR-2000), Barcelona, Spain, 3–7 September 2000; pp. 543–546.
12. Khorsheed, M. A lexicon based system with multiple hmms to recognise typewritten and handwritten Arabic words. In Proceedings of the 17th National Computer Conference, Madinah, Saudi Arabia, 5–8 April 2004; pp. 613–621.
13. Khorsheed, M.S. Offline recognition of omnifont Arabic text using the HMM ToolKit (HTK). *Pattern Recognit. Lett.* **2007**, *28*, 1563–1571.
14. Krayem, A.; Sherkat, N.; Evett, L.; Osman, T. Holistic Arabic whole word recognition using HMM and block-based DCT. In Proceedings of the 12th International Conference on Document Analysis and Recognition (DAR), Washington, DC, USA, 25–28 August 2013; pp. 1120–1124.
15. Nasrollahi, S.; Ebrahimi, A. Printed persian subword recognition using wavelet packet descriptors. *J. Eng.* **2013**, *2013*, doi:10.1155/2013/465469.
16. Slimane, F.; Kanoun, S.; El Abed, H.; Alimi, A.M.; Ingold, R.; Hennebert, J. ICDAR2013 competition on multi-font and multi-size digitally represented arabic text. In Proceedings of the 12th International Conference on Document Analysis and Recognition (ICDAR), Washington, DC, USA, 25–28 August 2013; pp. 1433–1437.
17. Slimane, F.; Ingold, R.; Kanoun, S.; Alimi, A.M.; Hennebert, J. A new arabic printed text image database and evaluation protocols. In Proceedings of the 10th International Conference on Document Analysis and Recognition (ICDAR), Barcelona, Spain, 26–29 July 2009; pp. 946–950.
18. Zagoris, K.; Ergina, K.; Papamarkos, N. A document image retrieval system. *Eng. Appl. Artif. Intell.* **2010**, *23*, 1563–1571.
19. Linde, Y.; Buzo, A.; Gray, R. An algorithm for vector quantizer design. *IEEE Trans. Commun.* **1980**, *28*, 84–95.
20. Arabic Gigaword Fifth Edition. Available online: https://catalog.ldc.upenn.edu/LDC2003T12 (accessed on 1 March 2014).
21. Rashwan, M.A.A.; Al-Badrashiny, M.A.; Attia, M.; Abdou, S.M.; Rafea, A. A stochastic Arabic diacritizer based on a hybrid of factorized and un-factorized textual features. *IEEE Trans. Audio Speech Lang. Proc.* **2011**, *19*, 166–175.

Journal of **Imaging**

MDPI

Article

Efficient Query Specific DTW Distance for Document Retrieval with Unlimited Vocabulary

Gattigorla Nagendar [1,*], Viresh Ranjan [2], Gaurav Harit [3] and C. V. Jawahar [1]

[1] Center for Visual Information Technology, IIIT Hyderabad, Hyderabad 500 032, India; jawahar@iiit.ac.in
[2] CSE Department, Stony Brook University, Stony Brook, NY 11794, USA; viresh.ranjan@stonybrook.edu
[3] Department of Computer Science and Engineering, IIT Jodhpur, Jodhpur 342037, India; gharit@iitj.ac.in
*   Correspondence: nagendar.g@research.iiit.ac.in

Received: 31 October 2017; Accepted: 2 February 2018; Published: 8 February 2018

Abstract: In this paper, we improve the performance of the recently proposed Direct Query Classifier (DQC). The (DQC) is a classifier based retrieval method and in general, such methods have been shown to be superior to the OCR-based solutions for performing retrieval in many practical document image datasets. In (DQC), the classifiers are trained for a set of frequent queries and seamlessly extended for the rare and arbitrary queries. This extends the classifier based retrieval paradigm to an unlimited number of classes (words) present in a language. The (DQC) requires indexing cut-portions (n-grams) of the word image and DTW distance has been used for indexing. However, DTW is computationally slow and therefore limits the performance of the (DQC). We introduce query specific DTW distance, which enables effective computation of global principal alignments for novel queries. Since the proposed query specific DTW distance is a linear approximation of the DTW distance, it enhances the performance of the (DQC). Unlike previous approaches, the proposed query specific DTW distance uses both the class mean vectors and the query information for computing the global principal alignments for the query. Since the proposed method computes the global principal alignments using n-grams, it works well for both frequent and rare queries. We also use query expansion (QE) to further improve the performance of our query specific DTW. This also allows us to seamlessly adapt our solution to new fonts, styles and collections. We have demonstrated the utility of the proposed technique over 3 different datasets. The proposed query specific DTW performs well compared to the previous DTW approximations.

Keywords: DTW distance; query classifiers; word spotting; indexing; retrieval

1. Introduction

Retrieving relevant documents (pages, paragraphs or words) is a critical component in information retrieval solutions associated with digital libraries. The problem has been looked at in two settings: recognition based [1,2] like OCR and recognition free [3,4]. Most of the present day digital libraries use Optical Character Recognizers (OCR) for the recognition of digitized documents and thereafter employ a text based solution for the information retrieval. Though OCRs have become the de facto preprocessing for the retrieval, they are realized as insufficient for degraded books [5], incompatible for older print styles [6], unavailable for specialized scripts [7] and very hard for handwritten documents [8]. Even for printed books, commercial OCRs may provide highly unacceptable results in practice. The best commercial OCRs can only give word accuracy of 90% on printed books [4] in modern digital libraries. This means that every 10th word in a book is not searchable. Recall of retrieval systems built on such erroneous text is thus limited. Recognition free approaches have gained interest in recent years. Word spotting [3] is a promising method for recognition free retrieval. In this method, word images are represented using different features (e.g., Profiles, SIFT-BOW), and the features are compared with the help of appropriate distance measures (Euclidean, Earth Movers [9], DTW [10]). Word spotting has the advantage that it does not require prior learning due to its appearance-based matching. These techniques have been popularly used in document image retrieval.

Konidaris et al. [5] retrieve words from a large collection of printed historical documents. A search keyword typed by the user is converted into a synthetic word image which is used as a query image. Word matching is based on computing the L_1 distance metric between the query feature and all the features in the database. Here the features are calculated using the density of the character pixels and the area that is formed from the projections of the upper and lower profile of the word. The ranked results are further improved by relevance feedback. Sankar and Jawahar [7] have suggested a framework of probabilistic reverse annotation for annotating a large collection of images. Word images were segmented from 500 Telugu books. Matching of the word images is done using the DTW approach [11]. Hierarchical agglomerative clustering was used to cluster the word images. Exemplars for the keywords are generated by rendering the word to form a keyword-image. Annotation involved identifying the closest word cluster to each keyword cluster. This involves estimating the probability that each cluster belongs to the keyword. Yalniz and Manmatha [4] have applied word spotting to scanned English and Telugu books. They are able to handle noise in the document text by the use of SIFT features extracted on salient corner points. Rath and Manmatha [11] used projection profile and word profile features in a DTW based matching technique.

Recognition free retrieval was attempted in the past for printed as well as handwritten document collections [4,7,12,13]. Since most of these methods were designed for smaller collections (few handwritten documents as in [12]), computational time was not a major concern. Methods that extended this to a larger collection [14–16] used mostly (approximate) nearest neighbor retrieval. For searching complex objects in large databases, SVMs have emerged as the most popular and accurate solution in the recent past [12]. For linear SVMs, both training and testing have become very fast with the introduction of efficient algorithms and excellent implementations [17]. However, there are two fundamental challenges in using a classifier based solution for word retrieval (i) A classifier needs a good amount of annotated training data (both positive and negative) for training. Obtaining annotated data for every word in every style is practically impossible. (ii) One could train a set of classifiers for a given set of frequent queries. However, they are not applicable for rare queries.

In [18], Ranjan et al. proposed a one-shot classifier learning scheme (Direct query classifier). The proposed one shot learning scheme enables direct design of a classifier for novel queries, without having any access to the annotated training data, i.e., classifiers are trained for a set of frequent queries, and seamlessly extended for the rare and arbitrary queries, as and when required. The authors hypothesize that word images, even if degraded, can be matched and retrieved effectively with a classifier based solution. A properly trained classifier can yield an accurate ranked list of words since the classifier looks at the word as a whole, and uses a larger context (say multiple examples) for matching. The results of this method are significant since (i) It does not use any language specific post-processing for improving the accuracy. (ii) Even for a language like English, where OCRs are fairly advanced and engineering solutions were perfected, the classifier based solution is as good, if not superior to the best available commercial OCRs .

In the direct query classifier (DQC) scheme [18], the authors used DTW distance for indexing the frequent mean vectors. Since the DTW distance is computationally slow, the authors do not use all the frequent mean vectors for indexing. For comparing two word images, DTW distance typically takes one second [3]. This limits the efficiency of DQC. To overcome this limitation, the authors used Euclidean distance for indexing. The authors use the top 10 (closest in terms of Euclidean distance) frequent mean vectors for indexing. Since the DTW distance better captures the similarities compared to Euclidean distance for word image retrieval, this restricts the performance of DQC.

For speed-up, DTW distance has been previously approximated [19,20] using different techniques. In [20], the authors proposed a fast approximate DTW distance, in which, the DTW distance is approximated as a sum of multiple weighted Euclidean distances. For a given set of sequences, there are similarities between the top alignments (least cost alignments) of different pairs of sequences. In [20], the authors explored these similarities by learning a small set of global principal alignments from the given data, which captures all the possible correlations in the data. These global principal alignments are then used to compute the DTW distance for the new test sequences. Since these methods [19,20] avoid the computation of optimal alignments, these are computationally efficient compared to naive DTW distance. The fast approximate DTW distance can be used for efficient indexing in DQC classifier. However, it gives sub-optimal results. For best results, it needs query specific global principal alignments. In this paper, we introduce query specific DTW distance, which enables the direct design of global principal alignments for novel queries. Global principal alignments are computed for a set of frequent classes and seamlessly extended for the rare and arbitrary queries, as and when required, without using language specific knowledge. This is a distinct advantage over an OCR engine, which is difficult to adapt to varied fonts and noisy images and would require language specific knowledge to generate possible hypotheses for out of vocabulary words. Moreover, an OCR engine can respond to a word image query only by first converting it into text, which is again prone to recognition errors. In [21,22], deep learning frameworks are used for word spotting. In [23], a attribute based learning model PHOC is presented for word spotting. In training phase, each word image is to be given with its transcription. Both word image feature vectors and its transcriptions are used to create the PHOC representation. An SVM is learned for each attribute in this representation. Our approach bears similarity with the PHOC representation based word spotting [23]. In this sense, both the approaches are designed for handling out-of-vocabulary queries. Our work takes advantage of granular description at ngrams (cut-portion) level. This somewhat resembles the arrangement of characters used in the PHOC encoding. However, training efforts for PHOC are substantial with a large number of classifiers (604 classifiers) being trained and requires complete data for training, which is huge for large datasets. In our work, the amount of training data is restricted to only frequent classes, which is much less compared to PHOC. Further, PHOC requires labels in the form of transcriptions, whereas in our work the labels need not be transcriptions. In addition, PHOC is language dependent [24] and it is very difficult to apply over different languages. The method proposed in this paper is language independent; it can be applied to any language.

The paper is organized as follows. The next section describes the Direct query classifier (DQC). Fast approximation of (DTW) distance is discussed in Section 3. The query specific DTW distance is presented in Section 4. Experimental settings and results are discussed in Section 5, followed by concluding remarks in Section 6.

2. Direct Query Classifier (DQC)

In [18], Ranjan et al. proposed Direct Query Classifier (DQC), which is a one-shot learning scheme for dynamically synthesizing classifiers for novel queries. The main idea is to compute an SVM classifier for the query class using the classifiers obtained from the frequent classes of the database. The number of possible words in a language could be very large and it would be practically difficult to build a classifier for each of the words. However, all these words come from a small set of *n-grams*. The words corresponding to the frequent queries are expected to contain the n-grams that cover the full vocabulary. Exemplar SVM classifiers are computed for the frequent queries (word classes) and then appropriately concatenated to create novel classifiers for the rare queries. However, this process has its challenges due to

(i)   Variations due to nature of script and writing style,
(ii)   Classifiers for smaller ngrams could be noisy.

The authors address these limitations by building the SVM classifiers for most frequent queries and use classifier synthesis only for rare queries. This improves its overall performance. They use Query Expansion (QE) for further improving the performance. An overview of the direct query classifier is given in the following sections.

2.1. DP DQC: Design of DQC Using Dynamic Programming

Given a set of classifiers for frequent classes $\mathcal{W}_w = \{w_1, w_2, \ldots, w_N\}$ and a query vector X_q, the query classifier w_q is designed as a piecewise fusion of parts (*n-grams*) from the available classifiers from $\mathcal{W}_w$. Let p be the number of portions to be selected for computing the query classifier w_q. These portions are characterized by the sequence of indices $a_1, \ldots, a_{p+1}$. The classifier synthesis problem is formulated as that of picking up the optimal set of classifiers $\{c_i\}$ and the set of segment indices $\{a_i\}$ such that $\{a_i\}$ form a monotonically increasing sequence of indices. This involves the following optimization:

$$\max_{\{a_i\},\{c_i\}} \sum_{i=1}^{p} \sum_{k=a_i}^{a_{i+1}} w_{c_i}^k X_q^k \tag{1}$$

where w_{c_i} corresponds to the weight vector of the c_i^{th} classifier that we choose and the inner summation applies the index k in the range (a_i, a_{i+1}) to use the kth component $w_{c_i}^k$ from the classifier c_i. The index i in the outer summation refers to the cut portions, and p is the total number of portions we need to consider.

In [12], Malisiewicz et al. proposed the idea of exemplar SVEN (ESVM) where a separate (SVM) is learned for each example. Almazan et al. [25] use ESVMs for retrieving word images. ESVMs are inherently highly tuned to its corresponding example. Given a query, it can retrieve highly similar word images. This constrains the recall, unless one has large variations of the query word available. Another demerit of ESVM is the large overall training time since a separate SVM needs to be trained for each exemplar. One approach to reducing training time is to make the negative example mining step offline and selecting a common set of negative examples [26]. Gharbi et al. [27] provide another alternative for fast training of exemplar SVM in which the hyperplane between a single positive point and a set of negative points can be seen as finding the tangent to the manifold of images at the positive point.

Given a query q, the similar vectors in the dataset are identified by adopting the ESVM formulation proposed by Gharbi et al. [27] which yields an approach equivalent to Linear Discriminant Analysis. It involves a fast computation of the weight vector by adopting a parametric representation of the negative examples approximated as a Gaussian model on the complete set of training points. The normal to the Gaussian at the query point q is computed using the covariance matrix to yield the weight vector w_q as follows:

$$w_q = \Sigma^{-1}(\mu_q - \mu_0) \tag{2}$$

where Σ and μ_0 are the covariance and mean computed over the entire dataset. Since Σ and μ_0 are common for all data, finding w_q requires finding the mean vector μ_q of the class to which the query q belongs to. Let us define the set of class mean vectors for the frequent classes as $\mathcal{W}_\mu = \{\mu_1, \ldots, \mu_N\}$. The mean vector μ_q for the class of the query q is computed by making use of appropriate cut portions from the mean vectors of the frequent classes. Optimizing (1) for variable length cut portions entails high computational complexity. Therefore, instead of matching variable-length *n-grams*, the method divides X_q into p number of fixed length portions.

1. The class mean vectors of the most frequent 1000 classes are concatenated.
2. Now, each query cut portion X_q^k is searched in the concatenated mean vector using subsequence dynamic time warping [28]
3. The most similar segment in the concatenated mean vector is taken as the corresponding portion of the query class mean μ_q^k.
4. The concatenation of these query class mean cut portions μ_q^k synthesizes the query class mean $\mu_q = [\mu_q^1, \ldots, \mu_q^p]$.

Since DTW is computationally slow, applying subsequence DTW, in this case, is computationally expensive.

2.2. NN DQC: Design of DQC Using Approximate Nearest Neighbour

A speed-up is obtained by using approximate nearest neighbor search instead of using DTW.

- Instead of concatenating the class mean vectors, now each class mean vector is divided into same p number of fixed length portions. An index is built over frequent class means cut portions using FLANN.
- Each cut portion of X_q is compared with frequent class means cut portions using nearest neighbor search with Euclidean distance.
- The best matching cut portions of the mean vectors are used to synthesize the mean vector for the query class.

However, using nearest neighbor (NN DQC) instead of subsequence DTW based scheme (DP DQC) compromises the optimality of the classifier synthesis.

Few qualitative examples for the two versions of DQC are given in Figure 1. We have shown the retrieval results for frequent queries and rare queries. For each case, we have compared the retrieval results for NN DQC and DP DQC. For rare query, we have also shown the results for Query expansion (QE).

Query		Method	Retrieved Results				
			Rank 1	Rank 2	Rank 3	Rank 4	Rank 5
Frequent Query	but	NN DQC	but	but	but	but	but
		DP DQC	but	but	but	but	but
Rare Query	money	NN DQC	money	money	every	money	makes
		DP DQC	money	money	money	month	lovely
		QE with NN DQC	money	money	money	money	money.

Figure 1. Figure shows few query words and their corresponding retrieval results. The first column shows the query image and the corresponding images in each row are its retrieval results. First two rows show frequent query results. The first row shows the results for NN DQC and second row show the results for DP DQC. Row 3 to Row 5 show the retrieval results for a rare query. Row 3 shows the results for NN DQC and Row 4 show the results for DP DQC and Row 5 show the results for query expansion.

3. Approximating the DTW Distance

In general, DTW distance has quadratic complexity in the length of the sequence. Nagender et al. [20] proposed Fast approximate DTW distance (Fast Apprx DTW), which is a linear approximation to the DTW distance. For a pair of given sequences, DTW distance is computed using the optimal alignment from all the possible alignments. This optimal alignment gives a similarity between the given sequences by ignoring local shifts. Computation of optimal alignment is the most expensive operation in finding the DTW distance.

For a given set of sequences, there are similarities between the optimal alignments of different pairs of sequences. For example, if we take two different classes, the top alignments (optimal alignments/least cost alignments) between the samples of class 1 and the samples of class 2 always have some similarity. For a small dataset, the top alignments between few class 1 samples and few class 2 samples are plotted in Figure 2. It can be observed that the top alignments are in harmony. Based on this idea, we compute a set of global principal alignments from the training data such that the computed global principal alignments should be good enough for approximating the DTW distance between *any* new pair of sequences. For new test sequences, instead of finding the optimal alignments, the global principal alignments are used for computing the DTW distance. This avoids the computation of optimal alignments. Now, the DTW distance is approximated as the sum of the Euclidean distances over the global principal alignments.

$$FastApprxDTW(x_1, x_2) = \sum_{\pi \in G_{\mathcal{X}}} Euclid_{\pi}(x_1, x_2) \tag{3}$$

where $G_{\mathcal{X}}$ is the set of global principal alignments for the given data $\mathcal{X}$ and $Euclid_{\pi}(x_1, x_2)$ is the Euclidean distance between x_1 and x_2 over the alignment π. Notice that the DTW distance between two samples is the Euclidean distance (ground distance) over the optimal alignment.

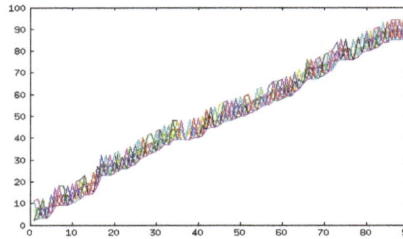

Figure 2. The top alignments between few samples from 2 different classes. Here, X-axis is the length of the samples from class 1 and Y-axis is the length of the samples from class 2.

To show the performance of Fast Apprx DTW [20], we have compared with naive DTW distance and Euclidean distance for word retrieval problem. Here, these distance measures are used for comparing word image representations. The dataset contains images from three different word classes. The results are given in Table 1. Nearest neighbor is used for retrieving the similar samples. The performance is measured by mean Average Precision (mAP). From the results, we can observe that Fast Apprx DTW is comparable to naive DTW distance and it performs better than Euclidean distance.

Table 1. The comparison of the performance of DTW distance, Fast Apprx DTW and Euclidean distance as a similarity measure for a word retrieval problem.

	DTW Distance	Fast Apprx DTW	Euclidean
mAP score	0.96	0.94	0.82

4. Query Specific Fast DTW Distance

In Fast approximate DTW distance [20] (Section 3), the global principal alignments are computed from the given data. Here, no class information is used while computing the alignments and also these alignments are query independent, i.e., query information is not used while computing the global principal alignments. In this section, we introduce Query specific DTW distance, which is computed using query specific (global) principal alignments. The proposed Query specific DTW distance has been found to give a much better performance when used with the direct query classifier.

Let $\mathcal{X}$ be the given data and all the samples are scaled to a fixed size. Let $\{C_1, C_2, \ldots, C_N\}$ be the most frequent N classes from the data and $\mu_1, \ldots, \mu_N$ be their corresponding class means. The matching process using the query specific principal alignments is as follows:

(i) Divide each sample from the frequent classes to a fixed number p of equal size portions. Let $x_{i_1}, \ldots, x_{i_{|c_i|}}$ be the samples (sequences) from the ith class c_i, where $|c_i|$ is the number of samples in the class c_i. The cut portions for the class means μ^i are denoted as $\mu_1^i, \ldots, \mu_p^i$, where each cut portion is of length d. Similarly, divide the query X_q into same number p of fixed length portions.

(ii) For each class, compute the global principal alignments for each cut portion separately. These are the cut specific principal alignments for the class. For ith class and jth cut portion the cut specific principal alignments are computed from $\{x_{i_1}^j, \ldots x_{i_{|c_i|}}^j\}$ and these are denoted as G_i^j. These alignments are computed for all the cut portions for each class.

(iii) The final step computes the cut specific principal alignments for the given query X_q as follows. For each cut portion of X_q, we compute the DTW distance (Euclidean distance over the cut specific principal alignments) with the corresponding cut portions of all the class means using their corresponding cut specific principal alignments. The distance between the jth cut portion of X_q i.e., X_q^j and the jth cut portion of the ith class mean i.e., μ_i^j is denoted as

$$Dis_i^j = \sum_{\pi \in G_i^j} Euclid_\pi(X_q^j, \mu_i^j) \tag{4}$$

For each cut portion of X_q, we compute the minimum distance mean cut portion over all the class mean vectors. The corresponding cut specific principal alignments of the closest matching mean cut portions are taken as the cut specific principal alignments of the query cut portion. In addition, the corresponding class mean cut portion is taken as the matching cut portion for constructing the query mean. Let the jth cut portion of the query have the best match with the jth cut-portion of the class with index c.

$$c = \arg\min_i \ Dis_i^j \tag{5}$$

Here the minimum distance is computed over all the frequent classes. We thus have

$$G_{X_q}^j \longleftarrow G_c^j \quad \text{and} \quad \mu_q^j \longleftarrow \mu_c^j \tag{6}$$

Here $G_{X_q}^j$ is the cut specific principal alignments for the jth cut portion of X_q.

Together, all these query mean cut portions give the query class mean. The query class mean μ_q is given as $\mu_q = (\mu_q^1, \mu_q^2, \ldots, \mu_q^p)$. This query class mean μ_q is then used as in Equation (2) to compute the LDA weight w_q (query classifier weight).

The query specific (QS) DTW distance between the query X_q and a sample X from the data is given as

$$\underset{qs}{dtw}(X_q, X) = \sum_{i=1}^p dtw_{G_{X_q}^i}(X_q^i, X^i) \tag{7}$$

where p is the number of cut portions.

Figure 3 shows all the processing stages of the nearest neighbor DQC. To summarize, we generate query specific principal alignments on the fly by selecting and concatenating the global principal alignments corresponding to the smaller *n grams* (cut portions). Our strategy is to build cut-specific principal alignments for the most frequent classes; these are the word classes that will be queried more frequently. These cut-specific principal alignments are then used to synthesize the query specific principal alignments (see Figure 4). The results demonstrate that our strategy gives good performance for queries from both the frequent word classes and rare word classes.

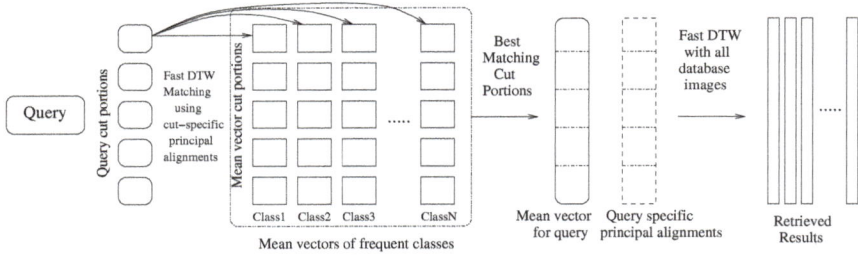

Figure 3. Overall Scheme for NN DQC. In an offline phase, the mean vectors for the frequent word classes are computed and their cut-specific principal alignments are computed. To process a query word image, it is divided into cut portions and FastDTW matching is used to get the best matching cut-portions from the frequent class mean vectors with the cut-portions of the query image. These best matching cut-portions are used to construct the mean vector for the query class and the query specific principal alignments. FastDTW [20] matching between the query image and the database images is done using the query specific principal alignments.

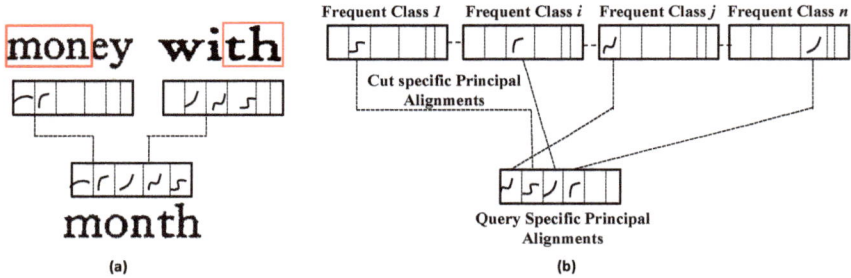

Figure 4. Synthesis of query specific principal alignments. (**a**) Cut specific principal alignments corresponding to "ground" and "leather" are joined to form the principal alignments for "great". Note that the appropriate cut portions are automatically found. (**b**) In a general setting, query specific principal alignments gets formed from multiple constituent cut specific principal alignments computed for frequent classes.

To ensure wider applicability of our approach, we consider that the alignments trained on one dataset may not work well on another dataset. This is mainly due to the print and style variations. For adapting to different styles, we use query expansion (QE), a popular approach in the information retrieval domain in which the query is reformulated to further improve the retrieval performance. An index is built over the given sample vectors from the database and using approximate nearest neighbor search, the top 10 similar vectors to the given query are computed. These top 10 similar vectors are then averaged to get the new reformulated query. This reformulated query is expected to better capture the variations in the query class. In our experiments, this further improves the retrieval performance. Approximate nearest neighbors are obtained using FLANN [29].

5. Results and Discussions

In this section, we validate the DQC classifier using query specific Fast DTW distance for efficient indexing on multiple word image collections and also demonstrate its quantitative and qualitative performance.

5.1. Data Sets and Evaluation Protocols

In this subsection, we discuss datasets and the experimental settings that we follow in the experiments. Our datasets, given in Table 2, comprise scanned English books from a digital library collection. We manually created ground truth at word level for the quantitative evaluation of the methods. The first collection (D1) of words is from a book which is reasonably clean. Second dataset (D2) is larger in size and is used to demonstrate the performance in case of heterogeneous print styles. Third dataset (D3) is a noisy book and is used to demonstrate the utility of the performance of our method in degraded collections. We have also given the results over the popular George Washington dataset. For the experiments, we extract profile features [11] for each of the word images. In this, we divide the image horizontally into two parts and the following features are computed: (i) vertical profile i.e the number of ink pixels in each column (ii) location of lowermost ink pixel, (ii) location of uppermost ink pixel and (iv) number of ink to background transitions. The profile features are calculated on binarized word images obtained using the Otsu thresholding algorithm. The features are normalized to [0, 1], so as to avoid dominance of any specific feature.

To evaluate the quantitative performance, multiple query images were generated. The query images are selected such that they have multiple occurrences in the database and are mostly functional words and do not include the stop words. The performance is measured by mean Average Precision (mAP), which is the mean of the area under the precision-recall curve for all the queries.

Table 2. Details of the datasets considered in the experiments. The first collection (D1) of words is from a book which is reasonably clean. The second dataset (D2) is obtained from 2 books and is used to demonstrate the performance in case of heterogeneous print styles. The third dataset (D3) is a noisy book.

Dataset	Source	Type	# Images	# Queries
D1	1 Book	Clean	14,510	100
D2	2 Books	Clean	32,180	100
D3	1 Book	Noisy	4100	100

5.2. Experimental Settings

For representing word images, we prefer a fixed length sequence representation of the visual content, i.e., each word image is represented as a fixed length sequence of vertical strips. A set of features $f_1, \ldots, f_L$ are extracted, where $f_i \in \mathbb{R}^M$ is the feature representation of the ith vertical strip and L is the number of vertical strips. This can be considered as a single feature vector $F \in \mathbb{R}^d$ of size $d = LM$. We implement the query specific alignment based solution as discussed in Section 4. For query expansion based solution, we identify the five most similar samples to the query using approximate nearest neighbor search and compute their mean.

Each dataset contains certain words which are more frequent than others. The number of samples in the frequent word classes are more compared to the rare classes. The retrieval results for frequent queries give better performance because the number of relevant samples available in the dataset is greater. It is worth emphasizing that for the method proposed in this paper (QS DTW), the degradation in the performance for rare queries is much less compared to other methods.

5.3. Results for Frequent Queries

Table 3 compares the retrieval performance of the direct query classifier DQC with the nearest neighbor classifier using different options for distance measures. The performance is shown in terms of mean average precision (mAP) values on three datasets. For the nearest neighbor classifier, we experimented with five distance measures: naive DTW distance, Fast approximate DTW distance [20], query specific DTW (QS DTW) distance, FastDTW [30] and Euclidean distance. We see that DTW performs comparably with DTW for all the datasets. It performs superior compared to the Fast DTW, Fast approximate DTW distance [20] and performs significantly better compared to Euclidean distance.

For DQC, we experimented with four options for indexing the frequent class mean vectors: subsequence DTW [18] (sDTW), approximate nearest neighbor NN DQC [18] (aNN), FastDTW, and QS DTW. We use the cut-portions obtained from the mean vectors of the most frequent 1000 word classes for (i) computing the cut-specific principal alignments in case of QS DTW, (ii) computing the closest matching cut-portion (i.e., one with the smallest distance, which can be Euclidean or DTW) with a cut-portion from the query vector, in case of aNNor FastDTW.

However, since sDTW has computational complexity $O(n^2)$, we restrict the number of frequent words used for indexing to 100. The QS DTW distance improves the performance of the DQC classifier. This is mainly due to the improved alignments involved in the QS DTW distance. The query specific alignments better capture the variations in the query class. Moreover, unlike the case of sDTW distance, the QS DTW distance has linear complexity and therefore we are able to index all the frequent mean vectors in the DQC classifier. Thus, the proposed method of QS DTW enhances the performance of the DQC classifier [18].

For frequent queries, the experiments revealed that the QS DTW gets the global principal alignments from the mean vector of the same (query) class. Since the alignments are coming from the query class, it gives minimum distance only for the samples which belong to its own class. Therefore, the retrieved samples largely belong to the query class. The performance is therefore improved compared to sDTW distance. In contrast, the Fast approximate DTW distance [20] computes the global principal alignments using all samples in the database, without exploiting any class information. The computed global principal alignments, therefore, include alignments from classes that may be different from the query class. For this reason, it performs inferior to the proposed DTW distance.

Table 3. Retrieval performance of various methods for frequent queries.

Dataset	Retrieval Results (mAP) for Frequent Queries								
	Using Nearest Neighbour Classifier					Using DQC (Exemplar SVM)			
	DTW	Fast Apprx DTW [20]	QS DTW	Euclidean	FastDTW [30]	sDTW	aNN	FastDTW	QS DTW
D1	0.94	0.92	0.92	0.81	0.91	0.98	0.98	1	**1**
D2	0.91	0.89	0.9	0.75	0.87	0.96	0.95	0.97	**0.99**
D3	0.83	0.79	0.81	0.67	0.76	0.91	0.92	0.93	**0.96**

5.4. Results for Rare Queries

The faster indexing offered by the use of QS DTW with DQC allows us to make use of the mean vectors of all the 1000 frequent classes. This gives us a much improved performance of the DQC on rare queries, compared to sDTW [18] which uses mean vectors from 100 frequent classes. Table 4 shows the retrieval performance of DQC with a nearest neighbour classifier using different options for distance measures. The performance is showed in terms of mean average precision (mAP) values on rare queries from three datasets. For the nearest neighbor classifier, we experimented with five distance measures: naive DTW distance, Fast approximate DTW distance [20], query specific DTW (QS DTW) distance, FastDTW [30] and Euclidean distance. We see that QS DTW performs comparably with DTW distance for all the datasets. It performs superior compared to the Fast approximate DTW distance [20], FastDTW and significantly better compared to Euclidean distance.

For DQC, we observe that QS DTW improves the performance compared to sDTW. This improvement of QS DTW over sDTWis more for rare queries compared to that for frequent queries. This shows that QS DTW can be used for faster indexing for both frequent and rare queries.

For rare queries, the query specific DTW distance outperforms Fast approximate DTW [20] distance. This happens because the Fast approximate DTW computes the global principal alignments from the database and its performance depends on the number of samples. Also, these alignments are query independent, i.e., they do not use any query information for computing the global principal alignments.

For a given query, it needs enough samples from the query class for getting novel global principal alignments. However, in any database, the number of samples for frequent classes dominate the number of samples for rare classes. The global principal alignments for frequent queries are likely to dominate the rare queries. Therefore, the precomputed global principal alignments in Fast approximate DTW may not capture all the correlations for rare query classes. In the proposed QS DTW distance, the global principal alignments are learned from the *ngrams* (cut-portions) of frequent classes. These *n-grams* are in abundance and also shared with rare queries, thus there are enough *n-gram* samples for learning the cut-specific alignments. The computed query specific alignments for the cut-portions outperform the alignments obtained from Fast approximate DTW.

Table 4. Retrieval performance of various methods for rare queries.

Dataset	Retrieval Results (mAP) for Rare Queries									
	Using Nearest Neighbour Classifier					Using DQC (Exemplar SVM)				
	DTW	Fast Apprx DTW [20]	QS DTW	Euclidean	FastDTW [30]	sDTW	aNN	FastDTW	QS DTW	QE
D1	0.82	0.77	0.83	0.69	0.75	0.91	0.90	0.91	0.95	**0.98**
D2	0.81	0.74	0.80	0.65	0.74	0.89	0.90	0.90	0.94	**0.95**
D3	0.73	0.66	0.71	0.59	0.62	0.80	0.78	0.80	0.91	**0.96**

It is worth mentioning that FastDTW [30], which is an approximation method, attempts to compute the DTW distance in an efficient way. It does not consider cut portion similarities, which may be influenced by various printing styles. Hence, these approaches are not applicable in our setting where the dataset can have words printed in varied printing styles, and thus can result in a marked degradation of performance for rare queries. Since query specific DTW finds the approximate DTW distance using cut specific principal alignments, it can exploit properties which cannot be used by other DTW approximation methods.

To summarize, the experiments demonstrate that the proposed query specific DTW performs well for both frequent and rare queries. Since it is learning the alignments from *ngrams*, it performs comparable to sDTWdistance for rare queries. For some queries, it performed better than the DTW distance.

5.5. Results for Rare Query Expansion

The results for QS DTW enhanced with query expansion (QE) using five best matching samples are also given in Table 4. It is observed that QE further improves the performance of our proposed method. To show the effectiveness of query expansion, we have computed the average of the DTW distance between the given query and all database samples that belonged to the query class. Likewise, we computed the average of the DTW distance for the reformulated query. Table 5 shows a comparison of the averaged DTW distance for the given query and the reformulated query using 2, 5, 7, and 10 most similar (to the query) samples from the database. From the results, we can observe that compared to the given query, the reformulated query using five best matching samples gives the lowest averaged DTW distance to the samples from the query class. This means the reformulated query is a good representative for the given query. However, using nine best matching samples for reformulating the query leads to a higher average of DTW distances. This means some irrelevant samples to the query are coming in the top similar samples.

Table 5. The table gives the average sum of DTW distance for the given query and the reformulated query with varying number of samples n from the query class.

Average of DTW Distance				
For given query	For Reformulated Query			
	$n = 2$	$n = 5$	$n = 7$	$n = 10$
2.67 ± 0.19	2.69 ± 0.23	**2.52 ± 0.13**	2.58 ± 0.21	2.94 ± 0.29

5.6. Results on George Washington Dataset

The George Washington (GW) dataset [31] contains 4894 word images from 1471 word classes. This is one of the popular dataset for word images. We applied our proposed method of DQC using QS DTW for word retrieval on the GW dataset. Table 6 provides comparative results for seven methods. Experiments are repeated for 100 random queries and the average over these results are reported in the table. We can observe that for the DQC the proposed QS DTW gives better performance than DTW. We can also observe that for the nearest neighbor classifier, QS DTW distance is performing slightly superior to the DTW distance and Fast approximate DTW distance. The superiority is because of the principal alignments which are query specific.

Table 6. Retrieval performance on the George Washington (GW) dataset. The DQC makes use of top 800 frequent classes for indexing the cut-portions.

Dataset	mAP Using Nearest Neighbour			mAP Using DQC			
	DTW	Fast Apprx DTW [20]	QS DTW	Euclidean	sDTW	FastDTW [30]	QS DTW
GW	0.51	0.50	0.52	0.32	0.62	0.63	**0.70**

5.7. Setting the Hyperparameters

The proposed method has few hyperparameters, like the length of the cut portion and the number of cut specific principal alignments. For tuning these parameters, we randomly choose 100 queries for each dataset and validate the performance over these queries. Queries included in the validation set are not used for reporting the final results.

In Table 7, we report the effect of varying the cut portion length on retrieval performance. The mAP score is less for smaller cut portion length. In this case, the learned alignments are not capturing the desired correlations. This happens because the occurrence of smaller cut portions is very frequent in the word images. For length more than 30, the mAP is again decreased. This is because the occurrences of larger cut portions are rare. Cut portion lengths in the range of 10 to 20 give better results. In this case, the cut portions are good enough to yield global principal alignments that can distinguish the different word images.

Table 7. The table shows the change in retrieval performance with the change in the length of cut portion over all the datasets (D1, D2, D3). Here l is the length of the cut portion.

l	D1	D2	D3
1	0.81	0.78	0.7
10	**0.86**	**0.83**	0.74
20	0.86	0.82	**0.75**
30	0.82	0.77	0.72

We assessed the effect of varying the number of cut-specific principal alignments on the retrieval performance on the three datasets and the results are given in Table 8. It is seen that the performance degrades for all the datasets when the number of alignments is chosen as 30. This can be attributed to some redundant alignments getting included in the set of principal alignments. Increasing the number of alignments from 10 to 20 improves performance for dataset D1, but has no effect on the performance for datasets D2 and D3. Therefore, we can conclude that restricting the number of principal alignments in the range 10 to 20 would give good results. In all our experiments, we set the number of cut-specific principal alignments as 10.

Table 8. Retrieval performance on the 3 datasets D1, D2 and D3 for varying number of cut specific principal alignments.

Number of Cut Specific	mAP for Different Datasets		
Principal Alignments	D1	D2	D3
10	0.92	**0.89**	**0.81**
20	**0.93**	0.89	0.81
30	0.91	0.88	0.78

5.8. Computation Time

Table 9 gives the computational time complexity for the methods based on DTW. The main computation involved in the use of QS DTW is that of computing the cut specific principal alignments for the frequent classes. Figure 5 shows the time for computing the cut specific principal alignments for the three datasets. The computation of these cut specific principal alignments can be carried out independently for all the classes. Since we can compute these principal alignments in parallel with each other, the proposed QS DTW scales well with the number of samples compared to Fast Apprx DTW [20].

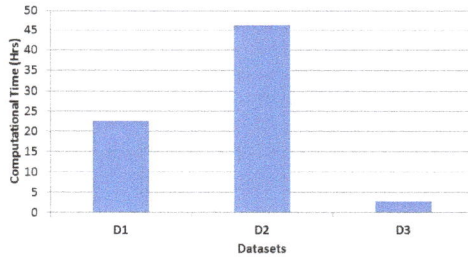

Figure 5. Computation time for computing the cut specific principal alignments for all the datasets. It includes the computation of cut specific principal alignments for all the frequent classes over all the cut portions.

Table 9. Computational complexities of DTW-based methods for distance computation. Here n is the length of the cut-portion of the feature vector.

Methods	SDTW	Fast Apprx DTW [20]	FastDTW [30]	QS DTW
Computational Complexity	$O(n^2)$	$O(n)$	$O(n)$	$O(n)$

Unlike the case of QS DTW, where the principal alignments are computed for the small cut portions, in Fast Apprx DTW, the principal alignments are computed for the full word image representation. Further, in Fast Apprx DTW, the principal alignments are computed from the entire dataset, unlike the case of QS DTW in which the principal alignments are computed for the individual classes. For these reasons, Fast Apprx DTW is computationally slower compared to the QS DTW.

For a given dataset, computing the cut specific principal alignments for the frequent classes is an offline process. When performing retrieval for a given query, DQC involves computing the query mean by composing together the nearest cut portions from the mean vectors of frequent classes. Further, the query specific principal alignments are not explicitly computed but rather constructed using the cut-specific principal alignments corresponding to the nearest cut portions. Once the query specific principal alignments are obtained, computation of QS DTW involves computing the Euclidean distance (using the query specific principal alignments) with the database images.

For the given two samples x and y of length N, FastDTW [30] is computed in the following way. First, these two samples are reduced to smaller length (1/8 times) and the naive DTW distance is applied over the reduced length samples to find the optimal warp path. Next, both the optimal path and the reduced length samples from the previous step are projected to higher (two times) resolution. Instead of filling all the entries in the cost matrix in the higher resolution, only the entries around a neighborhood of the projected warp path, governed by a parameter called radius r, are filled up. This projection step is continued until the original resolution was obtained. The time complexity of FastDTW is $N(8r + 14)$, where r is the radius. The performance of FastDTW depends on the radius r. The higher the value of r, the better the performance is. The time complexity of QSDTW/Fast Apprx DTW is $N * p$, where p is the number of principal alignments. In general, $p << 8r + 14$, for getting the similar performance in both the methods.

6. Conclusions

We have proposed query specific DTW distance for faster indexing in the direct query classifier DQC [18]. The benefit of deploying QS DTW with DQC is that it results in linear time complexity. Therefore, we are able to index all the frequent mean vectors of the database for constructing the mean vector for the query class in the DQC classifier. Since QS DTW distance performs equally well as DTW distance and because we consider all the frequent mean vectors for indexing, the proposed method enhances the performance of the DQC. Unlike previous approaches, the proposed QS DTW distance uses both the class mean vectors and the query information for computing the global principal alignments for the query. The use of *ngrams* for computing the global principal alignments makes the method perform well for rare queries, which are query word images that belong to non-frequent word classes for which mean vectors are not computed for the database. The query expansion (QE) further improves the performance of QS DTW. We have demonstrated the utility of the proposed technique over three different datasets. The proposed query specific DTW performs well compared to the previous DTW approximations.

Acknowledgments: This work was supported from the grant received for the IMPRINT project titled "Information access from document images of Indian languages," from MHRD, Government of India.

Author Contributions: Gattigorla Nagendar and Viresh Ranjan performed the experiments. Gaurav Harit and C.V Jawahar wrote the paper.

Conflicts of Interest: The authors declare no conflict of interest.

References

1. Nagy, G. Twenty Years of Document Image Analysis in PAMI. *PAMI* **2008**, *22*, 38–62, doi:10.1109/34.824820.
2. Sivic, J.; Zisserman, A. Video Google: A Text Retrieval Approach to Object Matching in Videos. In Proceedings of the Ninth IEEE International Conference on Computer Vision, Nice, France, 13–16 October 2003; pp. 1470–1477.
3. Rath, T.M.; Manmatha, R. Word spotting for historical documents. *IJDAR* **2007**, *9*, 139–152, doi:10.1109/SIU.2008.4632567.
4. Zeki, Y.I.; Manmatha, R. An Efficient Framework for Searching Text in Noisy Document Images. In Proceedings of the 2012 10th IAPR International Workshop on Document Analysis Systems (DAS), Gold Cost, QLD, Australia, 27–29 March 2012; pp. 48–52.
5. Konidaris, T.; Gatos, B.; Ntzios, K.; Pratikakis, I.; Theodoridis, S.; Perantonis, S.J. Keyword-guided word spotting in historical printed documents using synthetic data and user feedback. *IJDAR* **2007**, *9*, 167–177, doi:10.1007/s10032-008-0067-3.
6. Basilios, G.; Nikolaos, S.; Georgios, L. ICDAR 2009 Handwriting Segmentation Contest. In Proceedings of the 10th International Conference on Document Analysis and Recognition, Barcelona, Spain, 26–29 July 2009; pp. 1393–1397.

7.   Sankar, K.P.; Jawahar, C.V. Probabilistic Reverse Annotation for Large Scale Image Retrieval. In Proceedings of the IEEE Conference on Computer Vision and Pattern Recognition, Minneapolis, MN, USA, 17–22 June 2007.

8.   Almazán, J.; Fernández, D.; Fornés, A.; Lladós, J.; Valveny, E. A Coarse-to-Fine Approach for Handwritten Word Spotting in Large Scale Historical Documents Collection. In Proceedings of the 2012 International Conference on Frontiers in Handwriting Recognition (ICFHR), Bari, Italy, 18–20 September 2012; pp. 455–460.

9.   Yossi, R.; Carlo, T.; Guibas, J.L. The Earth Mover's Distance As a Metric for Image Retrieval. *IJCV* **2000**, *40*, 99–121, doi:10.1023/A:1026543900054.

10. David, S.; Kruskal, J.B. *Time Warps, String Edits, and Macromolecules: The Theory and Practice of Sequence Comparison*; Addison-Wesley: Reading, MA, USA, 1983; pp. 1–44, ISBN 0-201-07809-0.

11. Rath, T.M.; Manmatha, R. Word Image Matching Using Dynamic Time Warping. In Proceedings of the 2003 IEEE Computer Society Conference on Computer Vision and Pattern Recognition, Madison, WI, USA, 18–20 June 2003; pp. 521–527.

12. Tomasz, M.; Abhinav, G.; Efros, A.A. Ensemble of exemplar-SVMs for Object Detection and Beyond. In Proceedings of the 2011 International Conference on Computer Vision, Barcelona, Spain, 6–13 November 2011; pp. 89–96.

13. Cao, H.; Govindaraju, V. Vector Model Based Indexing and Retrieval of Handwritten Medical Forms. In Proceedings of the Ninth International Conference on Document Analysis and Recognition, Parana, Brazil, 23–26 September 2007; pp. 88–92.

14. Rath, T.M.; Manmatha, R. Features for Word Spotting in Historical Manuscripts. In Proceedings of the Seventh International Conference on Document Analysis and Recognition, Edinburgh, UK, 6 August 2003.

15. Balasubramanian, A.; Million, M.; Jawahar, C.V. Retrieval from Document Image Collections. In Proceedings of the 7th International Workshop, DAS 2006, Nelson, New Zealand, 13–15 February 2006; pp. 1–12.

16. Kovalchuk, A.; Wolf, L.; Dershowitz, N. A Simple and Fast Word Spotting Method. In Proceedings of the 2014 14th International Conference on Frontiers in Handwriting Recognition (ICFHR), Heraklion, Greece, 1–4 September 2014; pp. 3–8.

17. Shai, S.; Yoram, S.; Nathan, S. Pegasos: Primal Estimated sub-GrAdient Solver for SVM. In Proceedings of the 24th International Conference on Machine Learning, Corvalis, OR, USA, 20–24 June 2007; pp. 807–814.

18. Ranjan, V.; Harit, G.; Jawahar, C.V. Document Retrieval with Unlimited Vocabulary. In Proceedings of the 2015 IEEE Winter Conference on Applications of Computer Vision (WACV), Waikoloa, HI, USA, 5–9 January 2015; pp. 741–748.

19. Nagendar, G.; Jawahar, C.V. Fast Approximate Dynamic Warping Kernels. In Proceedings of the Second ACM IKDD Conference on Data Sciences, Bangalore, India, 18–21 March 2015; pp. 30–38.

20. Nagendar, G.; Jawahar, C.V. Efficient word image retrieval using fast DTW distance. In Proceedings of the 2015 13th International Conference on Document Analysis and Recognition (ICDAR), Tunis, Tunisia, 23–26 August 2015; pp. 876–880.

21. Sudholt, S.; Fink, G.A. PHOCNet: A Deep Convolutional Neural Network for Word Spotting in Handwritten Documents. In Proceedings of the 2016 15th International Conference on Frontiers in Handwriting Recognition (ICFHR), Shenzhen, China, 23–26 October 2016; pp. 686–690.

22. Krishnan, P.; Dutta, K.; Jawahar, C.V. Deep Feature Embedding for Accurate Recognition and Retrieval of Handwritten Text. In Proceedings of the 2016 15th International Conference on Frontiers in Handwriting Recognition (ICFHR), Shenzhen, China, 23–26 October 2016; pp. 686–690.

23. Jon, A.; Albert, G.; Alicia, F.; Ernest, V. Word Spotting and Recognition with Embedded Attributes. *IEEE Trans. Pattern Anal. Mach. Intell.* **2014**, *36*, 2552–2566.

24. Sfikas, G.; Giotis, A.P.; Louloudis, G.; Gatos, B. Using attributes for word spotting and recognition in polytonic greek documents. In Proceedings of the 2015 13th International Conference on Document Analysis and Recognition (ICDAR), Tunis, Tunisia, 23–26 August 2015; pp. 686–690.

25. Almazán , J.; Gordo, A.; Fornés , A.; Valveny, E. Segmentation-free Word Spotting with Exemplar SVMs. *Pattern Recognit.* **2014**, *47*, 3967–3978, doi:10.1016/j.patcog.2014.06.005.

26. Takami, M.; Bell, P.; Ommer, P. Offline learning of prototypical negatives for efficient online Exemplar SVM. In Proceedings of the 2014 IEEE Winter Conference on Applications of Computer Vision (WACV), Steamboat Springs, CO, USA, 24–26 March 2014; pp. 377–384.

27. Gharbi, M.; Malisiewicz, T.; Paris, S.; Durand, F. *A Gaussian Approximation of Feature Space for Fast Image Similarity*; *MIT CSAIL Technical Report*; CSAIL Publications: Cambridge, MA, USA, 2012.
28. Meinard, M. *Information Retrieval for Music and Motion*; Springer: Secaucus, NJ, USA, 2007; pp. 32–58, ISBN 3540740473.
29. Muja, M.; Lowe, D.G. Fast approximate nearest neighbours with automatic algorithm configuration. In Proceedings of the 4th International Conference on Computer Vision Theory and Applications, Lisboa, Portugal, 5–8 February 2009; pp. 331–340.
30. Stan, S.; Philip, C. FastDTW: Toward accurate dynamic time warping in linear time and space. In Proceedings of the KDD Workshop on Mining Temporal and Sequential Data, Seattle, WA, USA, 22 August 2004.
31. Fischer, A.; Keller, A.; Frinken, V.; Bunke, H. Lexicon-free handwritten word spotting using character HMMs. *Pattern Recognit. Lett.* **2012**, *33*, 934–942.

Journal of
Imaging

MDPI

Article

Handwritten Devanagari Character Recognition Using Layer-Wise Training of Deep Convolutional Neural Networks and Adaptive Gradient Methods

Mahesh Jangid [1],* and Sumit Srivastava [2]

[1] Department of Computer Science and Engineering, School of Computing & Information Technology, Manipal University Jaipur, Rajasthan 303007, India
[2] Department of Information Technology, School of Computing & Information Technology, Manipal University Jaipur, Rajasthan 303007, India; sumit.310879@gmail.com
* Correspondence: mahesh_seelak@yahoo.co.in; Tel.: +91-0141-3999279

Received: 6 December 2017; Accepted: 12 February 2018; Published: 13 February 2018

Abstract: Handwritten character recognition is currently getting the attention of researchers because of possible applications in assisting technology for blind and visually impaired users, human–robot interaction, automatic data entry for business documents, etc. In this work, we propose a technique to recognize handwritten Devanagari characters using deep convolutional neural networks (DCNN) which are one of the recent techniques adopted from the deep learning community. We experimented the ISIDCHAR database provided by (Information Sharing Index) ISI, Kolkata and V2DMDCHAR database with six different architectures of DCNN to evaluate the performance and also investigate the use of six recently developed adaptive gradient methods. A layer-wise technique of DCNN has been employed that helped to achieve the highest recognition accuracy and also get a faster convergence rate. The results of layer-wise-trained DCNN are favorable in comparison with those achieved by a shallow technique of handcrafted features and standard DCNN.

Keywords: handwritten character recognition; deep learning; Devanagari characters; convolutional neural network; adaptive gradient methods

---

1. Introduction

In the last few years, deep learning approaches [1] have been successfully applied to various areas such as image classification, speech recognition, cancer cell detection, video search, face detection, satellite imagery, recognizing traffic signs and pedestrian detection, etc. The outcome of deep learning approaches is also prominent, and in some cases the results are superior to human experts [2,3] in the past years. Most of the problems are also being re-experimented with deep learning approaches with the view to achieving improvements in the existing findings. Different architectures of deep learning have been introduced in recent years, such as deep convolutional neural networks, deep belief networks, and recurrent neural networks. The entire architecture has shown the proficiency in different areas. Character recognition is one of the areas where machine learning techniques have been extensively experimented. The first deep learning approach, which is one of the leading machine learning techniques, was proposed for character recognition in 1998 on MNIST database [4]. The deep learning techniques are basically composed of multiple hidden layers, and each hidden layer consists of multiple neurons, which compute the suitable weights for the deep network. A lot of computing power is needed to compute these weights, and a powerful system was needed, which was not easily available at that time. Since then, the researchers have drawn their attention to finding the technique which needs less power by converting the images into feature vectors. In the last few decades, a lot of feature extraction techniques have been proposed such as HOG (histogram of oriented gradients) [5], SIFT (scale-invariant feature transform) [6,7], LBP (local binary pattern) [8] and SURF (speeded up robust features) [9]. These are prominent feature extraction methods, which have been experimented for many problems like image recognition, character recognition, face detection, etc. and the corresponding models are called shallow learning models, which are still popular for the pattern recognition. Feature extraction [10] is one type of dimensionality reduction technique that represents the important parts of a large image into a feature vector. These features are handcrafted and explicitly designed by the research community. The robustness and performance of these features depend on the skill and the knowledge of each researcher. There are the cases where some vital features may be unseen by the researchers while extracting the features from the image and this may result in a high classification error.

Deep learning inverts the process of handcrafting and designing features for a particular problem into an automatic process to compute the best features for that problem. A deep convolutional neural network has multiple convolutional layers to extract the features automatically. The features are extracted only once in most of the shallow learning models, but in the case of deep learning models, multiple convolutional layers have been adopted to extract discriminating features multiple times. This is one of the reasons that deep learning models are generally successful. The LeNet [4] is an example of deep convolutional neural network for character recognition. Recently, many other examples of deep learning models can be listed such as AlexNet [3], ZFNet [11], VGGNet [12] and spatial transformer networks [13]. These models have been successfully applied for image classification and character recognition. Owing to their great success, many leading companies have also introduced deep models. Google Corporation has made a GoogLeNet having 22 layers of convolutional and pooling layers alternatively. Apart from this model, Google has also developed an open source software library named Tensorflow to conduct deep learning research. Microsoft also introduced its own deep convolutional neural network architecture named ResNet in 2015. ResNet has 152-layer network architectures which made a new record in detection, localization, and classification. This model introduced a new idea of residual learning that makes the optimization and the back-propagation process easier than the basic DCNN model.

Character recognition is a field of image processing where the image is recognized and converted into a machine-readable format. As discussed above, the deep learning approach and especially deep convolutional neural networks have been used for image detection and recognition. It has also been successfully applied on Roman (MNIST) [4], Chinese [14], Bangla [15] and Arabic [16] languages. In this work, a deep convolutional neural network is applied for handwritten Devanagari characters recognition.

The main contributions of our work can be summarized in the following points:

1. This work is the first to apply the deep learning approach on the database created by ISI, Kolkata. The main contribution is a rigorous evaluation of various DCNN models.
2. Deep learning is a rapidly developing field, which is bringing new techniques that can significantly ameliorate the performance of DCNNs. Since these techniques have been published in the last few years, there is even a validation process for establishing their cross-domain utility. We explored the role of adaptive gradient methods in deep convolutional neural network models, and we showed the variation in recognition accuracy.
3. The proposed handwritten Devanagari character recognition system achieves a high classification accuracy, surpassing existing approaches in literature mainly regarding recognition accuracy.
4. A layer-wise technique of DCNN technique is proposed to achieve the highest recognition accuracy and also get a faster convergence rate.

The remainder of this paper is organized as follows. Section 2 discusses previous work in handwritten Devanagari character recognition, Section 3 presents the introduction of deep convolutional neural network and adaptive gradient methods, Section 4 outlines the experiments and discussions and, finally, Section 5 concludes the paper.

2. Previous Work

Devanagari handwritten character recognition has been investigated by different feature extraction methods and different classifiers. Researchers have used structural, statistical and topological features. Neural networks, KNN (K-nearest neighbors), and SVM (Support vector machine) are primarily used for classification. However, the first research work was published by I. K. Sethi and B. Chatterjee [17] in 1976. The authors recognized the handwritten Devanagari numerals by a structured approach which found the existence and the positions of horizontal and vertical line segments, D-curve, C-curve, left slant and right slant. A directional chain code based feature extraction technique was used by N. Sharma [18]. A bounding box of a character sample was divided into blocks and computed 64-D direction chain code features from each divided block, and then a quadratic classifier was applied for the recognition of 11,270 samples. The authors reported an accuracy of 80.36% for handwritten Devanagari characters. Deshpande et al. [19] used the same chain code features with a regular expression to generate an encoded string from characters and improved the recognition accuracy by 1.74%. A two-stage classification approach for handwritten characters was reported by S. Arora [20] where she used structural properties of characters like shirorekha and spine in the first stage and in another stage used intersection features. These features further fed into a neural network for the classification. She also defined a method for finding the shirorekha properly. This approach has been tested on 50,000 samples and obtained 89.12% accuracy. In [21], S. Arora combined different features such as chain codes, four side views, and shadow based features. These features were fed into a multilayer perceptron neural network to recognize 1500 handwritten Devanagari characters and obtain 89.58% accuracy.

A fuzzy model-based recognition approach has reported by M. Hanmandlu [22]. The features are extracted by the box approach which divided the character into 24 cells (6 × 4 grid), and a normalized vector distance for each box was computed except the empty cells. A reuse policy is also used to enhance the speed of the learning of 4750 samples and obtained 90.65% accuracy. The work presented in [23] computed shadow features, chain code features and classified the 7154 samples using two multilayer perceptrons and a minimum edit distance method for handwritten Devanagari characters. They reported 90.74% accuracy. Kumar [24] has tested five different features named Kirsch directional edges, chain code, directional distance distribution, gradient, and distance transform on the 25,000 handwritten Devanagari characters and reported 94.1% accuracy. During the experiment, he found the gradient feature outperformed the remaining four features with the SVM classifier, and the Kirsch directional edges feature was the weakest performer. A new kind of feature was also created that computed total distance in four directions after computing the gradient map and neighborhood pixels' weight from the binary image of the sample. In the paper [25], Pal applied the mean filter four times before extracting the direction gradient features that have been reduced using the Gaussian filter. They used modified quadratic classifier on 36,172 samples and reported 94.24% accuracy using cross-validation policy. Pal [26] has further extended his work with SVM and MIL classifier on the same database and obtained 95.13% and 95.19% recognition accuracy respectively.

Despite the higher recognition rate achieved by existing methods, there is still room for improvement of the handwritten Devanagari character recognition.

3. Deep Convolutional Neural Networks (DCNN)

The deep convolutional neural network can be broadly segregated into two major parts as shown in Figure 1, the first part contains the sequence of alternative convolutional with max-pooling layers, and another part contains the sequence of fully connected layers. An object can be recognized by its features which are directly dependent on the distributions of color intensity in the image. The Gaussian, Gabor, etc. filters are used to record these color intensity distributions. The values of a kernel for these filters are predefined, and they record only the specific distribution of color intensity. The kernel values are not going to change as per the response of the applied model. However, in DCNN, the values of the kernel are being updated according to the response of the model. That helps to find the best kernel values for the model. The alternative convolutional and max-pooling layers do this job perfectly. Another part of DCNN is fully connected layers which contain multiple neurons, like the simple neural network in each layer that gets a high-level feature from the previous convolutional-pooling layer and computes the weights to classify the object properly.

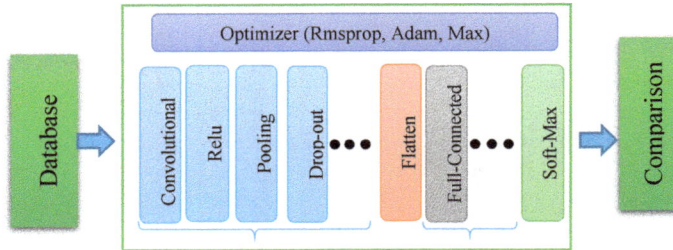

Figure 1. The schematic diagram of deep convolutional neural network (DCNN) architecture.

3.1. DCNN Notation

The deep convolutional neural network is a specially designed neural network for the image processing work. The most of the color images are being represented in three dimensions $h \times w \times c$, where h represents height, w represents the width of the image and c represents the number of channels of the image. However, the DCNN can only take an image which has the same height and width. So before feeding the image in DCNN, a normalization process has to follow to convert the image from $h \times w \times c$ size to $m \times m \times c$ size where m represents height and width of an image. The DCNN directly takes the three-dimensional normalized image/matrix X as an input and supplies to convolutional layer which has k kernels of size $n \times n \times p$, where $n < m$ and $p \leq c$. The convolutional layer performs the multiplication between the neighbors of a particular element of X with the weights provided by the kernel to generate the k different feature maps of size $l(m - n + 1)$. The convolutional layer is often followed by the activation functions. Rectified linear unit (Relu) was selected as activation function

$$Y_l^k = f\left(\sum_{i=1}^{n} X_i * W_{il}^k + B_l^k\right) \tag{1}$$

where k denotes the feature map layer, Y is a map of size $l \times l$ and W_{il} is a kernel weight of size $n \times n$, B_l^k represents the bias value and * represents the 2D convolution.

The next pooling layer works to reduce the feature maps by applying mean, max or min operation over $pl \times pl$ local region of feature map, where pl can vary from 2 to 5 generally. DCNNs have multiple consecutive layers of convolutional followed by pooling layers and each convolutional layer introduces a lot of unknown weight. The back-propagation algorithm—one of the famous techniques used in the simple neural network to find weight automatically—has been used to find the unknown weights during the training phase. The back-propagation updates the weights to minimize a loss $j(w)$ or error with an iterative process of gradient descent that can be expressed as

$$W_{t+1} = W_t - \alpha \nabla E|j(W_t)| + \mu v_t \tag{2}$$

Back-propagation algorithm helps to follow a direction towards where the cost function gives the minimum loss or error by updating the weights. The value α, called learning rate, helps to determine the step size or change in the previous weight. The back-propagation can be stuck at local minimum sometimes, which can be overcome by momentum μ which accumulates a velocity vector v in the direction of continuous reduction of loss function. The error or loss of a network can be found by various functions. The sum of squares function used to calculate the loss or error that can be expressed as

$$j(w) = \sum_{n=1}^{N} (y_n - \hat{y}_n)^2 + \lambda \sum_{l=1}^{L} W_l^2 \tag{3}$$

An L2 regularization λ was applied during the computation of loss to avoid the large progress of the parameters at the time of the minimization process.

The entire network of DCNN involves the multiple layers of convolutional, pooling, relu, fully connected and Softmax. These layers have a different specification to express them in a particular network. In this paper, we used a special convention to express network of DCNN.

- xINy: An input layer where x represents the width and height of the image and y represent the number of channels.
- xCy: A convolutional layer where x represents a number of kernels and y represents the size of kernel y*y.
- xPy: A pooling layer where x represents pooling size x*x, and y represents pooling stride.
- Relu: Represents rectified layer unit.
- xDrop: A dropout layer where x represents the probability value.
- xFC: A fully connected or dense layer where x represents a number of neurons.
- xOU: A output layer where x represents classes or labels.

3.2. Different Adaptive Gradient Methods

Basically, the neural network training updates the weights in each iteration, and the final goal of training is to find the perfect weight that gives the minimum loss or error. One of the important parameters of the deep neural network is learning rate, which decides the change in the weights. The selection of value for learning rate is a very challenging task because if the value of the learning rate selects low, then the optimization can be very slow and a network will take time to reach the minimum loss or error. On the other hand, if the value of learning rate selects higher, then the optimization can deviate and the network will not reach the minimum loss or error. This problem can be solved by the adaptive gradient methods that help in faster training and better convergence. The Adagrad [27] (adaptive gradient) algorithm was introduced by Duchi in 2011. It automatically incorporates low and high update for frequent and infrequent occurring features respectively. This method gives an improvement in convergence performance as compared to standard stochastic gradient descent for the sparse data. It can be expressed as,

$$W_{t+1} = W_t - \frac{\alpha}{\sqrt{\sum_t Av_t^2 + \epsilon}} \odot g_t \tag{4}$$

where Av_t is the previous adjustment gradient and ϵ is used to avoid divide by zero problems.

The Adagrad method divides the learning rate by the sum of the squared gradient that produces a small learning rate. This problem is solved by the Adadelta method [28] that can only accumulate a few past gradients in spite of entire past gradients. The equation of the Adadelta method can be expressed as

$$W_{t+1} = W_t - \frac{\alpha}{\sqrt{E[Av]^2 + \epsilon}} \odot g_t \tag{5}$$

where $E[Av]^2$ represents entire past gradients. It depends on current gradient and the previous average of the gradient. The problem of Adagrad is solved by Hinton [29] by the technique called RMSProp, which was designed for stochastic gradient descent. RMSProp is an updated version of Rprop which did not work with mini-batches. Rprop is same as the gradient, but it also divides by the size of the gradient. RMSProp keeps a moving average of the squared gradient for each weight and, further, it divides the gradient by square root of the mean square value. The first moving average of the squared gradient is given by,

$$Av_t \gamma Av_{t-1} + (1 - \gamma)(\nabla Qw)^2 \tag{6}$$

where γ is the forgetting factor, ∇Qw is the derivative of the error and Av_{t-1} is the previous adjustment value. The weights are updated as per following equation,

$$w_{t+1} w_t - \frac{\alpha}{\sqrt{Av_t}} \nabla Qw \tag{7}$$

where w is the previous weight and w_{t+1} is the updated weight whereas α is the global learning rate.

Adam (adaptive moment estimation) [30] is another optimizer for DCNN that needs the first-order gradient with small memory and computes adaptive learning rate for different parameters. This method has proven better than the RMSprop and rprop optimizers. The rescaling of the gradient is dependent on the magnitudes of parameter updates. The Adam does not need a stationary object and works with sparse gradients. It also contains a decaying average of past gradients M_t.

$$M_t = B_1 M_{t-1} + (1 - B_1) G_t \tag{8}$$

$$V_t = B_2 V_{t-1} + (1 - B_2) G_t^2 \tag{9}$$

where M_t and V_t are calculated first and the second moment of the gradients and these values are biased towards zero when the decay rates are small, and thereby bias-correction has done first and second moments estimates:

$$\check{M}_t = \frac{M_t}{1 - B_1^t} \tag{10}$$

$$\check{V}_t = \frac{V_t}{1 - B_2^t} \tag{11}$$

As per the authors of Adam, the default values of B_1 and B_2 were fixed at 0.9 and 0.999 empirically. They have shown its work in practice as a best choice as an adaptive learning method. Adamax is an extension of Adam, where in place of L^2 norm, an L^P norm-based update rule has been followed.

3.3. Layerwise Training DCNN Model

The work of training is to find the best weight for the deep neural network at which the network produces high accuracy or a very small error rate. The outcome of any deep model neural network somehow depends on how the model was trained and the number of layers. Usually, the model is created with the certain number of layers, and entire layers are being involved in the training phase. In this work, we proposed a layer-wise training model of DCNN in spite of involving entire layers during the training phase to recognize the handwritten Devanagari characters. The layer-wise training model starts with adding one layer of convolutional and pooling layer, followed by fully connected layer and applies the back-propagation algorithm to find the weights. In the next phase of the layer-wise training model, the next layer of convolutional, pooling layer is added and the back propagation algorithm is applied with previously found weights to calculate weights for the added layer.

After adding entire layers, a fine tuning was performed with the complete network to adjust the entire weights of the network on a very low learning rate. The back-propagation algorithm starts with some random weights, and during training it sharpens the weighs by updating them in each epoch. The layer-wise training model provides nice rough weights initially as the network starts with first layers and, further, it adds remaining layers to find the weights for remaining layers. The layer-wise training model is clearly shown in Figure 2. The training starts with only one pair of convolutional and pooling layer and further another pair is being added. Algorithm 1 shows the stepwise procedure to create the layer-wise DCNN model.

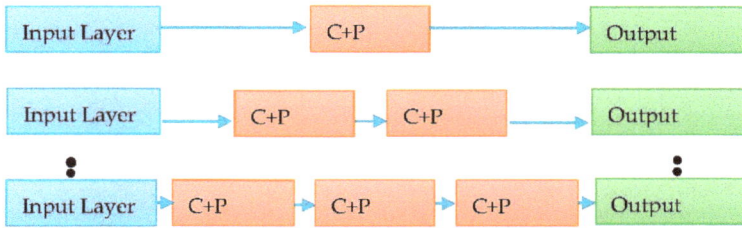

Figure 2. Layer-wise training of deep convolutional neural network.

Algorithm 1. Layer wise training of deep convolutional neural network

---

INPUT: Model, T, t, α_1, α_2, n \\ **T=** (TrainData), t = (TestData),
OUTPUT: TM \\ TrainedModel
Begin \\ Add first layer of convolutional layer and pooling layer
     Model.*add* (**xCy, T, Relu**)
     Model.*add* (**xPy**)
     Model.*add* (**xFC**)
     Model.*add* (**xOU**)
     Model.*compile* (**optimizer**)
     Model.*fit* (**T**, t, α_1)
       for all I := 1: n-1 **step 1 do**
         \\ Remove the last two layers (FC & OU)
         of existing model to add next layer of convolutional and pooling
         Model.layer.pop()
         Model.layer.pop()
         Model.*add* (**xCy, T, Relu**)
         Model.*add* (**xPy**)
         \\ Again added fully connected and output layer
         Model.*add* (**xFC**)
         Model.*add* (**xOU**)
         Model.*compile* (optimizer)
         Model.*fit* (**T**, t, α_1) \\ Trained the model with high learning rate
       end for
       Model.*fit* (**T**, t, α_2) \\ Perform fine tuning with low learning rate
end

---

4. Experiments and Discussions

Experiments were carried out on two databases: ISIDCHAR and V2DMDCHAR using the DCNN, layer-wise DCNN and different adaptive gradient methods. As it is hard to delineate the number of layers of DCNN that can produce the best result, we considered six different network architectures (NA) of DCNN as shown in Table 1. NA-1 contains only single convolutional-pooling layer and 500 fully connected neurons to observe the first response of DCNN. The next, NA-2 has double the number of fully connected neurons. The aim is to observe the impact of enhancement. Further, NA-3 and NA-4 have two C-P layers with variation in the number of kernels to analysis the impact of two C-P layers. The last, NA-5 and NA-6 have three C-P layers.

Initially, the different network architectures of DCNN were applied on each database to find out the best model for that particular database and then the proposed layer-wise DCNN was applied to observe the impact of that model. The models have also been tested with different adaptive gradient methods to these methods; they are also under experiment to observe their performance. Our work also shows the impact of different adaptive gradient methods on recognition accuracy.

Table 1. Various network architectures of deep convolutional neural network used.

Network	Model Architectures
NA-1	64IN64-64C2-**Relu**-4P2-500FC-47OU
NA-2	64IN64-64C2-**Relu**-4P2-1000FC-47OU
NA-3	64IN64-32C2-**Relu**-4P2-32C2-**Relu**-4P2-1000FC-47OU
NA-4	64IN64-64C2-**Relu**-4P2-64C2-**Relu**-4P2-1000FC-47OU
NA-5	64IN64-32C2-**Relu**-4P2-32C2-**Relu**-4P2-32C2-**Relu**-4P2-1000FC-47OU
NA-6	64IN64-64C2-**Relu**-4P2-64C2-**Relu**-4P2-64C2-**Relu**-4P2-1000FC-47OU

The experiments were all executed on the ParamShavak supercomputer system having two multicore CPUs with each CPU consisting of 12 cores along with two accelerator cards. This system has 64 GB RAM with CentOs 6.5 operating system. The deep neural network model was coded in Python using Keras—a high-level neural network API that uses Theano Python library. The basic pre-processing tasks like background elimination, gray-normalization and image resizing were done in Matlab. ISIDCHAR and V2DMDCHAR databases.

The ISIDCHAR [26] was prepared by researchers of the Indian Statistical Institute, Kolkata. They collected the samples from persons of different age groups to accommodate the maximum variation of written characters. Apart from that, the samples are also collected from the filled job forms and post-cards that makes this database so realistic. This database consists of 36,172 grayscale images of 47 different Devanagari characters. Owing to the assemblage of samples from many authors, this database delivers a variety of samples in each class, and the background of the samples is also highly uninformed. V2DMDCHAR [31] has been prepared by Vikas J. Dongre and Vijay H. Mankar's in 2012. This database has 20,305 samples of handwritten Devanagari characters.

4.1. Experimental Setup

The experiments were performed to investigate the effects of different network architectures, optimizers, and layer-wise trainings. The first phase of experiments was performed to observe the best network architecture for the database, and then the best-observed network architecture was tested with six different optimizers to find the best optimizer. A total of 12 (6 + 6) different experiments were performed on the database. The second phase of experiments aimed to observe the effect of layer-wise training. The layer-wise training was only performed with the best network architecture and best optimizer selected in the first phase.

Each optimizer had its own set of parameters. In our experiments, the optimizer parameters were kept as per their default values or as suggested by the author. The rectified linear activation function was used for entire experiments to mitigate the gradient vanishing problem. The sum of squares of the difference between target and observed values was calculated to estimate the loss of the deep network. Each network was trained for 100 epochs using mini-batches of size 200.

4.2. Results

The first phase of experiments was performed on ISIDCHAR to examine the best deep network architecture. We recorded the recognition accuracy at different network architecture using the Adam optimizer during each of the 50 epochs. The results in terms of the maximum, minimum, mean, and standard deviation values of recognition accuracy are reported in Table 2.

The best recognition accuracy was obtained with the network architecture NA-6, and the least recognition accuracy was obtained with the network architecture NA-1. Figure 3 shows the obtained recognition accuracy at each epoch. The network NA-1 produced 85% recognition accuracy because it has only one convolutional layer. The network NA-3 and NA-5 produced higher recognition accuracies of 91.53% and 93.24% respectively because these networks have a more convolutional layer. This enhancement signifies that the increment of the convolutional layer in deep convolutional neural network produced best results. In our experiments, we observed the enhancement in the recognition accuracy by increasing the number of kernels of convolutional layer. The network architectures NA-2, NA-4 and NA-6 had more kernels than NA-1, NA-3 and NA-5 and they produced higher recognition accuracy as observed in Table 2. The number of trainable parameters for each network architecture is shown in Table 3. The entire network architecture was also tested using the RMSProp optimizer, and the results have reported in Table 4. The NA-6 network produced 96.02% recognition accuracy with RMSProp while 95.58% with Adam. The behavior of NA-6 with RMSProp at each epoch can be seen in Figure 4.

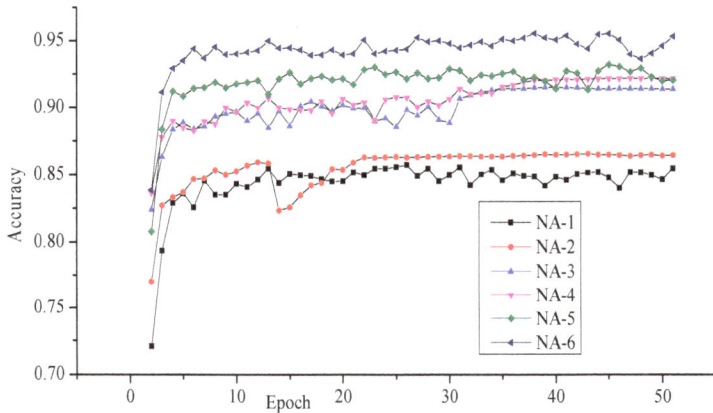

Figure 3. In this figure, we draw the recognition accuracy obtained with different network architectures on ISIDCHAR database at each epoch. The Adam optimizer was used.

Table 2. In this table, we report the results in term of maximum, minimum, mean, and standard deviation recognition accuracy obtained with different network architectures on ISIDCHAR when the system trained for 50 epochs with the Adam optimizer. The best scores are in bold.

Recognition Accuracy	Different Network Architectures					
	NA-1	**NA-2**	**NA-3**	**NA-4**	**NA-5**	**NA-6**
Maximum	0.8571	0.8654	0.9153	0.9224	0.9324	**0.9558**
Minimum	0.7208	0.7701	0.8237	0.8363	0.8077	**0.8385**
Average	0.8436	0.8549	0.9000	0.9058	0.9190	0.9427
Std. Deviation	0.0204	0.0169	0.0165	0.0158	0.0178	0.0168

Table 3. List of trainable parameters in each network architecture.

Network Architectures	Layer Type	Layer Size	Trainable Parameters	Total Parameters
NA-1	Conv1 layer	$64 \times 64 \times 64$	1088	34,873,135
	Dense layer	500	34,848,500	
	Output layer	47	23,547	
NA-2	Conv1 layer	$64 \times 64 \times 64$	1088	61,553,135
	Dense layer	1000	61,505,000	
	Output layer	47	47,047	
NA-3	Conv1 layer	$32 \times 64 \times 64$	544	7,265,007
	Conv2 layer	$32 \times 33 \times 33$	16,416	
	Dense layer	1000	7,201,000	
	Output layer	47	47,047	

Network Architectures	Layer Type	Layer Size	Trainable Parameters	Total Parameters
NA-4	Conv1 layer	$64 \times 64 \times 64$	1088	
	Conv2 layer	$64 \times 33 \times 33$	65,600	
	Dense layer	1000	14,401,000	14,514,735
	Output layer	47	47,047	
NA-5	Conv1 layer	$32 \times 64 \times 64$	544	
	Conv2 layer	$32 \times 33 \times 33$	16,416	
	Conv3 layer	$32 \times 17 \times 17$	16,416	1,649,423
	Dense layer	1000	1,569,000	
	Output layer	47	47,047	
NA-6	Conv1 layer	$64 \times 64 \times 64$	1088	
	Conv2 layer	$64 \times 33 \times 33$	65,600	
	Conv3 layer	$64 \times 17 \times 17$	65,600	3,316,335
	Dense layer	1000	3,137,000	
	Output layer	47	47,047	

Table 4. In this table, we report the results in term of maximum, minimum, mean, and standard deviation recognition accuracy obtained with different network architectures on ISIDCHAR when the system trained for 50 epochs with the RMSProp optimizer. The best scores are in bold.

Recognition Accuracy	Different Network Architectures					
	NA-1	NA-2	NA-3	NA-4	NA-5	NA-6
Maximum	0.8572	0.8641	0.903	0.9079	0.9311	**0.9602**
Minimum	0.7093	0.7475	0.7711	0.7788	0.7422	**0.8067**
Average	0.8383	0.8501	0.8927	0.8941	0.9150	0.9463
Std. Deviation	0.0321	0.0232	0.0210	0.0197	0.0308	0.0252

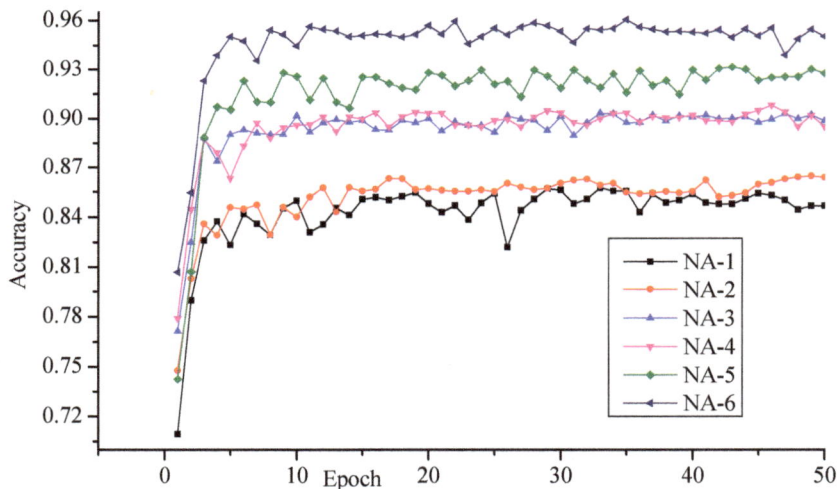

Figure 4. In this figure, we draw the recognition accuracy obtained with different network architectures on the ISIDCHAR database at each epoch. The RMSProp optimizer was used.

The best recognition accuracy of the ISIDCHAR database was obtained with NA-6 network architecture with RMSProp optimizer. However, it may be possible that this network could perform better with other optimizers. To further investigate, we performed experiments with six different optimizers. Table 5 shows the recognition accuracy obtained with NA-6 at different optimizers. The highest recognition accuracy 96.02% was recorded with NA-6 at RMSProp optimizer. The Adam optimizer outperformed the SGD and Adagrad optimizers. The AdaDelta, AdaMax, and RMSProp optimizers outperformed the Adam optimizer. Figure 5 shows the performance of individual optimizer.

Table 5. In this table, we report the results in term of maximum, minimum, mean, and standard deviation recognition accuracy obtained with NA-6 on ISIDCHAR when the system trained for 50 epochs with the different optimizers. The best scores are in bold.

Recognition Accuracy	Different Optimizers					
	SGD	Adagrad	Adam	AdaDelta	AdaMax	RMSProp
Maximum	0.931	0.9364	**0.9558**	0.9565	0.9579	**0.9602**
Minimum	0.6933	0.7703	**0.7585**	0.7605	0.7851	**0.8067**
Mean	0.9168	0.9280	0.9411	0.9457	0.9448	0.9463
Std. Deviation	0.0365	0.0252	0.0274	0.0311	0.0256	0.0252

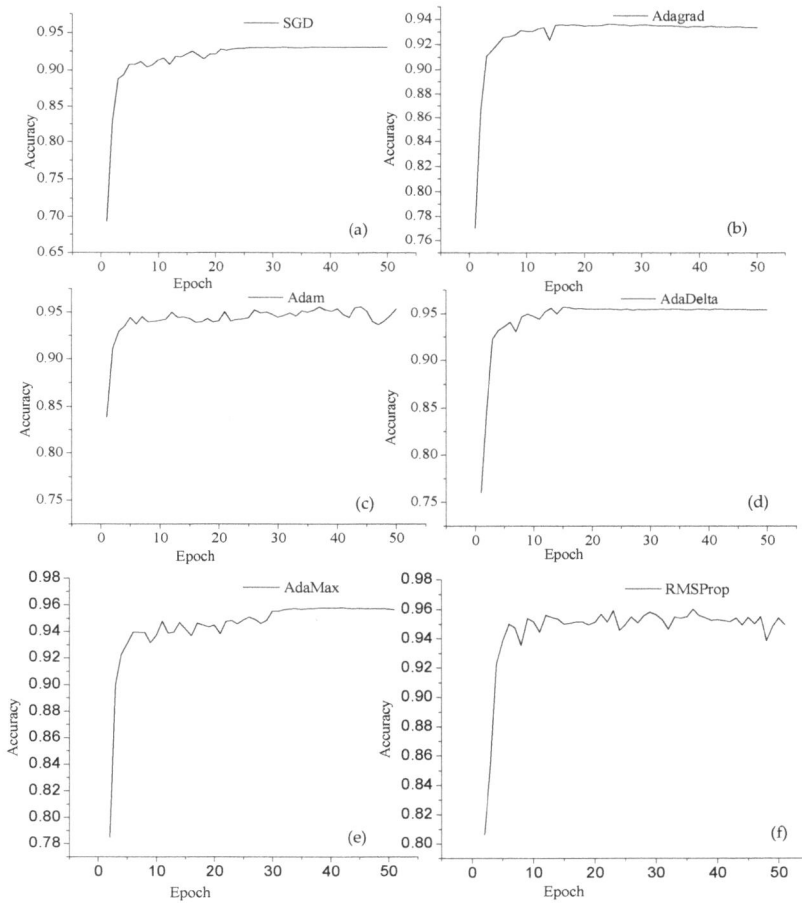

Figure 5. In this figure, we draw the recognition accuracy obtained with NA-6 network architecture at different optimizers (**a**) SGD; (**b**) Adagrad; (**c**) Adam; (**d**) AdaDelta; (**e**) AdaMax; (**f**) RMSProp; on the ISIDCHAR database at each epoch.

We found that the NA-6 network architecture with RMSProp optimizer produced the highest recognition accuracy. This network was again trained by layer-wise model as described in Section 3.3.

This network was tested with ISIDCHAR, V2DMDCHAR, and combined databases. The results are reported in Table 6. It has been seen that a nice enhancement in the recognition accuracy was recorded by the layer-wise training model. The 97.30% recognition accuracy was obtained on ISIDCHAR database and 97.65% recognition accuracy obtained on V2DMDCHAR database. The layer-wise training model was also applied after combining both the databases and obtained 98% recognition accuracy when 70% of the samples were used for training and the rest used for testing. The current work is compared to previous works on ISIDCHAR database in Table 7.

Table 6. In this table, we reported the maximum recognition accuracy obtained with NA-6 and RMSProp optimizer on ISIDCHAR, V2DMDCHAR and combined both when the model was trained layer-wise.

Database	No. of Samples	Recognition Accuracy	
		DCNN	Layer-Wise DCNN
ISIDCHAR	36,172	96.02%	97.30%
V2DMDCHAR	20,305	96.45%	97.65%
ISIDCHAR+V2DMDCHAR	56,477	96.53%	98.00%

Table 7. Comparison of recognition accuracy by other researchers.

S. No.	Accuracy Obtained	Feature; Classifier	Method Proposed by	Data Size
1	95.19	Gradient; MIL	U. Pal [26]	36,172
2	95.24	GLAC; SVM	M. Jangid [32]	36,172
3	96.58	Masking, SVM	M. Jangid [33]	36,172
4	96.45	DCNN	Proposed work	36,172
5	97.65	SL-DCNN	Proposed work	36,172
6	98	SL-DCNN	Proposed work	56,477

5. Conclusions

Deep learning is one of the prominent technologies that have been experimentally studied with entire major areas of computer vision and document analysis. In this paper, we experimentally developed a deep convolutional neural network (DCNN) and adaptive gradient methods to recognize the unconstrained handwritten Devanagari characters. The deep convolutional neural network helped us to find the best features automatically and also classify them. We experimented with a handwritten Devanagari character database with six different DCNN network architectures as well as six different optimizers. The highest recognition accuracy 96.02% was obtained using NA-6 network architecture and RMSProp—an adaptive gradient method (optimizer). Further, we again trained DCNN layer-wise, which is also adopted by many researchers to enhance the recognition accuracy, using NA-6 network architecture and the RMSProp adaptive gradient method. Using DCNN layer-wise training model, our database obtained 98% recognition accuracy, which is the highest recognition accuracy of the database.

Acknowledgments: The authors are thankful to the ISI, Kolkata to provide a database and the Manipal University Jaipur to provide the supercomputing facility without this facility deep learning concept was not possible for the handwritten Devanagari characters.

Author Contributions: Mahesh Jangid has envisaged the study, designed the experiments, and wrote the manuscript. Sumit Srivastava performed some part of experiments and corrected the manuscript. Both authors read and approved the final manuscript.

Conflicts of Interest: The authors declare that they have no competing interests.

References

1. Jürgen, S. Deep learning in neural networks: An overview. *Neural Netw.* **2015**, *61*, 85–117.
2. Ciregan, D.; Meier, U.; Schmidhuber, J. Multi-column deep neural networks for image classification. In Proceedings of the IEEE Conference on Computer Vision and Pattern Recognition (CVPR), Providence, RI, USA, 16–21 June 2012.
3. Krizhevsky, A.; Sutskever, I.; Hinton, G.E. Imagenet classification with deep convolutional neural networks. In Proceedings of the Advances in Neural Information Processing Systems, Lake Tahoe, NV, USA, 3–8 December 2012.
4. Lecun, Y.; Bottou, L.; Bengio, Y.; Haffner, P. Gradient-based learning applied to document recognition. *Proc. IEEE* **1998**, *86*, 2278–2324. [CrossRef]
5. Navneet, D.; Triggs, B. Histograms of oriented gradients for human detection. In Proceedings of the CVPR 2005 IEEE Computer Society Conference on Computer Vision and Pattern Recognition, San Diego, CA, USA, 20–25 June 2005; Volume 1.
6. Lowe, D.G. Distinctive image features from scale-invariant keypoints. *Int. J. Comput. Vis.* **2004**, *60*, 91–110. [CrossRef]
7. Surinta, O.; Karaaba, M.F.; Schomaker, L.R.B.; Wiering, M.A. Recognition of handwritten characters using local gradient feature descriptors. *Eng. Appl. Artif. Intell.* **2015**, *45*, 405–414. [CrossRef]
8. Ojala, T.; Pietikäinen, M.; Harwood, D. A comparative study of texture measures with classification based on featured distributions. *Pattern Recognit.* **1996**, *29*, 51–59. [CrossRef]
9. Bay, H.; Tuytelaars, T.; van Gool, L. Surf: Speeded up robust features. In Proceedings of the 9th European Conference on Computer Vision, Graz, Austria, 7–13 May 2006; pp. 404–417.
10. Wang, X.; Paliwal, K.K. Feature extraction and dimensionality reduction algorithms and their applications in vowel recognition. *Pattern Recognit.* **2003**, *36*, 2429–2439. [CrossRef]
11. Zeiler, M.D.; Rob, F. Visualizing and understanding convolutional networks. In Proceedings of the European Conference on Computer Vision, Zurich, Switzerland, 6–12 September 2014.
12. Simonyan, K.; Andrew, Z. Very deep convolutional networks for large-scale image recognition. *arXiv*, 2004.
13. Jaderberg, M.; Simonyan, K.; Zisserman, A. Spatial transformer networks. In Proceedings of the Advances in Neural Information Processing Systems, Montreal, QC, Canada, 11–12 December 2015.
14. Cireşan, D.; Ueli, M. Multi-column deep neural networks for offline handwritten Chinese character classification. In Proceedings of the 2015 International Joint Conference on Neural Networks (IJCNN), Killarney, Ireland, 12–17 July 2015.
15. Sarkhel, R.; Das, N.; Das, A.; Kundu, M.; Nasipuri, M. A Multi-scale Deep Quad Tree Based Feature Extraction Method for the Recognition of Isolated Handwritten Characters of popular Indic Scripts. *Pattern Recognit.* **2017**, *71*, 78–93. [CrossRef]
16. Ahranjany, S.S.; Razzazi, F.; Ghassemian, M.H. A very high accuracy handwritten character recognition system for Farsi/Arabic digits using Convolutional Neural Networks. In Proceedings of the 2010 IEEE Fifth International Conference on Bio-Inspired Computing: Theories and Applications (BIC-TA), Changsha, China, 23–26 September 2010.
17. Sethi, I.K.; Chatterjee, B. Machine Recognition of Hand-printed Devnagri Numerals. *IETE J. Res.* **1976**, *22*, 532–535. [CrossRef]
18. Sharma, N.; Pal, U.; Kimura, F.; Pal, S. Recognition of off-line handwritten devnagari characters using quadratic classifier. In *Computer Vision, Graphics and Image Processing*; Springer: Berlin/Heidelberg, Germany, 2006; pp. 805–816.
19. Deshpande, P.S.; Malik, L.; Arora, S. Fine Classification & Recognition of Hand Written Devnagari Characters with Regular Expressions & Minimum Edit Distance Method. *JCP* **2008**, *3*, 11–17.
20. Arora, S.; Bhatcharjee, D.; Nasipuri, M.; Malik, L. A two stage classification approach for handwritten Devnagari characters. In Proceedings of the International Conference on Computational Intelligence and Multimedia Applications, Sivakasi, Tamil Nadu, India, 13–15 December 2007; Volume 2, pp. 399–403.
21. Arora, S.; Bhattacharjee, D.; Nasipuri, M.; Basu, D.K.; Kundu, M.; Malik, L. Study of different features on handwritten Devnagari character. In Proceedings of the 2009 2nd International Conference on Emerging Trends in Engineering and Technology (ICETET), Nagpur, India, 16–18 December 2009; pp. 929–933.

22. Hanmandlu, M.; Murthy, O.V.R.; Madasu, V.K. Fuzzy Model based recognition of handwritten Hindi characters. In Proceedings of the 9th Biennial Conference of the Australian Pattern Recognition Society on Digital Image Computing Techniques and Applications, Glenelg, Australia, 3–5 December 2007; pp. 454–461.

23. Arora, S.; Bhattacharjee, D.; Nasipuri, M.; Basu, D.K.; Kundu, M. Recognition of non-compound handwritten devnagari characters using a combination of mlp and minimum edit distance. *arXiv*, 2010.

24. Kumar, S. Performance comparison of features on Devanagari handprinted dataset. *Int. J. Recent Trends* **2009**, *1*, 33–37.

25. Pal, U.; Sharma, N.; Wakabayashi, T.; Kimura, F. Off-line handwritten character recognition of devnagari script. In Proceedings of the Ninth International Conference on Document Analysis and Recognition, Curitiba, Parana, Brazil, 23–26 September 2007; Volume 1, pp. 496–500.

26. Pal, U.; Wakabayashi, T.; Kimura, F. Comparative study of Devnagari handwritten character recognition using different feature and classifiers. In Proceedings of the 10th International Conference on Document Analysis and Recognition, Catalonia, Spain, 26–29 July 2009; pp. 1111–1115.

27. Duchi, J.; Hazan, E.; Singer, Y. Adaptive subgradient methods for online learning and stochastic optimization. *J. Mach. Learn. Res.* **2011**, *12*, 2121–2159.

28. Zeiler, M.D. ADADELTA: an adaptive learning rate method. *arXiv*, 2012.

29. Hinton, G. Slide 6, Lecture Slide 6 of Geoffrey Hinton's Course. Available online: http://www.cs.toronto. edu/$\sim$tijmen/csc321/slides/lecture_slides_lec6.pdf (accessed on 19 July 2017).

30. Kingma, D.; Jimmy, B. Adam: A method for stochastic optimization. *arXiv*, 2014.

31. Dongre, V.J.; Vijay, H.M. Devnagari handwritten numeral recognition using geometric features and statistical combination classifier. *arXiv*, 2013.

32. Jangid, M.; Sumit, S. Gradient local auto-correlation for handwritten Devanagari character recognition. In Proceedings of the 2014 International Conference on High Performance Computing and Applications (ICHPCA), Bhubaneswar, India, 22–24 December 2014.

33. Jangid, M.; Sumit, S. Similar handwritten Devanagari character recognition by critical region estimation. In Proceedings of the 2016 International Conference on Advances in Computing, Communications and Informatics (ICACCI), Jaipur, India, 21–24 September 2016.

Journal of
Imaging

MDPI

Article

Benchmarking of Document Image Analysis Tasks for Palm Leaf Manuscripts from Southeast Asia

Made Windu Antara Kesiman [1,2,*], **Dona Valy** [3,4], **Jean-Christophe Burie** [1], **Erick Paulus** [5], **Mira Suryani** [5], **Setiawan Hadi** [5], **Michel Verleysen** [2], **Sophea Chhun** [4] and **Jean-Marc Ogier** [1]

1   Laboratoire Informatique Image Interaction (L3i), Université de La Rochelle, 17042 La Rochelle, France; jean-christophe.burie@univ-lr.fr (J.-C.B.); jean-marc.ogier@univ-lr.fr (J.-M.O.)
2   Laboratory of Cultural Informatics (LCI), Universitas Pendidikan Ganesha, Singaraja, Bali 81116, Indonesia; michel.verleysen@uclouvain.be
3   Institute of Information and Communication Technologies, Electronic, and Applied Mathematics (ICTEAM), Université Catholique de Louvain, 1348 Louvain-la-Neuve, Belgium; dona.valy@student.uclouvain.be
4   Department of Information and Communication Engineering, Institute of Technology of Cambodia, Phnom Penh, Cambodia; sophea.chhun@itc.edu.kh
5   Department of Computer Science, Universitas Padjadjaran, Bandung 45363, Indonesia; erick_paulus@yahoo.com (E.P.); mira.suryani@unpad.ac.id (M.S.); setiawanhadi@unpad.ac.id (S.H.)
*   Correspondence: made_windu_antara.kesiman@univ-lr.fr

Received: 15 December 2017; Accepted: 18 February 2018; Published: 22 February 2018

Abstract: This paper presents a comprehensive test of the principal tasks in document image analysis (DIA), starting with binarization, text line segmentation, and isolated character/glyph recognition, and continuing on to word recognition and transliteration for a new and challenging collection of palm leaf manuscripts from Southeast Asia. This research presents and is performed on a complete dataset collection of Southeast Asian palm leaf manuscripts. It contains three different scripts: Khmer script from Cambodia, and Balinese script and Sundanese script from Indonesia. The binarization task is evaluated on many methods up to the latest in some binarization competitions. The seam carving method is evaluated for the text line segmentation task, compared to a recently new text line segmentation method for palm leaf manuscripts. For the isolated character/glyph recognition task, the evaluation is reported from the handcrafted feature extraction method, the neural network with unsupervised learning feature, and the Convolutional Neural Network (CNN) based method. Finally, the Recurrent Neural Network-Long Short-Term Memory (RNN-LSTM) based method is used to analyze the word recognition and transliteration task for the palm leaf manuscripts. The results from all experiments provide the latest findings and a quantitative benchmark for palm leaf manuscripts analysis for researchers in the DIA community.

Keywords: document image analysis; binarization; character recognition; text line segmentation; word recognition; transliteration; palm leaf manuscript; dataset; benchmark; experimental test

1. Introduction

Since the world entered the digital age in the early 20th century, the need for a document image analysis (DIA) system is increasing. This is due to the dramatic increase in efforts to digitize the various types of document collections available, especially the ancient documents of historical relics found in various parts of the world. Some very interesting projects on a wide variety of heritage document collections can be mentioned here: for example, the tranScriptorium project (http://transcriptorium.eu/) [1]; the READ (Recognition and Enrichment of Archival Documents) project (https://read.transkribus.eu/) [2], which works on documents from the Middle Ages to today, and also focuses on different languages ranging from Ancient Greek to modern English; the

IAM Historical Document Database (IAM-HistDB) (http://www.fki.inf.unibe.ch/databases/iam-historical-document-database) [3], which includes handwritten historical manuscript images from the Saint Gall Database from the 9th century in Latin; the Parzival Database from the 13th century in German; the Washington Database from the 18th century in English; the Ancient Lives Project (https://www.ancientlives.org/) [4], which asks volunteers to transcribe Ancient Greek text fragments from the Oxyrhynchus Papyri collection; and many other projects.

To accelerate the process of accessing, preserving, and disseminating the contents of the heritage documents, a DIA system is needed. Besides aiming to preserve the existence of such ancient documents physically, the DIA system is expected to enable open access to the contents of the documents and provide opportunities for a wider audience to access all the important information stored in the document. DIA is the process of using various technologies to extract text, printed or handwritten, and graphics from digitized document files (http://www.cvisiontech.com/library/pdf/pdf-document/document-image-analysis.html) [5]. DIA systems generally have a major role in identifying, analyzing, extracting, structuring, and transferring document contents more quickly, effectively, and efficiently. This system is able to work semi-automatically or even fully automatically without human intervention. The DIA system is expected to save time, cost, and effort at many points in the heritage document preservation process.

However, although the DIA research develops rapidly, it is undeniable that most of the document collections used in the initial step are from developed regions such as America and European countries. The document samples from these countries are mostly written in English or old English with Latin/Roman script. Several important document collections were finally used as standard benchmarks for the evaluation of the latest DIA research results. The next wave of DIA research finally began to deal with documents from non-English-speaking areas with non-Latin scripts, such as Arabic, Chinese, and Japanese documents. During the evolution of DIA research in the last two decades, DIA researchers have proposed and achieved satisfactory solutions for many complex problems of document analysis for these types of documents. However, the DIA research challenge is ongoing. The latest challenge is documents from Asia, with new languages and more complex scripts to explore, such as Devanagari script [6], Gurmukhi script [7–10], Bangla script [11], and Malayalam script [12], and the case of multiple languages and scripts in documents from India. Optical character recognition (OCR) for Indian languages is considered more difficult in general than for European languages because of the large number of vowels, consonants, and conjuncts (combinations of vowels and consonants) [13].

This work was part of exploring DIA research for a palm leaf manuscripts collection from Southeast Asia. This collection offers a new challenge for DIA researchers because palm leaves are used as the writing medium and the language and script have never been analyzed before. In this paper, we did a comprehensive benchmark experimental test of some principal tasks in the DIA system, starting with binarization, text line segmentation, isolated character/glyph recognition, word recognition, and transliteration. To the best of our knowledge, this work is the first comprehensive study of the DIA researchers' community and the first to perform a complete series of experimental benchmarking analyses of palm leaf manuscripts. The results of this research will be very useful in accelerating, evaluating, and improving the performance of existing DIA systems for a new type of document.

This paper is organized as follow. Section 2 gives a brief description of the palm leaf manuscripts collection from Southeast Asia, especially the Khmer palm leaf manuscript corpus from Cambodia and two palm leaf manuscript corpuses, the Balinese and Sundanese manuscripts from Indonesia. The challenges of DIA for this manuscript corpus are also presented in this section. Section 3 describes the DIA tasks that need to be developed for the palm leaf manuscript collections, followed by a description of the methods investigated for those tasks. The datasets and evaluation methods for each DIA task used in the experimental studies for this work are presented in Section 4. Section 5 reports and analyzes the detailed results of the experiments. Finally, conclusions are given in Section 6.

2. Palm Leaf Manuscripts from Southeast Asia

Regarding the use of writing materials and tools, history records the discovery of important documents written on stone plates, clay plates or tablets, bark, skin, animal bones, ivory, tortoiseshell, papyrus, parchment (form of leather made of processed sheepskin or calfskin) (http://www.casepaper. com/company/paper-history) [14], copper and bronze plates, bamboo, palm leaves, and other materials [15]. The choice of natural materials that can be used as a medium for document writing is strongly influenced by the geographical condition and location of a nation. For example, because bamboo and palm trees are easily found in Asia, both types of materials were the first choice of writing material in Asia. In Southeast Asia, most ancient manuscripts were written on palm leaves. For example, in Cambodia, palm leaves have been used as a writing material dating back to the first appearance of Buddhism in the country. In Thailand, dried palm leaves have also been used as one of the most popular written documents for over 500 years [16]. Palm leaves were also historically used as writing supports in manuscripts from the Indonesian archipelago. The leaves of sugar, or toddy, palm (*Borassus flabellifer*) are known as *lontar*. The existence of ancient palm leaf manuscripts in Southeast Asia is very important both in terms of the quantity and variety of historical contents.

2.1. Balinese Palm Leaf Manuscripts—Collection from Bali, Indonesia

2.1.1. Corpus

Apart from the collection at the museum (Museum Gedong Kertya Singaraja and Museum Bali Denpasar), it is estimated that there are more than 50,000 *lontar* collections that are owned by private families (Figure 1). For this research, in order to obtain a large variety of manuscript images, sample images have been collected from 23 different collections, which come from five different locations (regions): two museums and three private families. They consist of 10 randomly selected collections from Museum Gedong Kertya, City of Singaraja, Regency of Buleleng, North Bali, Indonesia, four collections from manuscript collections of Museum Bali, City of Denpasar, South Bali, seven collections from a private family collection from the village of Jagaraga, Regency of Buleleng, and two other private family collections from the village of Susut, Regency of Bangli and the village of Rendang, Regency of Karangasem [17].

Figure 1. Balinese palm leaf manuscripts.

2.1.2. Balinese Script and Language

Although the official language of Indonesia, Bahasa Indonesia, is written in the Latin script, Indonesia has many local, traditional scripts, most of which are ultimately derived from Brahmi [18]. In Bali, palm leaf manuscripts were written in the Balinese script in the Balinese language, in the ancient literary texts composed in the old Javanese language of Kawi and Sanskrit. Balinese language is a Malayo-Polynesian language spoken by more than 3 million people, mainly in Bali, Indonesia (www.omniglot.com/writing/balinese.htm) [19]. Balinese is the native language of the people of Bali, known locally as Basa Bali [18]. The alphabet and numbers of Balinese script are composed of ±100 character classes including consonants, vowels, and some other special compound characters. According to the Unicode Standard 9.0, the Balinese script actually has the Unicode table from 1B00 to 1B7F.

2.2. Khmer Palm Leaf Manuscripts—Collection from Cambodia

2.2.1. Corpus

In Cambodia, Khmer palm leaf manuscripts (Figure 2) are still seen in Buddhist establishments and are traditionally used by monks as reading scriptures. Various libraries and institutions have been collecting and digitizing these manuscripts and have even shared the digital images with the public. For instance, the École Française d'Extrême-Orient (EFEO) has launched an online database (http://khmermanuscripts.efeo.fr) [20] of microfilm images of hundreds of Khmer palm leaf manuscript collections. Some digitized collections are also obtained from the Buddhist Institute, which is one of the biggest institutes in Cambodia responsible for research on Cambodian literature and language related to Buddhism, and also from the National Library (situated in the capital city, Phnom Penh), which is home to a large collection of palm leaf manuscripts. Moreover, a standard digitization campaign was conducted in order to collect palm leaf manuscript images found in Buddhist temples in different locations throughout Cambodia: Phnom Penh, Kandal, and Siem Reap [21].

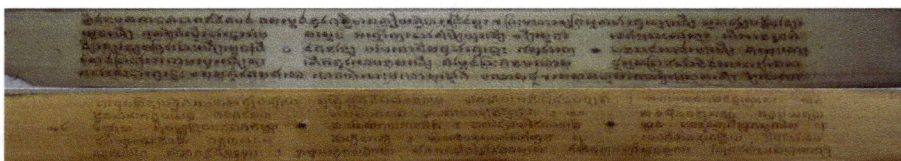

Figure 2. Khmer palm leaf manuscript.

2.2.2. Khmer Script and Language

According to the era during which the documents were created, slightly different versions of Khmer characters are used in the writing of Khmer palm leaf manuscripts. The Khmer alphabet is famous for its numerous symbols (~70), including consonants, different types of vowels, diacritics, and special characters. Certain symbols even have multiple shapes and forms depending on what other symbols are combined with them to create words. The languages written on palm leaf documents vary from Khmer, the official language of Cambodia, to Pali and Sanskrit, by which the modern Khmer language was considerably influenced. Only a minority of Cambodian people, such as philologists and Buddhist monks, are able to read and understand the latter languages.

2.3. Sundanese Palm Leaf Manuscripts—Collection from West Java, Indonesia

2.3.1. Corpus

The collection of Sundanese palm leaf manuscripts (Figure 3) comes from Situs Kabuyutan Ciburuy, Garut, West Java, Indonesia. The Kabuyutan Ciburuy is a complex cultural heritage from Prabu Siliwangi and Prabu Kian Santang, the king and the son of the Padjadjaran kingdom. The cultural complex consists of six buildings. One of them is Bale Padaleuman, which is used to store the Sundanese palm leaf manuscripts. The oldest Sundanese palm leaf manuscript in Situs Kabuyutan Ciburuy came from the 15th century. In Bale Padaleuman, there are 27 collections of Sundanese manuscripts. Each collection contains 15 to 30 pages, with dimensions of 25–45 cm in length × 10–15 cm in width [22].

2.3.2. Sundanese Script and Language

The Sundanese palm leaf manuscripts were written in the ancient Sundanese language and script. The characters consist of numbers, vowels (such as a, i, u, e, and o), basic characters (such as *ha*, *na*, *ca*, *ra*, etc.), punctuation, diacritics (such as *panghulu, pangwisad, paneuleung, panyuku,* etc.), and many special compound characters.

Figure 3. Sundanese palm leaf manuscript.

2.4. Challenges of Document Image Analysis for Palm Leaf Manuscripts

There are two main technical challenges to assessing palm leaf manuscripts in a DIA system. The first challenge is the physical condition of the palm leaf manuscript, which will strongly influence the quality of the document images captured. For the image capturing process for DIA research, data in a paper document are usually captured by optical scanning, but when the document is on a different medium such as microfilm, palm leaves, or fabric, photographic methods are often used to capture the images [13]. Nowadays, due to the specific characteristics of the physical support of the manuscripts, the development of DIA methods for palm leaf manuscripts in order to extract relevant information is considered a new research problem in handwritten document analysis. Ancient palm leaf manuscripts contain artifacts due to aging, foxing, yellowing, strain, local shading effects, low intensity variations or poor contrast, random noises, discolored parts, fading, and other types of degradation.

The second challenge is the complexity of the script. The Southeast Asian manuscripts with different scripts and languages provide real challenges for document analysis methods, not only because of the different forms of characters in the script, but also because the writing style of each script (e.g., how to join or separate a character in a text line) differs. It ranges widely from a binarization process [23–25], text line segmentation [26,27], and character and text recognition tasks [25,28,29], to the word spotting methods [30].

In the domain of DIA, handwritten character and text recognition has been the subject of intensive research during the last three decades. Some methods have already reached a satisfactory performance, especially for Latin, Chinese, and Japanese scripts. However, the development of handwritten character and text recognition methods for other various Asian scripts presents many issues. In the OCR task and development for palm leaf manuscripts from Southeast Asia, several deformations in the character shapes are visible due to the merges and fractures of the use of nonstandard fonts. The similarities of distinct character shapes, overlaps, and interconnection of the neighboring characters further complicate the OCR system [31]. One of the main problems faced when dealing with segmented handwritten character recognition is the ambiguity and illegibility of the characters [32]. These characteristics provide suitable conditions to test and evaluate the robustness of feature extraction methods that were proposed for character recognition.

3. Document Image Analysis Tasks and Investigated Methods

Heritage document preservation is not just about converting physical documents into document images. With many physical documents being digitized and stored in large document databases, and then sent and received via digital machines, the interest and demand grew to require more functionalities than simply viewing and print the images [33]. Further treatment is required before the collection of document images can be explored more extensively. For example, a more specific research field needed to be developed to add machine capabilities for extracting information from these images, reading text on a document page, finding sentences, and locating paragraphs, lines, words, and symbols on a diagram [33].

In this work, the methods for each DIA task were investigated for palm leaf manuscripts. The binarization task is evaluated using the latest methods from binarization competitions. The seam carving method is evaluated for the text line segmentation task, compared to a recent text line segmentation method for palm leaf manuscripts [27]. For the isolated character/glyph recognition task, the evaluation is reported from the handcrafted feature extraction method, the neural network with unsupervised learning feature to the CNN based method. Finally, the RNN-LSTM based method is used to analyze the word recognition and transliteration task for palm leaf manuscripts.

3.1. Binarization

Binarization is widely applied as the first pre-processing step in image document analysis [34]. Binarization is a common starting point for document image analysis and converts gray image values into binary representation for background and foreground, or, more specifically, text and non-text, which is then fed into further document processing tasks such as text line segmentation and optical character recognition. The performance of binarization techniques has a great impact and directly affects the performance of the recognition task [35]. Non-optimal binarization methods produce unrecognizable characters with noise [16]. Many binarization methods have been reported. These methods have been tested and evaluated on different types of document collections. Based on the choice of the thresholding value, binarization methods can generally be divided into two types, global binarization and local adaptive binarization [16]. Some surveys and comparative studies of the performance of several binarization methods have been reported [35,36]. A binarization method that performs well for one document collection may not necessarily be applied to another document collection with the same performance [34]. For this reason, there is always a need to perform a comprehensive evaluation of the existing binarization methods for a new document collection that has different characteristics, for example the historical archive documents [36].

In this work, we compared several alternative binarization algorithms for palm leaf manuscripts. We tested and evaluated some well-known standard binarization methods, and some binarization methods that are experimentally promising for historical archive documents, though not specifically for images of palm leaf manuscripts. We also tested the binarization methods from the Document Image Binarization Competition (DIBCO) competition [37,38], for example Howe's method [39] and the ones from the International Conference on Frontiers in Handwriting Recognition (ICFHR) competition (amadi.univ-lr.fr/ICFHR2016_Contest) [25,40].

3.1.1. Global Thresholding

Global thresholding is the simplest technique and the most conventional approach for binarization [34,41]. A single threshold value was calculated from the global characteristics of the image. This value should be properly chosen based on a heuristic technique or a statistical measurement to be able to give promising optimal binarization results [36]. It is widely known that using a global threshold to process a batch of archive images with different illumination and noise variation is not a proper choice. The variation between images in the foreground and background colors on low-quality document images gives unsatisfactory results. It is difficult to choose one fixed threshold value that is adaptable for all images [36,42].

Otsu's method is a very popular global binarization technique [34,41]. Conceptually, Otsu's method tries to find an optimum global threshold on an image by minimizing the weighted sum of variances of the objects and background pixels [34]. Otsu's method is implemented as a standard binarization technique in a built-in Matlab function called *graythresh* (https://fr.mathworks.com/help/images/ref/graythresh.html) [43].

3.1.2. Local Adaptive Binarization

To overcome the weakness of the global binarization technique, many local adaptive binarization techniques were proposed, for example Niblack's method [34,36,41,42,44], Sauvola's method [34,36,41,42,44,45], Wolf's method [42,44,46], NICK method [44], and the Rais method [34]. The threshold value in local adaptive binarization technique is calculated in each smaller local image area, region, or window. Niblack's method proposed a local thresholding computation based on the local mean and local standard deviation of a rectangular local window for each pixel on the image. The rectangular sliding local window will cover the neighborhood for each pixel. Using this concept, Niblack's method was reported to outperform many thresholding techniques and gave optimal results for many document collections. However, there is still a drawback to this method. It was found that Niblack's method works optimally only on the text region, but is not well suited for large non-text regions of an image. The absence of text in local areas forces Niblack's method to detect noise as text. The suitable window size should be chosen based on the character and stroke size, which may vary for each image. Many other local adaptive binarization techniques were proposed to improve the performance of the basic Niblack method. For example, Sauvola's method is a modified version of Niblack's method. Sauvola's method proposes a local binarization technique to deal with light texture, large variations, and uneven illumination. The improvement over Niblack's method is in the use of adaptive contribution of standard deviation in determining the local threshold on the gray values of text and non-text pixels. Sauvola's method processes the image in $N \times N$ adjacent and non-overlapping blocks separately.

Wolf's method tried to overcome the problem of Sauvola's method when the gray values of text and non-text pixels are close to each other by normalizing the contrast and the mean gray value of the image to compute the local threshold. However, a sharp change in background gray values across the image decreases the performance of Wolf's method. Two other improvements to Niblack's method are NICK method and the Rais method. NICK method proposes a threshold computation derived from the basic Niblack's method and the Rais method proposes an optimal size of window for the local binarization.

3.1.3. Training-Based Binarization

The top two proposed methods in the Binarization Challenge for the ICFHR 2016 Competition on the Analysis of Handwritten Text in Images of Balinese Palm Leaf Manuscripts are training-based binarization methods [25]. The best method in this competition employs a Fully Convolutional Network (FCN). It takes a color subimage as input and outputs the probability that each pixel in the sub-image is part of the foreground. The FCN is pre-trained on normal handwritten document images with automatically generated "ground truth" binarizations (using the method of Wolf et al. [46]). The FCN is then fine-tuned using DIBCO and HDIBCO competition images and their corresponding ground truth binarizations. Finally, the FCN is fine-tuned again on the provided Balinese palm leaf images. Consequently, the pixel probabilities of foreground are efficiently predicted for the whole image at once and thresholded at 0.5 to create a binarized output image.

The second-best method uses two neural network classifiers, C_1 and C_2, to classify each pixel as background or not. Two binarized images, B_1 and B_2, are generated in this step. C_1 is a rough classifier that tries to detect all the foreground pixels, while probably making mistakes for some background pixels. C_2 is an accurate classifier that should not classify a background pixel as a foreground pixel but probably misses some foreground pixels. Secondly, these two binary images are joined to get the final classification result.

3.2. Text Line Segmentation

Text line segmentation is a crucial pre-processing step in most DIA pipelines. The task aims at extracting and separating text regions into individual lines. Most line segmentation approaches in the literature require that the input image be binarized. However, due to the degradation and noise often found in historical documents such as palm leaf manuscripts, the binarization task is not able to produce good enough results (see Section 5.1). In this paper, we investigate two line segmentation methods that are independent of the binarization task. These approaches work directly on color/grayscale images.

3.2.1. Seam Carving Method

Arvanitopoulos and Süsstrunk [47] proposed a binarization-free method based on a two-stage process: medial seam and separating seam computation. The approach computes medial seams by splitting the input page image into columns whose smoothed projection profiles are then calculated. The positions of the medial seams are obtained based on the local maxima locations of the profiles. The goal of the second stage of the approach is to compute separating seams with the application on the energy map within the area restricted by the medial seams of two neighboring lines found in the previous stage. The technique carves paths that traverse the image from left to right, accumulating energy. The path with the minimum cumulative energy is then chosen.

3.2.2. Adaptive Path Finding Method

This approach was proposed by Valy et al. [27]. The method takes as input a grayscale image of a document page. Connected components are extracted from the input image using the stroke width information by applying the stroke width transform (SWT) on the Canny edge map. The set of extracted components (filtered to remove components that come from noise and artifacts) is used to create a stroke map. Using column-wise projection profiles on the output map, estimated number and medial positions of text line can be defined. To adapt better to skew and fluctuation, an unsupervised learning called competitive learning is applied on the set of connected components found previously. Finally, a path finding technique is applied in order to create seam borders between adjacent lines by using a combination of two cost functions: one penalizing the path that goes through the foreground text (intensity difference cost function D) and another one favoring the path that stays close to the estimated medial lines (vertical distance cost function V). Figure 4 illustrates an example of an optimal path.

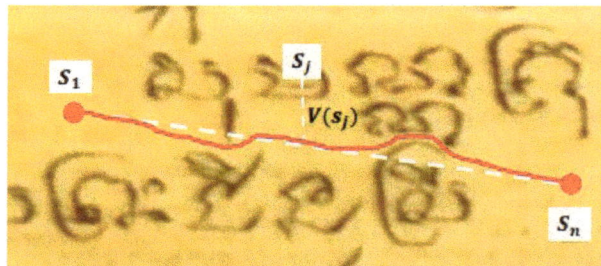

Figure 4. An example of an optimal path going from start state S_1 to goal state S_n.

3.3. Isolated Character/Glyph Recognition

In a DIA system, word or text recognition tasks are generally categorized into two different approaches: segmentation-based and segmentation-free methods. In segmentation-based methods, the isolated character recognition task is a very important process [9]. A proper feature extraction and a correct classifier selection can increase the recognition rate [48]. Although many methods for isolated character recognition have been developed and tested, especially for Latin-based scripts and alphabets, there is still a need for in-depth evaluation of those methods as applied to various other scripts. This includes the isolated character recognition task for many Southeast Asian scripts, and more specifically scripts that were written on ancient palm leaf manuscripts.

Previous studies on isolated character recognition in palm leaf manuscripts have already been reported, but only with the Balinese script as the benchmark dataset [28,29]. In that first work, an experimental study on feature extraction methods for character recognition of Balinese script was performed [28]. For the second work, a training-based method with neural network and unsupervised feature learning was used to increase the recognition rate [29]. In this paper, we will conduct a broader evaluation of the robustness of the methods previously tested on Balinese script, using the other two palm leaf manuscripts with Khmer and Sundanese scripts. In the next sub-sections, we provide a brief description of the methods. For a detailed description of each method, interested readers can refer to our previous works.

3.3.1. Handcrafted Feature Extraction Methods

Since the beginning of pattern recognition research, many feature extraction methods for character recognition have been presented in the literature. In our previous work [28], we investigated and evaluated the performance of 10 feature extraction methods with two classifiers, k-NN (k-Nearest Neighbor) and SVM (Support Vector Machine), in 29 different schemes for Balinese script on palm leaf manuscripts. After evaluating the performance of those individual feature extraction methods, we found that the Histogram of Gradient (HoG) features as directional gradient-based features [9,49] (Figure 5), the Neighborhood Pixels Weights (NPW) [50] (Figure 6), the Kirsch Directional Edges [50], and Zoning [12,32,50,51] (Figure 7) give very promising results. We then proposed a new feature extraction method applying NPW on Kirsch edge images (Figure 8) and concatenated the NPW–Kirsch with two other features, HoG and Zoning method, with k-NN as the classifier.

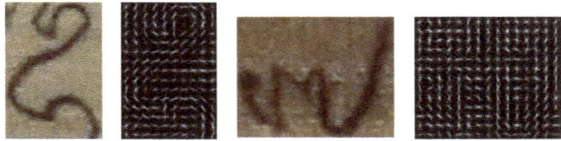

Figure 5. The representation of the array of cells in HoG [28].

Figure 6. Neighborhood pixels for NPW features [28].

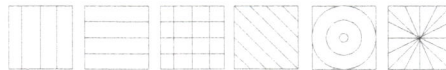

Figure 7. Type of Zoning (from left to right: vertical, horizontal, block, diagonal, circular, and radial zoning) [28].

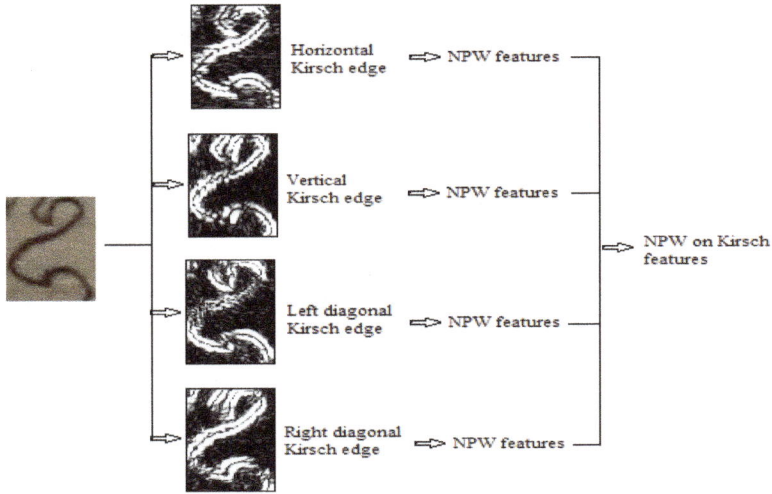

Figure 8. Scheme of NPW on Kirsch features [28].

3.3.2. Unsupervised Learning Feature and Neural Network

With the aim of improving the performance of our proposed feature extraction method, we continued our research on isolated character recognition by implementing the neural network as classifier. In this second step [29], the same combination of feature extraction methods was used and sent as the input feature vector to a single-layer neural network character recognizer. In addition to using only the neural network, we also applied an additional sub-module for the initial unsupervised learning based on K-Means clustering (Figure 9). This schema was inspired by the study of Coates et al. [52,53]. The unsupervised learning calculates the initial learning weight for the neural network training phase from the cluster centers of all feature vectors.

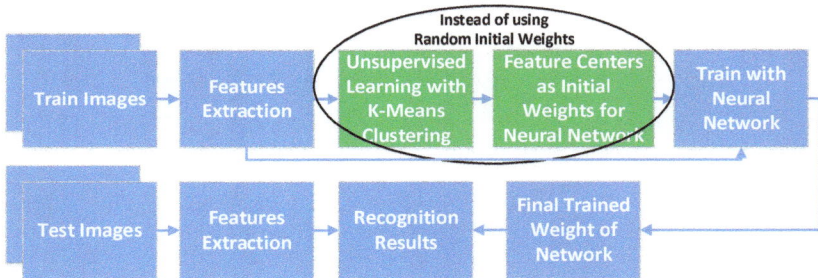

Figure 9. Schema of character recognizer with feature extraction method, unsupervised learning feature, and neural network [29].

3.3.3. Convolutional Neural Network

The multilayer convolutional neural networks (CNN) have proven very effective in areas such as image recognition and classification. In this evaluation experiment, a vanilla CNN is used. The architecture of the CNN (Figure 10) is described as follows (this architecture has also been reported in Khmer isolated character recognition baseline in [21]). The grayscale input images of isolated characters are rescaled to 48 × 48 pixels in size and normalized by applying histogram stretching.

The network consists of three sets of convolution and max pooling pairs. All convolutional layers use a stride of one and are zero padded so that the output is the same size as the input. The output of each convolutional layer is activated using the ReLu function and followed by a max pooling of 2×2 blocks. The numbers of feature maps (of size 5×5) used in the three consecutive convolutional layers are 8, 16, and 32, respectively. The output of the last layers is flattened, and a fully-connected layer with 1024 neurons (also activated with ReLu) is added, followed by the last output layer (softmax activation) consisting of N_{class} neurons, where N_{class} is the number of character classes. Dropout with probability $p = 0.5$ is applied before the output layer to prevent overfitting. We trained the network using an Adam optimizer with a batch size of 100 and a learning rate of 0.0001.

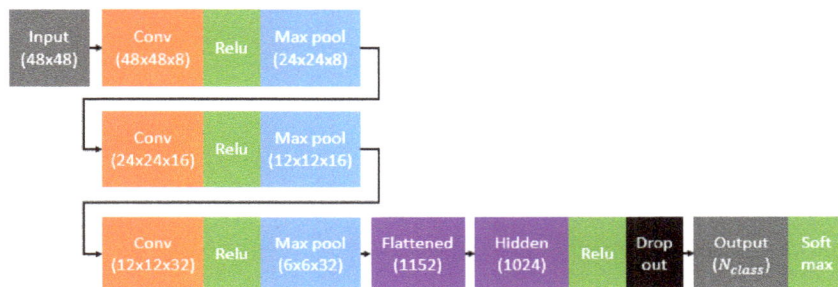

Figure 10. Architecture of the CNN.

3.4. Word Recognition and Transliteration

In order to make the palm leaf manuscripts more accessible, readable, and understandable to a wider audience, an optical character recognition (OCR) system should be developed. In many DIA systems, word or text recognition is the final task in the processing pipeline. However, normally in Southeast Asian script the speech sound of the syllable change is related to some certain phonological rules. In this case, an OCR system is not enough. Therefore, a transliteration system should also be developed to help transliterate the ancient scripts on these manuscripts. By definition, transliteration is defined as the process of obtaining the phonetic translation of names across languages [54]. Transliteration involves rendering a language from one writing system to another. In [54], the problem is stated formally as a sequence labeling problem from one language alphabet to another. It will help us to index and to quickly and efficiently access the content of the manuscripts. In our previous work [29], a complete scheme for segmentation-based glyph recognition and transliteration specific to Balinese palm leaf manuscripts was proposed. In this work, a segmentation-free method will be evaluated to recognize and transliterate the words from three different scripts of a palm leaf manuscript.

RNN/LSTM-Based Methods

From the last decade, sequence-analysis-based methods using a Recurrent Neural Network-Long Short-Term Memory (RNN-LSTM) type of learning network have been very popular among researchers in text recognition. RNN-LSTM-based method together with a Connectionist Temporal Classification (CTC) works as a segmentation-free learning-based method to recognize the sequence of characters in a word or text without any handcrafted feature extraction method. The raw image pixel can be sent directly as the input to the learning network and there is no requirement to segment the training data sequence. RNN is basically an extended version of the basic feedforward neural network. In a RNN, the neurons in the hidden layer are connected to each other. RNN offers very good context-aware processing to recognize patterns in a sequence or time series. One drawback of RNN is the vanishing gradient problem. To deal with this problem, the LSTM architecture was introduced. The LSTM network adds multiplicative gates and additive feedback. Bidirectional LSTM is an LSTM architecture with two-directional (forward and backward) context processing. LSTM architecture is widely evaluated as a generic and language-independent text recognizer [55]. In this work, the OCRopy (https://github.com/tmbdev/ocropy) [56] framework is used to test and evaluate the word recognition and transliteration tasks for the palm leaf manuscript collection. OCRopy provides the functional library of the OCR system by using RNN-LSTM architecture (http://graal.hypotheses. org/786) [57,58]. We evaluated the dataset with unidirectional LSTM and the (Bidirectional LTSM) BLSTM architecture.

4. Experiments: Datasets and Evaluation Methods

From the three manuscript corpuses (Khmer, Balinese, and Sundanese), the datasets for each DIA task were extracted and used in the experimental work for this research.

4.1. Binarization

4.1.1. Datasets

The palm leaf manuscript datasets for binarization task are presented in Table 1. For Khmer manuscripts, one ground truth binarized image is provided for each image, but for Balinese and Sundanese manuscripts, each image has two different ground truth binarized images [17,25]. The study of ground truth variability and subjectivity was reported in the previous work [24]. In this research, we only used the first binarized ground truth image for evaluation. The binarized ground truth images for Khmer manuscripts were generated manually with the help of photo editing software (Figure 11). A pressure-sensitive tip stylus is used to trace each text stroke by keeping the original size of the stroke width [59]. For the manuscripts from Bali, the binarized ground truth images have been created with a semi-automatic scheme [17,23–25] (Figure 12). The binarized ground truth images for Sundanese manuscripts were manually [22] generated using PixLabeler [60] (Figure 13). The training set is provided only for the Balinese dataset. We used all images of the Khmer and Sundanese corpuses as a test set because the training-based binarization method (ICFHR G1 method, see Section 5.1) was evaluated for the Khmer and Sundanese datasets by using only the pre-trained Balinese training set weighted model.

Table 1. Palm leaf manuscript datasets for binarization task.

Manuscripts	Train	Test	Ground Truth	Dataset
Balinese	50 pages	50 pages	2 × 100 pages	Extracted from AMADI_LontarSet [17,25,40]
Khmer	-	46 pages	1 × 46 pages	Extracted from EFEO [20,59]
Sundanese	-	61 pages	2 × 61 pages	Extracted from Sunda Dataset ICDAR2017 [22]

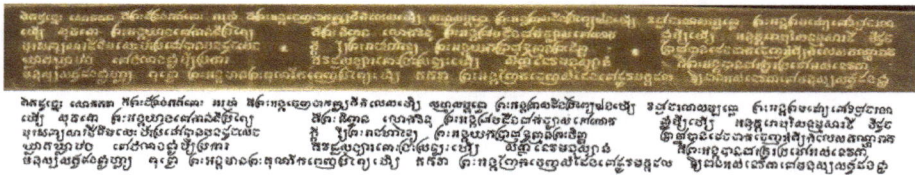

Figure 11. Khmer manuscript with binarized ground truth image.

Figure 12. Balinese manuscript with binarized ground truth image.

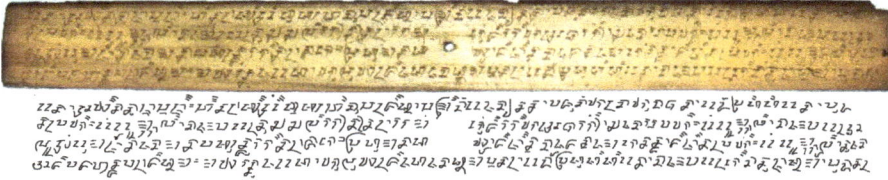

Figure 13. Sundanese manuscript with binarized ground truth image.

4.1.2. Evaluation Method

Following our previous work [24] and the evaluation method from the ICFHR competition [25], three metrics of binarization evaluation that were used in the DIBCO 2009 contest [37] are used in the binarization task evaluation for this work. Those three metrics are F-Measure (FM) (Equation (3)), *Peak SNR (PSNR)* (Equation (5)), and Negative Rate Metric (NRM) (Equation (8)).

F-Measure (FM): FM is defined from Recall and Precision.

$$Recall = \frac{TP}{FN + TP} \times 100 \tag{1}$$

$$Precision = \frac{TP}{FP + TP} \times 100 \tag{2}$$

TP, defined as true positive, occurs when the image pixel is labeled as foreground and the ground truth is also. *FP*, defined as false positive, occurs when the image pixel is labeled as foreground but the ground truth is labeled as background. *FN*, defined as false negative, occurs when the image pixel is labeled as background but the ground truth is labeled as foreground (Equations (1) and (2)).

$$FM = \frac{2 \times \text{Recall} \times \text{Precision}}{\text{Recall} + \text{Precision}} \tag{3}$$

A higher F-measure indicates a better match.

Peak SNR (PSNR): PSNR is calculated from *Mean Square Error (MSE)* (Equation (4)).

$$MSE = \sum_{x=1}^{M} \sum_{y=1}^{N} \frac{(I_1(x,y) - I_2(x,y))^2}{M * N} \tag{4}$$

$$PSNR = 10 \times \log_{10}(\frac{C^2}{MSE}), \tag{5}$$

where *C* is defined as 1, the difference between foreground and background colors in the case of a binary image. A higher PSNR indicates a better match.

Negative Rate Metric (NRM): NRM is defined from the negative rate of false negative (NR_{FN}) (Equation (6)) and the negative rate of false positive (NR_{FP}) (Equation (7)):

$$NR_{FN} = \frac{FN}{FN + TP} \tag{6}$$

$$NR_{FP} = \frac{FP}{FP + TN} \tag{7}$$

TN, defined as true negative, occurs when both the image pixel and ground truth are labeled as background. The definitions of *TP*, *FN*, and *FP* are the same as the ones given for the F-Measure.

$$NRM = \frac{NR_{FN} + NR_{FP}}{2} \tag{8}$$

A lower NRM indicates a better match.

4.2. Text Line Segmentation

4.2.1. Datasets

The palm leaf manuscript datasets for text line segmentation task are presented in Table 2. The text line segmentation ground truth data for Balinese and Sundanese manuscripts have been generated by hand based on the binarized ground truth images [17]. For Khmer 1, a semi-automatic scheme is used [26,59]. A set of medial points for each text is generated automatically on the binarization ground truth of the page image. Then those points can be moved up or down with a tool to fit the skew and fluctuation of the real text lines. We also note touching components spreading over multiple lines and the locations where they can be separated. For Khmer 2 and 3, an ID of the line it belongs to is associated with each annotated character. The region of a text line is the union of the areas of the polygon boundaries of all annotated characters composing it [21,27].

Table 2. Palm leaf manuscript datasets for text line segmentation task.

Manuscripts	Pages	Text Lines	Dataset
Balinese 1	35 pages	140 text lines	Extracted from AMADI_LontarSet [17,26,40]
Balinese 2	Bali-2.1: 47 pages	181 text lines	Extracted from AMADI_LontarSet [17]
	Bali-2.2: 49 pages	182 text lines	
Khmer 1	43 pages	191 text lines	Extracted from EFEO [20,26,59]
Khmer 2	100 pages	476 text lines	Extracted from SleukRith Set [21,27]
Khmer 3	200 pages	971 text lines	Extracted from SleukRith Set [21]
Sundanese 1	12 pages	46 text lines	Extracted from Sunda Dataset [26]
Sundanese 2	61 pages	242 text lines	Extracted from Sunda Dataset [22]

4.2.2. Evaluation Method

Following our previous work [26], we use the evaluation criteria and tool provided by ICDAR2013 Handwriting Segmentation Contest [61]. First, the one-to-one (o2o) match score is computed for a region pair based on the evaluator's acceptance threshold. In our experiments, we used 90% as the acceptance threshold. Let N be the count of ground truth elements, and M the count of result elements. With the o2o score, three metrics are calculated: detection rate (DR), recognition accuracy (RA), and performance metric (FM).

4.3. Isolated Character/Glyph Recognition

4.3.1. Datasets

The palm leaf manuscript datasets for isolated character/glyph recognition task are presented in Table 3. For the Balinese character dataset, Balinese philologists manually annotated the segment of connected components that represented a correct character in Balinese script from the word-level binarized images that were manually annotated [11,17,20] using Aletheia (http://www.primaresearch. org/tools/Aletheia) [62,63] (Figure 14). The Sundanese character dataset was annotated manually [22] (Figure 15). For the Khmer character dataset, a tool has been developed to annotate characters/glyphs on the document page. The polygon boundary of each character is traced manually by dotting out its vertex one by one. A label is given to each annotated character after its boundary has been constructed [21] (Figure 16).

Table 3. Palm leaf manuscript datasets for isolated character/glyph recognition task.

Manuscripts	Classes	Train	Test	Dataset
Balinese	133 classes	11,710 images	7673 images	AMADI_LontarSet [17,25,28]
Khmer	111 classes	113,206 images	90,669 images	SleukRith Set [21]
Sundanese	60 classes	4555 images	2816 images	Sunda Dataset [22]

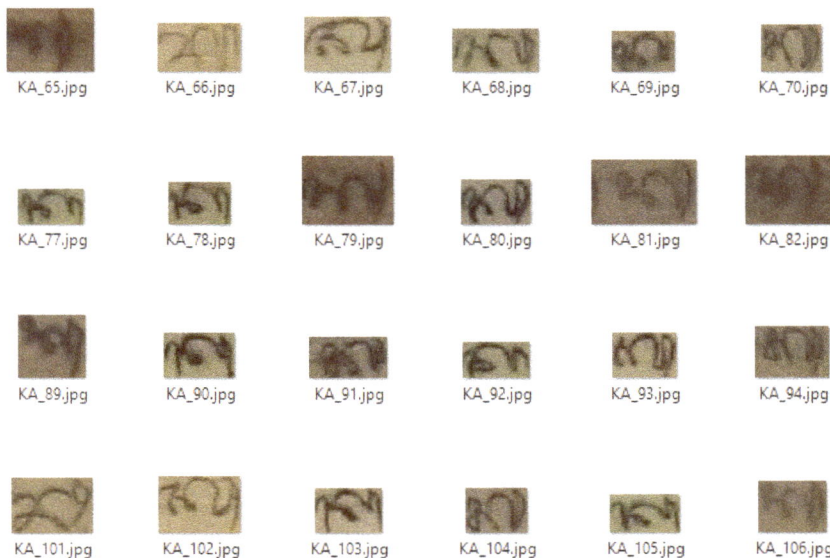

Figure 14. Balinese character dataset.

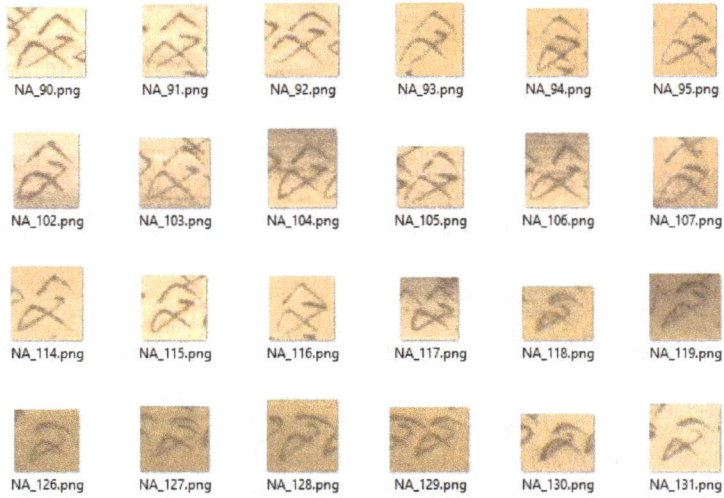

Figure 15. Sundanese character dataset.

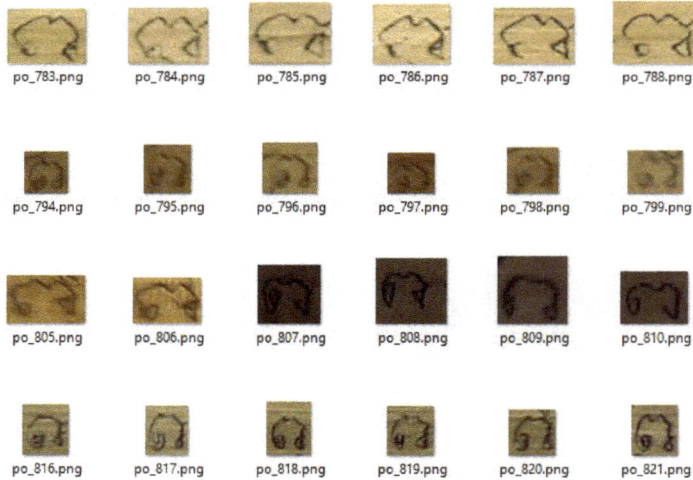

Figure 16. Khmer character dataset.

4.3.2. Evaluation Method

Following the evaluation method from the ICFHR competition [25], the recognition rate, i.e., the percentage of correctly classified samples over the test samples (C/N) is calculated, where C is the number of correctly recognized samples and N is the total number of test samples.

4.4. Word Recognition and Transliteration

4.4.1. Datasets

The palm leaf manuscript datasets for word recognition and transliteration task are presented in Table 4. For the Khmer dataset, all characters on the page have been annotated and grouped together into words (Figure 17). More than one label may be given to the created word. The order of how each character in the word is selected is also kept [21]. Balinese (Figure 18) and the Sundanese (Figure 19) word dataset was manually annotated using Aletheia [63].

Table 4. Palm leaf manuscript datasets for word recognition and transliteration tasks.

Manuscripts	Train	Test	Text	Published
Balinese	15,022 images from 130 pages	10,475 images from 100 pages	Latin	AMADI_LontarSet [17,25]
Khmer	16,333 images (part of 657 pages)	7791 images (part of 657 pages)	Latin and Khmer	SleukRith Set [21]
Sundanese	1427 images from 20 pages	318 images from 10 pages	Latin	Sunda Dataset [22]

Figure 17. Khmer word dataset.

Figure 18. Balinese word dataset.

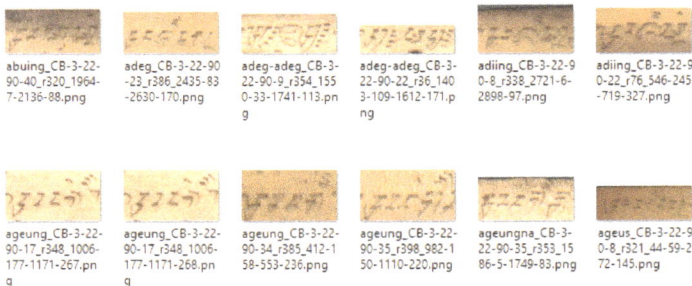

Figure 19. Sundanese word dataset.

4.4.2. Evaluation Method

The error rate is defined by edit distances between ground truth and recognizer output and is computed using the provided OCRopy function ocropus-errs (https://github.com/tmbdev/ocropy/blob/master/ocropus-errs) [56].

5. Experimental Results and Discussion

In this section, the performance of each method for the DIA tasks on palm leaf manuscript collections is presented.

5.1. Binarization

The experimental results for the binarization task are presented in Table 5. These results show that the performance of all methods on each dataset is still quite low. Most of the methods achieve less than a 50% FM score. This means that palm leaf manuscripts are still an open challenge for the binarization task. The different parameter values for the local adaptive binarization methods show significant improvement in performance, but still give unsatisfactory results. In these experiments, the ICFHR G1 method was evaluated for the Khmer and Sundanese datasets using the pre-trained Balinese training set weighted model. Based on these experiments, Niblack's method gives the highest FM score for Sundanese manuscripts (Figure 20), ICFHR G1 method gives the highest FM score for Khmer manuscripts (Figure 21), and ICFHR G2 gives the highest FM score for Balinese manuscripts (Figure 22). However, visually, there are still many broken and unrecognizable characters/glyphs, and noise is detected in the images.

Figure 20. Binarization of Sundanese manuscript with Niblack's method.

Figure 21. Binarization of Khmer manuscript with ICFHR G1 method.

Table 5. Experimental results for binarization task in F-Measure (FM), Peak SNR (PSNR), and Negative Rate Metric (NRM). A higher F-measure and PSNR, and a lower NRM, indicate a better result.

Methods	Parameter	Manuscripts	FM (%)	NRM	PSNR (%)
OtsuGray [34,41]	Otsu from gray image Using Matlab graythresh [43]	Balinese	18.98178	0.398894	5.019868
		Khmer	23.92159	0.313062	7.387765
		Sundanese	23.70566	0.326681	9.998433
OtsuRed [34,41]	Otsu from red image channel Using Matlab graythresh	Balinese	29.20352	0.300145	10.94973
		Khmer	21.15379	0.337171	5.907433
		Sundanese	21.25153	0.38641	12.60233
Sauvola [34,36,41,42,44,45]	$window = 50, k = 0.5, R = 128$	Balinese	13.20997	0.462312	27.69732
		Khmer	44.73579	0.268527	26.06089
		Sundanese	6.190919	0.479984	24.78595
Sauvola2 [34,36,41,42,44,45]	$window = 50, k = 0.2, R = 128$	Balinese	40.18596	0.274551	25.0988
		Khmer	47.55924	0.155722	21.96846
		Sundanese	43.04994	0.299694	23.65228
Sauvola3 [34,36,41,42,44,45]	$window = 50, k = 0.0, R = 128$	Balinese	35.38635	0.165839	17.05408
		Khmer	30.5562	0.190081	12.78953
		Sundanese	40.29642	0.181465	16.25056
Niblack [34,36,41,42,44]	$window = 50, k = -0.2$	Balinese	41.55696	0.175795	21.24452
		Khmer	38.01222	0.160807	16.84153
		Sundanese	46.79678	0.195015	20.31759
Niblack2 [34,36,41,42,44]	$window = 50, k = 0.0$	Balinese	35.38635	0.165839	17.05408
		Khmer	30.5562	0.190081	12.78953
		Sundanese	40.29642	0.181465	16.25056
NICK [44]	$window = 50, k = -0.2$	Balinese	37.85919	0.328327	27.59038
		Khmer	51.2578	0.176003	24.51998
		Sundanese	29.5918	0.390431	24.26187
Rais [34]	$window = 50$	Balinese	34.46977	0.171096	16.84049
		Khmer	31.59138	0.187948	13.52816
		Sundanese	40.65458	0.177016	16.35472
Wolf [42,44]	$window = 50, k = 0.5$	Balinese	27.94817	0.392937	27.1625
		Khmer	46.78589	0.23739	25.1946
		Sundanese	42.40799	0.299157	23.61075
Howe1 [39]	Default values [39]	Balinese	44.70123	0.267627	28.35427
		Khmer	40.20485	0.280604	25.59887
		Sundanese	45.90779	0.235175	21.90439
Howe2 [39]	Default values	Balinese	40.5555	0.273994	28.02874
		Khmer	32.35603	0.294016	25.96965
		Sundanese	35.35973	0.274865	22.36583
Howe3 [39]	Default values	Balinese	42.15377	0.304962	28.38466
		Khmer	30.7186	0.382087	26.36983
		Sundanese	25.77321	0.350349	23.66912
Howe4 [39]	Default values	Balinese	45.73681	0.273018	28.60561
		Khmer	36.48396	0.280519	25.83969
		Sundanese	38.98445	0.281118	22.83914
ICFHR G1	See ref. [25]	Balinese	63.32	0.15	31.37
		Khmer	52.65608	0.250503	28.16886
		Sundanese	38.95626	0.329042	24.15279
ICFHR G2	See ref. [25]	Balinese	68.76	0.13	33.39
		Khmer	-	-	-
		Sundanese	-	-	-
ICFHR G3	See ref. [25]	Balinese	52.20	0.18	26.92
		Khmer	-	-	-
		Sundanese	-	-	-
ICFHR G4	See ref. [25]	Balinese	58.57	0.17	29.98
		Khmer	-	-	-
		Sundanese	-	-	-

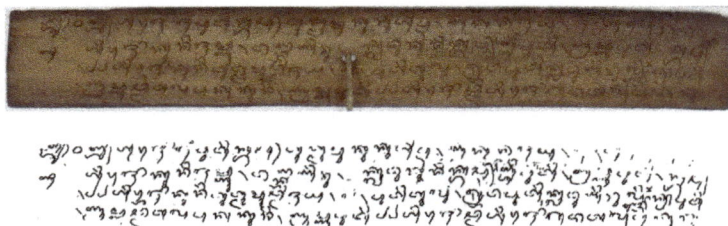

Figure 22. Binarization of Balinese manuscript with ICFHR G2 method.

5.2. Text Line Segmentation

The experimental results for text line segmentation task are presented in Table 6. According to these results, both methods perform sufficiently well for most datasets, except Khmer 1 (Figures 23–25). This is because all images in this set are of low quality due to the fact that they are digitized from microfilms. Nevertheless, the adaptive path finding method achieves better results than the seam carving method on all datasets of palm leaf manuscripts in our experiment. The main difference between these two approaches is that instead of finding an optimal separating path within an area constrained by medial seam locations of two adjacent lines (in the seam carving method), the adaptive path finding approach tries to find a path close to an estimated straight seam line section. These line sections already represent well the seam borders between two neighboring lines, so they can be considered a better guide for finding good paths, hence producing better results.

One common error that we encounter for both methods is in the medial position computation stage. Detecting correct medial positions of text lines is crucial for the path-finding stage of the methods. In our experiment, we noticed that some parameters play an important role. For instance, the number of columns/slices r of the seam carving method and the high and low thresholding values of the edge detection algorithm in the adaptive path finding approach are important. In order to select these parameters, a validation set consisting of five random pages is used. The optimal values of the parameters are then empirically selected based on the results from this validation set.

Table 6. Experimental results for text line segmentation task: the count of ground truth elements (N), and the count of result elements (M), the one-to-one (o2o) match score is computed for a region pair based on 90% acceptance threshold, detection rate (DR), recognition accuracy (RA), and performance metric (FM).

Methods	Manuscripts	N	M	o2o	DR (%)	RA (%)	FM (%)
Seam carving [47]	Balinese 1	140	167	128	91.42	76.64	83.38
	Bali-2.1	181	210	163	90.05	77.61	83.37
	Bali-2.2	182	219	161	88.46	73.51	80.29
	Khmer 1	191	145	57	29.84	39.31	33.92
	Khmer 2	476	665	356	53.53	74.79	62.40
	Khmer 3	971	1046	845	87.02	80.78	83.78
	Sundanese 1	46	43	36	78.26	83.72	80.89
	Sundanese 2	242	257	218	90.08	84.82	87.37
Adaptive Path Finding [27]	Balinese 1	140	143	132	94.28	92.30	93.28
	Bali-2.1	181	188	159	87.84	84.57	86.17
	Bali-2.2	182	191	164	90.10	85.86	87.93
	Khmer 1	191	169	118	61.78	69.82	65.55
	Khmer 2	476	484	446	92.15	93.70	92.92
	Khmer 3	971	990	910	93.71	91.91	92.80
	Sundanese 1	46	50	41	89.13	82.00	85.41
	Sundanese 2	242	253	222	91.73	87.74	89.69

Figure 23. Text line segmentation of Balinese manuscript with the Seam Carving method (green) and Adaptive Path Finding (red).

Figure 24. Text line segmentation of Khmer manuscript with the Seam Carving method (green) and Adaptive Path Finding (red).

Figure 25. Text line segmentation of Sundanese manuscript with the Seam Carving method (green) and Adaptive Path Finding (red).

5.3. Isolated Character/Glyph Recognition

The experimental results for isolated character/glyph recognition task are presented in Table 7. For handcrafted feature with k-NN, the Khmer set with 113,206 train images and 90,669 test images will need a considerable amount of time for one-to-one k-NN comparison, so we do not think it is reasonable to use it. For CNN 1, previous work only reported results for the Balinese set. For all ICFHR competition methods, the competition was proposed only for the Balinese set, so we only have the reported results for the Balinese set. According to these results, the handcrafted feature extraction combination of HoG-NPW-Kirsch-Zoning is a proper choice resulting in a good recognition rate for Balinese and Khmer characters/glyphs. The CNN methods also show satisfactory results, but the differences in recognition rates are not too significant with the handcrafted feature combinations. The unbalanced number of image samples for each character class means the CNN method did not perform optimally. For the Sundanese dataset, the handcrafted feature with NN slightly outperformed the CNN method. The UFL method slightly increased the recognition rate of the pure NN method for the Khmer and Balinese datasets.

Table 7. Experimental results for isolated character/glyph recognition tasks (in % recognition rate).

Methods	Balinese	Khmer	Sundanese
Handcrafted Feature (HoG-NPW-Kirsch-Zoning) with k-NN [28]	85.16	-	72.91
Handcrafted Feature (HoG-NPW-Kirsch-Zoning) with NN [29]	85.51	92.15	79.69
Handcrafted Feature (HoG-NPW-Kirsch-Zoning) with UFL + NN [29]	85.63	92.44	79.33
CNN 1 [28]	84.31	-	-
CNN 2	85.39	93.96	79.05
ICFHR G1: VCMF [25]	87.44	-	-
ICFHR G1: VMQDF [25]	88.39	-	-
ICFHR G3 [25]	77.83	-	-
ICFHR G5 [25]	77.70	-	-

5.4. Word Recognition and Transliteration

The experimental results for word recognition and transliteration task are presented in Table 8. The error rates for word recognition and transliteration tests set on each training model iteration are shown in Figures 26–28. The LSTM-based architecture of OCRopy seems very promising in terms of recognizing and directly transliterating Balinese words. For the Khmer and Sundanese datasets, the LSTM architecture seems to struggle to learn the training data. More synthetic data training with a more frequent word should be generated in order to support the training process. For the Balinese dataset, a sequence depth of 100 pixels with a neuron size of 200 gives a better result for both LSTM and BLTSM architecture. Most of the Southeast Asian scripts are syllabic scripts. One character/glyph in these scripts represents a syllable, with a sequence of letters in Latin script. In this case, word transliteration is not just word recognition with one-to-one glyph-to-letter association. This makes word transliteration more challenging than character/glyph recognition.

Table 8. Experimental results for word recognition and transliteration tasks (in % error rate for test).

Methods (with OCRopy [56] Framework)	Balinese	Khmer	Sundanese
BLSTM 1 (seq_depth 60, neuron size 100)	43.13	Latin text: 73.76 Khmer text: 77.88	75.52
LSTM 1 (seq_depth 100, neuron size 100)	42.88	-	-
BLSTM 2 (seq_depth 100, neuron size 200)	40.54	-	-
LSTM 2 (seq_depth 100, neuron size 200)	39.70	-	-

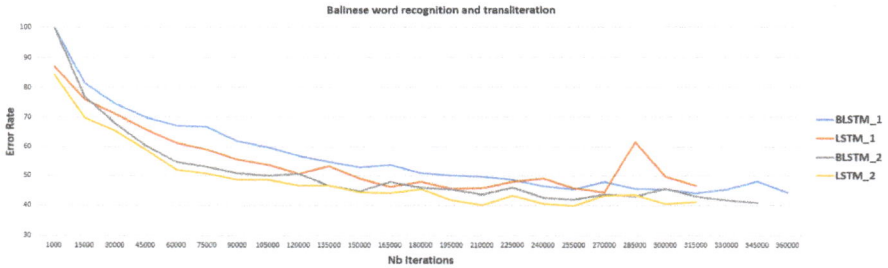

Figure 26. Error rate for Balinese word recognition and transliteration test set.

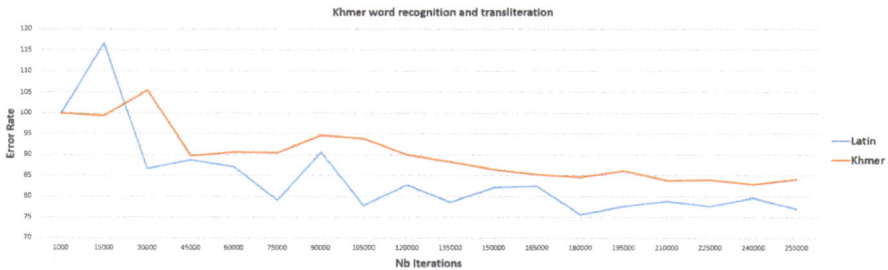

Figure 27. Error rate for Khmer word recognition and transliteration test set.

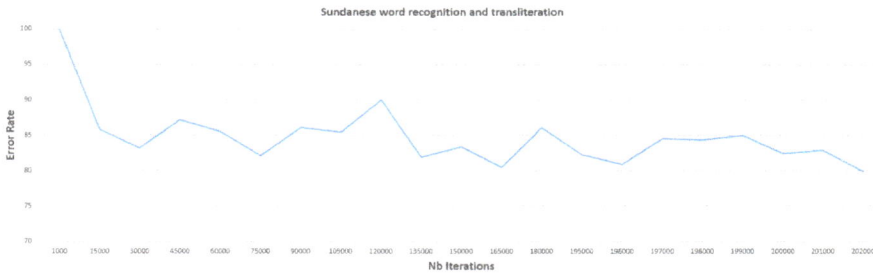

Figure 28. Error rate for Sundanese word recognition and transliteration test set.

6. Conclusions and Future Work

A comprehensive experimental test of the principal tasks in a DIA system, starting with binarization, text line segmentation, and isolated character/glyph recognition, and continuing on to word recognition and transliteration for a new collection of palm leaf manuscripts from Southeast Asia, is presented. The results from all experiments provide the latest findings and a quantitative benchmark of palm leaf manuscripts analysis for researchers in the DIA community. Binarizing the palm leaf manuscript images seems very challenging. Still, with many broken and unrecognizable characters/glyphs and noises detected in the images, binarization should be reconsidered the first step in the DIA process for palm leaf manuscripts. On the other hand, although there are already training-based DIA methods that do not require this binarization process, they usually require adequate training data. The problem of inadequate training data also influences glyph recognition and word transliteration. The unbalanced number of image samples for each character class means the CNN methods did not perform optimally in glyph recognition. The differences in the recognition rates of the CNN methods are not too significant with the handcrafted feature combinations. For future work, more synthetic data training for palm leaf manuscript images should be generated in order to support the training process. Especially for the word transliteration task, more synthetic data training with a more frequent word should be generated in order to improve the training process. Many examples of glyph-to-syllable association should be synthetically generated to transliterate syllabic scripts from Southeast Asia. The special characteristics and challenges posed by the palm leaf manuscript collections will require a thorough adaptation of the DIA system. Some specific adjustments need to be applied to the DIA methods for other types of documents. The adaptation of a DIA for palm leaf manuscripts is not unique and is not universal for all types of problem from different collections. However, among the DIA system's non-unique solutions, one specific solution can still be designed to deliver the most optimal DIA system performance while still taking into account the conditions of that collection.

Acknowledgments: The authors would like to thank Museum Gedong Kertya, Museum Bali, Undang Ahmad Darsa, the philologists from Sundanese Centre Studies of Universitas Padjadjaran, the Situs Kabuyutan Ciburuy Garut, all families in Bali, Indonesia, the EFEO team, the Buddhist Institute, and the National Library in Cambodia for providing us with samples of palm leaf manuscripts. We also thank the students from the Department of Informatics Education and the Department of Balinese Literature, University of Pendidikan Ganesha, the Institute of Technology of Cambodia, and the National Institute of Post, Telecommunication and ICT for helping us with the ground truthing process for this research project. This work is supported by the DIKTI BPPLN Indonesian Scholarship Program, the STIC Asia Program implemented by the French Ministry of Foreign Affairs and International Development (MAEDI), and ARES-CCD (program AI 2014-2019) under the funding of Belgian university cooperation, and DRPMI Universitas Padjadjaran, DIKTI International Collaboration and Publication grant 2017.

Author Contributions: The Balinese dataset was prepared by Made Windu Antara Kesiman. The Khmer dataset was prepared by Dona Valy and Sophea Chhun. The Sundanese dataset was prepared by Erick Paulus, Mira Suryani, and Setiawan Hadi. Jean-Christophe Burie, Michel Verleysen, and Jean-Marc Ogier contributed to designing a ground truth validation protocol. Made Windu Antara Kesiman and Dona Valy conceived, designed, and performed the experiments. Made Windu Antara Kesiman, Dona Valy, and Jean-Christophe Burie contributed to paper writing and editing.

Conflicts of Interest: The authors declare no conflict of interest. The founding sponsors had no role in the design of the study; in the collection, analyses, or interpretation of data; in the writing of the manuscript, and in the decision to publish the results.

References

1. tranScriptorium. Available online: http://transcriptorium.eu/ (accessed on 20 February 2018).
2. READ Project—Recognition and Enrichment of Archival Documents. Available online: https://read.transkribus.eu/ (accessed on 20 February 2018).
3. IAM Historical Document Database (IAM-HistDB)—Computer Vision and Artificial Intelligence. Available online: http://www.fki.inf.unibe.ch/databases/iam-historical-document-database (accessed on 20 February 2018).
4. Ancient Lives: Archive. Available online: https://www.ancientlives.org/ (accessed on 20 February 2018).
5. Document Image Analysis—CVISION Technologies. Available online: http://www.cvisiontech.com/library/pdf/pdf-document/document-image-analysis.html (accessed on 20 February 2018).
6. Ramteke, R.J. Invariant Moments Based Feature Extraction for Handwritten Devanagari Vowels Recognition. *Int. J. Comput. Appl.* **2010**, *1*, 1–5. [CrossRef]
7. Siddharth, K.S.; Dhir, R.; Rani, R. Handwritten Gurmukhi Numeral Recognition using Different Feature Sets. *Int. J. Comput. Appl.* **2011**, *28*, 20–24. [CrossRef]
8. Sharma, D.; Jhajj, P. Recognition of Isolated Handwritten Characters in Gurmukhi Script. *Int. J. Comput. Appl.* **2010**, *4*, 9–17. [CrossRef]
9. Aggarwal, A.; Singh, K.; Singh, K. Use of Gradient Technique for Extracting Features from Handwritten Gurmukhi Characters and Numerals. *Procedia Comput. Sci.* **2015**, *46*, 1716–1723. [CrossRef]
10. Lehal, G.S.; Singh, C.A. Gurmukhi script recognition system. In Proceedings of the 15th International Conference on Pattern Recognition, Barcelona, Spain, 3–7 September 2000; pp. 557–560.
11. Rothacker, L.; Fink, G.A.; Banerjee, P.; Bhattacharya, U.; Chaudhuri, B.B. Bag-of-features HMMs for segmentation-free Bangla word spotting. In Proceedings of the 4th International Workshop on Multilingual OCR, Washington, DC, USA, 24 August 2013; p. 5.
12. Ashlin Deepa, R.N.; Rao, R.R. Feature Extraction Techniques for Recognition of Malayalam Handwritten Characters: Review. *Int. J. Adv. Trends Comput. Sci. Eng.* **2014**, *3*, 481–485.
13. Kasturi, R.; O'Gorman, L.; Govindaraju, V. Document image analysis: A primer. *Sadhana* **2002**, *27*, 3–22. [CrossRef]
14. Paper History, Case Pap. Available online: http://www.casepaper.com/company/paper-history/ (accessed on 20 February 2018).
15. Doermann, D. *Handbook of Document Image Processing and Recognition*; Tombre, K., Ed.; Springer London: London, UK, 2014; p. 1055.
16. Chamchong, R.; Fung, C.C.; Wong, K.W. *Comparing Binarisation Techniques for the Processing of Ancient Manuscripts*; Nakatsu, R., Tosa, N., Naghdy, F., Wong, K.W., Codognet, P., Eds.; Springer Berlin: Berlin, Germany, 2010; pp. 55–64.
17. Kesiman, M.W.A.; Burie, J.-C.; Ogier, J.-M.; Wibawantara, G.N.M.A.; Sunarya, I.M.G. AMADI_LontarSet: The First Handwritten Balinese Palm Leaf Manuscripts Dataset. In Proceedings of the 15th International Conference on Frontiers in Handwriting Recognition (ICFHR), Shenzhen, China, 23–26 October 2016; pp. 168–172.
18. *The Unicode® Standard*; version 9.0—Core Specification; The Unicode Consortium: Mountain View, CA, USA, 2016.
19. Balinese Alphabet, Language and Pronunciation. Available online: http://www.omniglot.com/writing/balinese.htm (accessed on 20 February 2018).
20. Khmer Manuscript—Recherche. Available online: http://khmermanuscripts.efeo.fr/ (accessed on 20 February 2018).

21. Valy, D.; Verleysen, M.; Chhun, S.; Burie, J.-C. A New Khmer Palm Leaf Manuscript Dataset for Document Analysis and Recognition—SleukRith Set. In Proceedings of the 4th International Workshop on Historical Document Imaging and Processing, Kyoto, Japan, 10–11 November 2017; pp. 1–6.
22. Suryani, M.; Paulus, E.; Hadi, S.; Darsa, U.A.; Burie, J.-C. The Handwritten Sundanese Palm Leaf Manuscript Dataset From 15th Century. In Proceedings of the 14th IAPR International Conference on Document Analysis and Recognition (ICDAR), Kyoto, Japan, 9–15 November 2017; pp. 796–800.
23. Kesiman, M.W.A.; Prum, S.; Burie, J.-C.; Ogier, J.-M. An Initial Study on the Construction of Ground Truth Binarized Images of Ancient Palm Leaf Manuscripts. In Proceedings of the 13th International Conference on Document Analysis and Recognition (ICDAR), Nancy, France, 23–26 August 2015; pp. 656–660.
24. Kesiman, M.W.A.; Prum, S.; Sunarya, I.M.G.; Burie, J.-C.; Ogier, J.-M. An Analysis of Ground Truth Binarized Image Variability of Palm Leaf Manuscripts. In Proceedings of the 5th International Conference Image Processing Theory Tools Application (IPTA 2015), Orleans, France, 10–13 November 2015; pp. 229–233.
25. Burie, J.-C.; Coustaty, M.; Hadi, S.; Kesiman, M.W.A.; Ogier, J.-M.; Paulus, E.; Sok, K.; Sunarya, I.M.G.; Valy, D. ICFHR 2016 Competition on the Analysis of Handwritten Text in Images of Balinese Palm Leaf Manuscripts. In Proceedings of the 15th International Conference on Frontiers in Handwriting Recognition (ICFHR), Shenzhen, China, 23–26 October 2016; pp. 596–601.
26. Kesiman, M.W.A.; Valy, D.; Burie, J.-C.; Paulus, E.; Sunarya, I.M.G.; Hadi, S.; Sok, K.H.; Ogier, J.-M. Southeast Asian palm leaf manuscript images: A review of handwritten text line segmentation methods and new challenges. *J. Electron. Imaging.* **2016**, *26*, 011011. [CrossRef]
27. Valy, D.; Verleysen, M.; Sok, K. Line Segmentation for Grayscale Text Images of Khmer Palm Leaf Manuscripts. In Proceedings of the 7th International Conference Image Processing Theory Tools Application (IPTA 2017), Montreal, QC, Canada, 28 November–1 December 2017.
28. Kesiman, M.W.A.; Prum, S.; Burie, J.-C.; Ogier, J.-M. Study on Feature Extraction Methods for Character Recognition of Balinese Script on Palm Leaf Manuscript Images. In Proceedings of the 23rd International Conference Pattern Recognition, Cancun, Mexico, 4–8 December 2016; pp. 4017–4022.
29. Kesiman, M.W.A.; Burie, J.-C.; Ogier, J.-M. A Complete Scheme of Spatially Categorized Glyph Recognition for the Transliteration of Balinese Palm Leaf Manuscripts. In Proceedings of the 14th IAPR International Conference on Document Analysis and Recognition (ICDAR), Kyoto, Japan, 9–15 November 2017; pp. 125–130.
30. Bezerra, B.L.D. *Handwriting: Recognition, Development and Analysis*; Bezerra, B.L.D., Zanchettin, C., Toselli, A.H., Pirlo, G., Eds.; Nova Science Publishers, Inc.: Hauppauge, NY, USA, 2017; ISBN 978-1-53611-957-2.
31. Arica, N.; Yarman-Vural, F.T. Optical character recognition for cursive handwriting. *IEEE Trans. Pattern Anal. Mach. Intell.* **2002**, *24*, 801–813. [CrossRef]
32. Blumenstein, M.; Verma, B.; Basli, H. A novel feature extraction technique for the recognition of segmented handwritten characters. In Proceedings of the Seventh International Conference on Document Analysis and Recognition, Edinburgh, UK, 3–6 August 2003; pp. 137–141.
33. O'Gorman, L.; Kasturi, R. *Executive briefing: Document Image Analysis*; IEEE Computer Society Press: Los Alamitos, CA, USA, 1997; p. 107.
34. Naveed Bin Rais, M.S.H. Adaptive thresholding technique for document image analysis. In Proceedings of the 8th International Multitopic Conference, Lahore, Pakistan, 24–26 December 2004; pp. 61–66.
35. Ntirogiannis, K.; Gatos, B.; Pratikakis, I. An Objective Evaluation Methodology for Document Image Binarization Techniques. In Proceedings of the Eighth IAPR International Workshop Document Annual System 2008, Nara, Japan, 16–19 September 2008; pp. 217–224.
36. He, J.; Do, Q.D.M.; Downton, A.C.; Kim, J.H. A comparison of binarization methods for historical archive documents. In Proceedings of the Eighth International Conference on Document Analysis and Recognition (ICDAR'05), Seoul, South Korea, 31 August–1 September 2005; pp. 538–542.
37. Gatos, B.; Ntirogiannis, K.; Pratikakis, I. DIBCO 2009: Document image binarization contest. *Int. J. Doc. Anal. Recognit.* **2011**, *14*, 35–44. [CrossRef]
38. Pratikakis, I.; Gatos, B.; Ntirogiannis, K. ICDAR 2013 Document Image Binarization Contest (DIBCO 2013). In Proceedings of the 12th International Conference on Document Analysis and Recognition, Washington, DC, USA, 25–28 August 2013; pp. 1471–1476.
39. Howe, N.R. Document binarization with automatic parameter tuning. *Int. J. Doc. Anal. Recognit.* **2013**, *16*, 247–258. [CrossRef]

40. ICFHR2016 Competition on the Analysis of Handwritten Text in Images of Balinese Palm Leaf Manuscripts. Available online: http://amadi.univ-lr.fr/ICFHR2016_Contest/ (accessed on 20 February 2018).
41. Gupta, M.R.; Jacobson, N.P.; Garcia, E.K. OCR binarization and image pre-processing for searching historical documents. *Pattern Recognit.* **2007**, *40*, 389–397. [CrossRef]
42. Feng, M.-L.; Tan, Y.-P. Contrast adaptive binarization of low quality document images. *IEICE Electron. Express* **2004**, *1*, 501–506. [CrossRef]
43. Global Image Threshold Using Otsu's Method—MATLAB Graythresh—MathWorks France. Available online: https://fr.mathworks.com/help/images/ref/graythresh.html?requestedDomain=true (accessed on 20 February 2018).
44. Khurshid, K.; Siddiqi, I.; Faure, C.; Vincent, N. Comparison of Niblack Inspired Binarization Methods for Ancient Documents. In Proceedings of the Document Recognition and Retrieval XVI, 72470U, San Jose, CA, USA, 21 January 2009; p. 72470U. [CrossRef]
45. Sauvola, J.; Pietikäinen, M. Adaptive document image binarization. *Pattern Recognit.* **2000**, *33*, 225–236. [CrossRef]
46. Wolf, C.; Jolion, J.-M.; Chassaing, F. Text Localization, Enhancement and Binarization in Multimedia Documents. In Proceedings of the Object recognition supported by user interaction for service robots, Quebec City, QC, Canada, 11–15 August 2002; pp. 1037–1040.
47. Arvanitopoulos, N.; Susstrunk, S. Seam Carving for Text Line Extraction on Color and Grayscale Historical Manuscripts. In Proceedings of the 14th International Conference on Frontiers in Handwriting Recognition, Heraklion, Greece, 1–4 September 2014; pp. 726–731.
48. Hossain, M.Z.; Amin, M.A.; Yan, H. Rapid Feature Extraction for Optical Character Recognition. Available online: http://arxiv.org/abs/1206.0238 (accessed on 20 February 2018).
49. Fujisawa, Y.; Shi, M.; Wakabayashi, T.; Kimura, F. Handwritten numeral recognition using gradient and curvature of gray scale image. In Proceedings of the Fifth International Conference on Document Analysis and Recognition. ICDAR'99, Bangalore, India, 22 September 1999; pp. 277–280.
50. Kumar, S. Neighborhood Pixels Weights-A New Feature Extractor. *Int. J. Comput. Theory Eng.* **2009**, *2*, 69–77. [CrossRef]
51. Bokser, M. Omnidocument technologies. *Proc. IEEE.* **1992**, *80*, 1066–1078. [CrossRef]
52. Coates, A.; Lee, H.; Ng, A.Y. An Analysis of Single-Layer Networks in Unsupervised Feature Learning. In Proceedings of the Fourteenth International Conference on Artificial Intelligence and Statistics, Fort Lauderdale, FL, USA, 11–13 April 2011; pp. 215–223.
53. Coates, A.; Carpenter, B.; Case, C.; Satheesh, S.; Suresh, B.; Wang, T.; Wu, D.J.; Ng, A.Y. Text Detection and Character Recognition in Scene Images with Unsupervised Feature Learning. In Proceedings of the International Conference on Document Analysis and Recognition, Beijing, China, 18–21 September 2011; pp. 440–445.
54. Shishtla, P.; Ganesh, V.S.; Subramaniam, S.; Varma, V. A language-independent transliteration schema using character aligned models at NEWS 2009. In Proceedings of the Association for Computational Linguistics, Suntec, Singapore, 7 August 2009; p. 40. [CrossRef]
55. Ul-Hasan, A.; Breuel, T.M. Can we build language-independent OCR using LSTM networks? In Proceedings of the 4th International Workshop on Multilingual OCR, Washington, DC, USA, 24 August 2013.
56. Ocropy: Python-Based Tools for Document Analysis and OCR, 2018. Available online: https://github.com/tmbdev/ocropy (accessed on 20 February 2018).
57. Homemade Manuscript OCR (1): OCRopy, Sacré Grl. Available online: https://graal.hypotheses.org/786 (accessed on 20 February 2018).
58. Breuel, T.M.; Ul-Hasan, A.; Al-Azawi, M.A.; Shafait, F. High-Performance OCR for Printed English and Fraktur Using LSTM Networks. In Proceedings of the 12th International Conference on Document Analysis and Recognition, Washington, DC, USA, 25–28 August 2013; pp. 683–687. [CrossRef]
59. Valy, D.; Verleysen, M.; Sok, K. Line Segmentation Approach for Ancient Palm Leaf Manuscripts using Competitive Learning Algorithm. In Proceedings of the 15th International Conference on Frontiers in Handwriting Recognition (ICFHR), Shenzhen, China, 23–26 October 2016.
60. Saund, E.; Lin, J.; Sarkar, P. PixLabeler: User Interface for Pixel-Level Labeling of Elements in Document Images. In Proceedings of the 10th International Conference on Document Analysis and Recognition, Barcelona, Spain, 26–29 July 2009; pp. 646–650. [CrossRef]

61. Stamatopoulos, N.; Gatos, B.; Louloudis, G.; Pal, U.; Alaei, A. ICDAR 2013 Handwriting Segmentation Contest. In Proceedings of the 12th International Conference on Document Analysis and Recognition, Washington, DC, USA, 25–28 August 2013; pp. 1402–1406. [CrossRef]
62. PRImA. Available online: http://www.primaresearch.org/tools/Aletheia (accessed on 20 February 2018).
63. Clausner, C.; Pletschacher, S.; Antonacopoulos, A. Aletheia—An Advanced Document Layout and Text Ground-Truthing System for Production Environments. In Proceedings of the International Conference on Document Analysis and Recognition, Beijing, China, 18–21 September 2011; pp. 48–52. [CrossRef]

Journal of
Imaging

MDPI

Article

Transcription of Spanish Historical Handwritten Documents with Deep Neural Networks

Emilio Granell [1,*], Edgard Chammas [2], Laurence Likforman-Sulem [3], Carlos-D. Martínez-Hinarejos [1], Chafic Mokbel [2] and Bogdan-Ionuţ Cîrstea [3]

[1]   PRHLT Research Center, Universitat Politècnica de València, 46022 València, Spain; cmartine@dsic.upv.es
[2]   Department of Computer Engineering, University of Balamand, 2960 Balamand, Lebanon;
      edgard@balamand.edu.lb (E.C.); chafic.mokbel@balamand.edu.lb (C.M.)
[3]   Institut Mines-Télécom/Télécom ParisTech, Université Paris-Saclay, 75013 Paris, France;
      likforman@telecom-paristech.fr (L.L.-S.); bogdan-ionut.cirstea@telecom-paristech.fr (B.-I.C.)
*   Correspondence: egranell@dsic.upv.es

Received: 30 October 2017; Accepted: 2 January 2018; Published: 11 January 2018

Abstract: The digitization of historical handwritten document images is important for the preservation of cultural heritage. Moreover, the transcription of text images obtained from digitization is necessary to provide efficient information access to the content of these documents. Handwritten Text Recognition (HTR) has become an important research topic in the areas of image and computational language processing that allows us to obtain transcriptions from text images. State-of-the-art HTR systems are, however, far from perfect. One difficulty is that they have to cope with image noise and handwriting variability. Another difficulty is the presence of a large amount of Out-Of-Vocabulary (OOV) words in ancient historical texts. A solution to this problem is to use external lexical resources, but such resources might be scarce or unavailable given the nature and the age of such documents. This work proposes a solution to avoid this limitation. It consists of associating a powerful optical recognition system that will cope with image noise and variability, with a language model based on sub-lexical units that will model OOV words. Such a language modeling approach reduces the size of the lexicon while increasing the lexicon coverage. Experiments are first conducted on the publicly available *Rodrigo* dataset, which contains the digitization of an ancient Spanish manuscript, with a recognizer based on Hidden Markov Models (HMMs). They show that sub-lexical units outperform word units in terms of Word Error Rate (WER), Character Error Rate (CER) and OOV word accuracy rate. This approach is then applied to deep net classifiers, namely Bi-directional Long-Short Term Memory (BLSTMs) and Convolutional Recurrent Neural Nets (CRNNs). Results show that CRNNs outperform HMMs and BLSTMs, reaching the lowest WER and CER for this image dataset and significantly improving OOV recognition.

Keywords: historical handwritten transcription; out-of-vocabulary word recognition; character-level language model; word structure retrieval

1. Introduction

The digitization of historical handwritten document images is important for the preservation of cultural heritage. Moreover, the transcription of text images obtained from digitization is necessary to provide efficient information access to the content of these documents. Automatic transcription of these documents is performed by Handwriting Text Recognition (HTR) systems, which are traditionally composed of an optical model, a dictionary and a Language Model (LM). However, HTR systems face several challenges at both the image and language modeling levels. Historical document images may include defects due to age, manipulation and bleed-through of ink. They may also include calligraphic initial letters and long character strokes as ornaments. This is particularly true for Spanish documents from the 16th century as seen in Figure 1. Ancient texts also include rare characters, grammatical forms, word spellings and named entities distinct from modern ones. Such forms lead to Out-Of-Vocabulary (OOV) words, i.e., words that do not belong to the dictionary of the HTR system. Improving HTR systems at both image and language levels is an important issue for the recognition of such ancient historical documents. The main goal of this paper is to design efficient HTR systems that process document images written in Spanish and that can cope with ancient character forms and language.

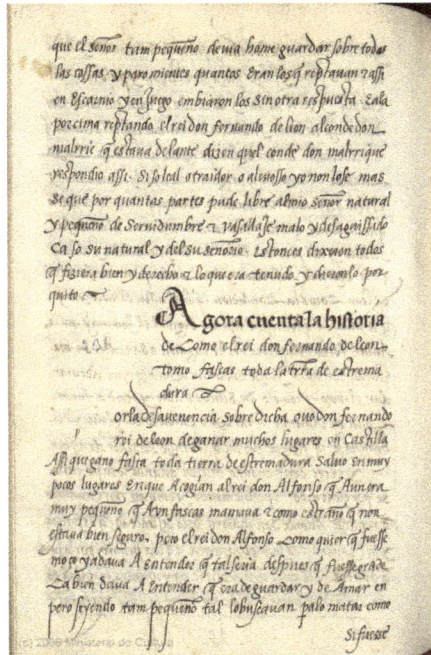

Figure 1. Sample image of a Spanish document from the 16th century.

Several approaches have been proposed to build optical models for handwriting recognition. Such approaches include Hidden Markov Models (HMMs) [1–4], Recurrent Neural Networks (RNNs) such as Long Short-Term Memory (LSTMs) and their variants: Bi-directional LSTMs (BLSTMs) and Multi-Dimensional LSTMs (MDLSTMs) [5]. HMMs enable embedded training and can be robust to noise and linear distortions. However, RNNs and their variants are generative models that perform better than HMMs in terms of accuracy. Nowadays, RNNs can be trained by using dedicated resources such as Graphic Processor Units (GPUs) that considerably reduce training time. By using GPUs, RNNs can be trained in a similar amount of time required to train HMMs with traditional Central Processing Units (CPUs).

Usually, the inputs of HMMs and RNNs are sequences of handcrafted features or pixel columns. However, deep learning approaches starting with convolutional layers as the first layers allow extracting learning-based features instead of handcrafted ones [6–8].

Generally, in HTR systems, the optical models are associated with dictionaries (lexical models) and Language Models (LMs), usually at the word level, in order to direct the recognition of real words and plausible word sequences (see Figure 2). In order to build open vocabulary systems, language models based on character units can be used [9]. Then, the dictionary is limited to the set of different characters, and the transition probabilities between the character models are given by a character LM. Character-based LMs are also useful for related tasks such as word spotting [10]. In the previous character LM approach or even in general word LM approaches, the optical models still model characters. However, in works such as [11,12], the optical models model strokes that are concatenated to form words.

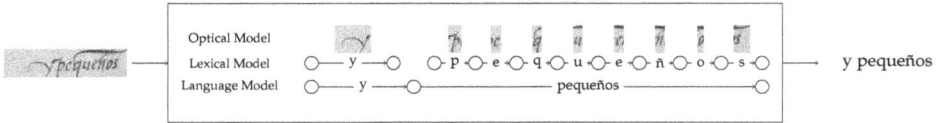

Figure 2. Scheme of a handwritten text recognition system.

When a word-based dictionary helps the recognition process, the handwriting recognition system can only transcribe a limited number of words. The size of the dictionary is a compromise between a too large size yielding word confusions and a too small one yielding many unknown words. Words of the test set that are not present in the HTR dictionary are denoted as Out-Of-Vocabulary (OOV) words. Several types of OOV words exist, such as common words using a less common grammatical form, misspellings, words attached to punctuation marks, hyphenated words or words containing rare characters (abbreviations, special signs, etc.).

An approach to cope with OOV words consists of extending the dictionary with external lexical resources, such as Wikipedia [13], or in the case of historical documents, with the transcription of other documents from the same period and topic [14]. From these resources, the language model can also be refined. However, in the general case, such resources may not be available, and a proportion of words (such as named entities and rare words) still remains as OOV. Another approach for coping with OOV words consists of modeling text at a sub-word level, as a sequence of characters, syllables or multi-grams [15]. Hybrid approaches [16,17] consist of using word-based language models for the most frequent words and character-based models for the less frequent ones. In sub-word approaches, the dictionary is considerably reduced to the number of lexical units, as well as the computational complexity. In addition, the language model can model unknown words by combining such lexical units.

In this work, we compare several HTR systems, based on HMMs, RNNs and convolutional RNNs (CRNNs). The CRNN is inspired from a very deep architecture presented in [18]. It consists of stacking BLSTMs and associating them with convolutional layers. Features are thus automatically extracted by the convolutional layers and processed by the BLSTM layers. We also model dictionaries and language models of our HTR systems with sub-word units. We apply this approach to the recognition of a publicly available Spanish historical documents dataset. We compare several HTR systems based on different types of sub-word units, and we show that sub-word units are more efficient than word units. We obtain, to our knowledge, the best recognition results on this Spanish dataset by associating sub-word units with the deepest HTR optical system, namely the CRNN. We also obtain high rates for the recognition of OOV words.

The rest of the paper is structured as follows: the Spanish historical manuscript used in the experimentation is presented in the next section (Section 2); the HTR systems and the experimental conditions are described in Section 3; our experiments and the obtained results are reported in Section 4; the conclusions and future work are drawn in Section 5; finally, in Appendix A, several recognition examples are shown.

2. The *Rodrigo* Dataset

The *Rodrigo* corpus [19] was obtained from the digitization of the book "Historia de España del arçobispo Don Rodrigo", written in ancient Spanish in 1545. It is a single writer book where most pages consist of a single block of well-separated lines of calligraphical text, as the examples presented in Figures 1 and 3. It is composed of 853 pages that were automatically divided into lines, giving a total number of 20,356 lines. In the standard training partition, the vocabulary size is of about 11,000 words with a set of 106 characters (the 105 different characters that appear in the text of the training partition and one extra character that appears in the text of the validation partition), including 10 numbers, 72 upper and lower case letters with and without accents, 5 punctuation marks, 1 blank space and 18 special symbols. The first 15,010 lines are publicly available on the website of the Pattern Recognition and Human Language Technology (PRHLT) research center [20]. In this work, we used this publicly available partition. The first 9000 lines were used for training the optical and language models, the next 1000 for validation and the last 5010 lines for testing.

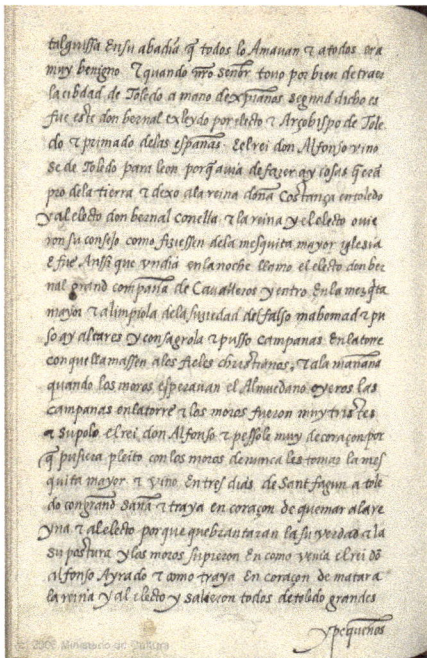

Figure 3. Page 515 of the *Rodrigo* dataset.

In the *Rodirgo* corpus, there are many rare words and words in their archaic forms yielding a large amount of OOV words. Moreover, this corpus contains scarce OOV characters (such as: \, $\acute{p}$, $\tilde{g}$, $\hbar$ and w) that do not belong to the training set. OOV words generally include words that appear in distinct form in the training and test sets (e.g., *portugal* and *portuğl*), abbreviations and words hyphenated differently in the training and test sets.

Table 1 presents a summary of the information contained in the partitions of the *Rodrigo* corpus used in this work at the three lexical units studied: words, sub-words and characters. This table presents for each lexical unit the total amount, the vocabulary size (different units), the amount of OOV units and the overlapping between the OOV contained in the validation and test partitions, i.e., the amount of OOV units contained in the test partition that are present in the validation partition.

Table 1. Description of the partitions of the *Rodrigo* corpus used in this work.

Partition	Lines	Words Total / Diff./ OOV (over.)	Sub-Words Total/Diff./OOV (over.)	Characters Total/Diff./OOV (over.)
Training	9000	98,232/12,650/-	148,070/3045/-	493,126/105/-
Validation	1000	10,899/3016/850	14,907/1074/7	54,936/82/1
Test	5010	55,195/7453/4918 (203)	73,660/1418/55 (11)	272,132/91/14 (1)

3. Handwritten Text Recognition Systems

This section presents our proposal, the feature extraction, the models used by the implemented HTR systems and the evaluation metrics used in the experimentation.

3.1. Proposal

The HTR problem can be formulated as finding the most likely word sequence $\hat{w}$ given a feature vector sequence $x = (x_1, x_2, \ldots, x_{|x|})$ that represents a handwritten text line image [21], that is:

$$\hat{w} = \arg\max_{w \in W} \Pr(w \mid x) = \arg\max_{w \in W} \frac{\Pr(x \mid w)\Pr(w)}{\Pr(x)} = \arg\max_{w \in W} \Pr(x \mid w)\Pr(w) \tag{1}$$

where W represents the set of all permissible word sequences, $\Pr(x)$ is the probability of observing x, $\Pr(w)$ is the probability of the word sequence $w = (w_1, w_2, \ldots, w_{|w|})$ and $\Pr(x \mid w)$ is the probability of observing x by assuming that w is the underlying word sequence for x. $\Pr(w)$ is approximated by the Language Model (LM), whereas $\Pr(x \mid w)$ is modeled by the optical model, which trains character models and concatenates them to build optical word or sub-word models.

Written words can be decomposed into small sub-word units such as characters, but they can also be decomposed into larger sub-word units such as graphemic syllables, hyphens or multigrams [15]. We choose here to compare character and hyphen word decompositions. In both cases, words are represented as a sequence of sub-word units $s = (s_1, s_2, \ldots, s_{|s|})$. Then, the HTR problem can be reformulated as finding the most likely sub-word sequence $\hat{s}$ given a feature vector sequence x that represents a handwritten text image. Therefore, Equation (1) becomes:

$$\hat{s} = \arg\max_{s \in S} \Pr(x \mid s)\Pr(s) \tag{2}$$

where $\Pr(s)$ is approximated by a sub-word LM, whereas $\Pr(x \mid s)$ can be modeled by the same optical model.

It should be noted that RNN-based systems directly provide in their outputs posterior distributions of character labels, at each time step, i.e., o_k^t for $k = 1, \ldots, L$ and $t = 1, \ldots, T$, T being the length of the observation sequence x and L the alphabet size. From these posteriors, the decoding can be constrained by a lexicon and a language model, in order to find the best output sequence $\hat{s}$. This can be done through Weighted Finite State Transducers (WFST) decoding (see Section 3.5), which can include several types of lexicon and language models (at word, hyphen or character levels).

Working at the sub-word level in HTR relaxes the restrictions imposed by the lexicon, allowing for a faster decoding, and given that the language model describes the relation between sub-word units, some OOV words can be decoded. Therefore, our proposal is to decode the handwritten text line images at the sub-word level and, then, from the obtained decoding output, reconstruct the words to build the final hypothesis.

First of all, the language model of sub-word units is trained using the transcription of the text lines of the training partition after a minimum preprocessing. This preprocessing consists of adding a new symbol (<SPACE>) for the separation between words and then splitting the words into sub-word sequences. In this way, the information of the separation between words is maintained.

As an example, the following text line from the training set:

```
Agora cuenta la historia
```

would be transformed into the following character sequence:

```
A g o r a~<SPACE> c u e n t a~<SPACE> l a~<SPACE> h i s t o r i a
```

or into the following sequence following the hyphenation rules for Spanish:

```
Ago ra <SPACE> cuen ta <SPACE> la <SPACE> his to ria
```

Then, these preprocessed transcriptions can be used to train the sub-word unit language model. Usually, *n*-gram language models of sub-word units are trained with a large *n* (large context). On the other side, the lexicon is reduced to match the list of sub-word units.

In the decoding process, the best hypothesis is processed to obtain the final hypothesis. This final process consists of collapsing the sub-word unit sequence to form words and to substitute the symbol used to mark the separation between words (<SPACE>) by a space. Figure 4 presents a text line example from the test partition whose reference transcription is:

```
vio e recognoscio el Astragamiento que perdiera de su gente
```

In this example, the words *recognoscio* and *Astragamiento* are OOV words. It is interesting to note their etymology. They are archaic forms from Early Modern Spanish (15th–17th century) that in Modern Spanish correspond to the forms *reconoció* and *Estragamiento*. For that reason, we could not find them in any external resource, not even in Google N-Grams [22].

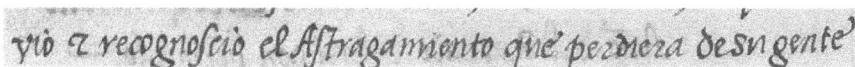

Figure 4. Text line sample. "Recognoscio" and "Astragamiento" are rare words; *recognoscio* is an archaic form of *reconoció* and *Astragamiento* an ancient form of *Estragamiento*.

The HMM decoding process with a traditional word-based approach offers the following best hypothesis:

```
vno & rea gustio el Astragar mando que perdona de lugar
```

which represents a Character Error Rate (CER) equal to 35.6% with respect to the reference text-line transcription. However, using a sub-word based approach, the following best hypothesis is obtained:

```
vio <SPACE> & <SPACE> re ca ges cio <SPACE> el <SPACE> As tra ga mien to
<SPACE> que <SPACE> per do na <SPACE> de <SPACE> lu gar <SPACE>
```

which is transformed into the improved hypothesis (CER = 22.0%):

```
vio & recagescio el Astragamiento que perdona de lugar
```

On the other hand, with a character-based approach, the following best hypothesis is obtained:

```
v i o <SPACE> & <SPACE> r e c e g e s c i o <SPACE> e l <SPACE> A s t r a~g a~m i e n t o
<SPACE> q u e <SPACE> p e r d i e r a~<SPACE> d e l <SPACE> s e g u n d o
```

which results in the next final best hypothesis (CER = 17.0%):

```
vio & recegescio el Astragamiento que perdiera del segundo
```

As can be observed, the final hypotheses obtained at sub-word levels (characters, hyphenation sub-word units) in HTR are considerably better than those obtained with the word-based approach. In addition, the OOV word *Astragamiento* has been fully recognized. The second OOV word is recognized as *recegescio* or *recagescio*, which also improves the word-based recognition *rea gustio*. In Section 4, word and sub-word language modeling approaches will be compared with several types of optical HTR systems.

3.2. Handcrafted Features

Features are computed in several steps from text line images. First, the image brightness is normalized, and a median filter of size 3×3 pixels is applied to the entire image. Next, slant correction is performed by using the maximum variance method with a threshold of 92% [23]. Then, size normalization is performed, and the final image is scaled to a height of 40 pixels. Finally, a sequence of 60-dimensional feature vectors is extracted by a sliding window, using the method described in [24].

3.3. Lexicon and Language Models

The lexicon and language models at the sub-word level were obtained by hyphenating the vocabulary words following the rules for modern Spanish by using the `testhyphens` package [25] for LaTeX. Lexicon models were in HTK lexicon format, where vocabulary words and sub-word units were modeled as a concatenation of symbols; however, characters were modeled as just the corresponding symbol.

Language Models (LM) were estimated as n-grams with Kneser–Ney back-off smoothing [26] by using the SRILM toolkit [27]. Different LMs were used in the experiments at word, sub-word and character levels. For the word-based system and the open-vocabulary case, the LM is trained directly from the text-line transcriptions of the training set. In the closed-vocabulary case, the LM is trained with the same transcriptions, plus the OOV words included as unigrams. For the character-based system, the closed-vocabulary case indicates that the character sequences that represent the OOV words are used for building the n-gram character LM. For both systems, word or character-based, "with validation" means that training and validation transcriptions are used for building the LM.

3.4. Optical Models

In this paper, three different approaches for optical modeling for HTR are used: traditional hidden Markov models and two deep network classifiers. The first one is based on recurrent neural networks with bi-directional long-short term memory, and the other one is based on convolutional recurrent neural networks.

3.4.1. Hidden Markov Models

The Hidden Markov Models (HMM) for optical modeling were trained with HTK [28]. The trained models are left-to-right character models including four states. The observation probabilities in each state are described by a mixture distribution of 64 Gaussians. The number of character models is 106, and words and sub-words are modeled by the concatenation of compound character HMMs. The HMM system uses as input sequences of handcrafted features. HMM HTR systems were implemented by using the iATROS recognizer [29].

3.4.2. Deep Models Based on BLSTMs

In this approach, we use an RNN to estimate the posterior probabilities of the characters at the frame level (features vector). Therefore, the size of the input layer corresponds to the size of the handcrafted feature vectors and the size of the output layer to the number of different characters. The frame-level labeling required to train this neural network was generated from a forced alignment decoding by a previously trained HMM recognition system [30]. This forced alignment decoding and the model training were repeated several times until the convergence of the assignment of the frame labels to the optical model.

Then, as presented in Figure 5, our RNN is formed by 60 neurones at the input layer, 500 BLSTM neurones at the hidden layer with a hyperbolic tangent activation function and 106 neurones at the output layer with a softmax function. The training was performed by using RNNLIB [31], and the main parameters (such as the size of the hidden layer) were tuned by using the validation partition. The Weighted Finite State Transducers (WFST) decoding (see Section 3.5) can be designed to output word, sub-word or character sequences. For each output type, the lexicon and language model have to be modified accordingly, and no additional modification is necessary in the system.

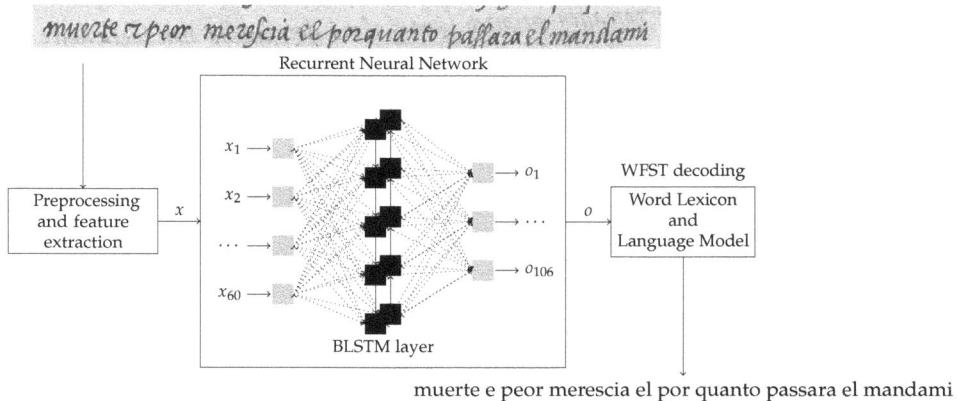

Figure 5. Bi-directional Long-Short Term Memory (BLSTM) system architecture. The BLSTM RNN outputs posterior distributions o at each time step. The decoding is performed with Weighted Finite State Transducers (WFST) using a lexicon and a language model at word level.

3.4.3. Deep Models Based on Convolutional Recurrent Neural Networks

The Convolutional Recurrent Neural Network (CRNN) [32] is inspired by the VGG16 architecture [33] that was developed for image recognition. We use a stack of 13 convolutional (3×3 filters, 1×1 stride) layers followed by three bi-directional LSTM layers with 256 units per layer (see Figure 6). Each LSTM unit has one cell with enabled peephole connections. Spatial pooling (max) is employed after some convolutional layers. To introduce non-linearity, the Rectified Linear Unit (ReLU) activation function was used after each convolution. It has the advantage of being resistant to the vanishing gradient problem while being simple in terms of computation and was shown to work better than sigmoid and hyperbolic tangent activation functions [34]. A square-shaped sliding window is used to scan the text-line image in the direction of the writing. The height of the window is equal to the height of the text-line image, which has been normalized to 64 pixels. The window overlap is equal to two pixels to allow continuous transition of the convolution filters. For each analysis window of 64×64 pixels in size, 16 feature vectors are extracted from the feature maps produced by the last convolutional layer and fed into the observation sequence. For each of the 16 columns of the last 512 feature maps, the columns of a height of two pixels are concatenated into a feature vector of size 1024 (512×2). Thanks to the CTCtranscription layer [35], the system is end-to-end trainable. The convolutional filters and the LSTM units weights are thus jointly learned using the back-propagation procedure. We combined the forward and backward outputs at the end of the BLSTM stack [36] rather than after each BLSTM layer, in order to decrease the number of parameters. We also chose not to add additional fully-connected layers since, by adding such layers, the network had more parameters, converged more slowly and performed worse. Hyper parameters such as the number of convolution layers and the number of BLSTM layers were set up on a validation set. The LSTM unit weights were initialized as per the method of [37], which proved to work well and helps the network to converge faster. This allows the network to maintain a constant variance across the network layers, which keeps the signal from exploding to a high value or vanishing to zero. The weight matrices were initialized with a uniform distribution.

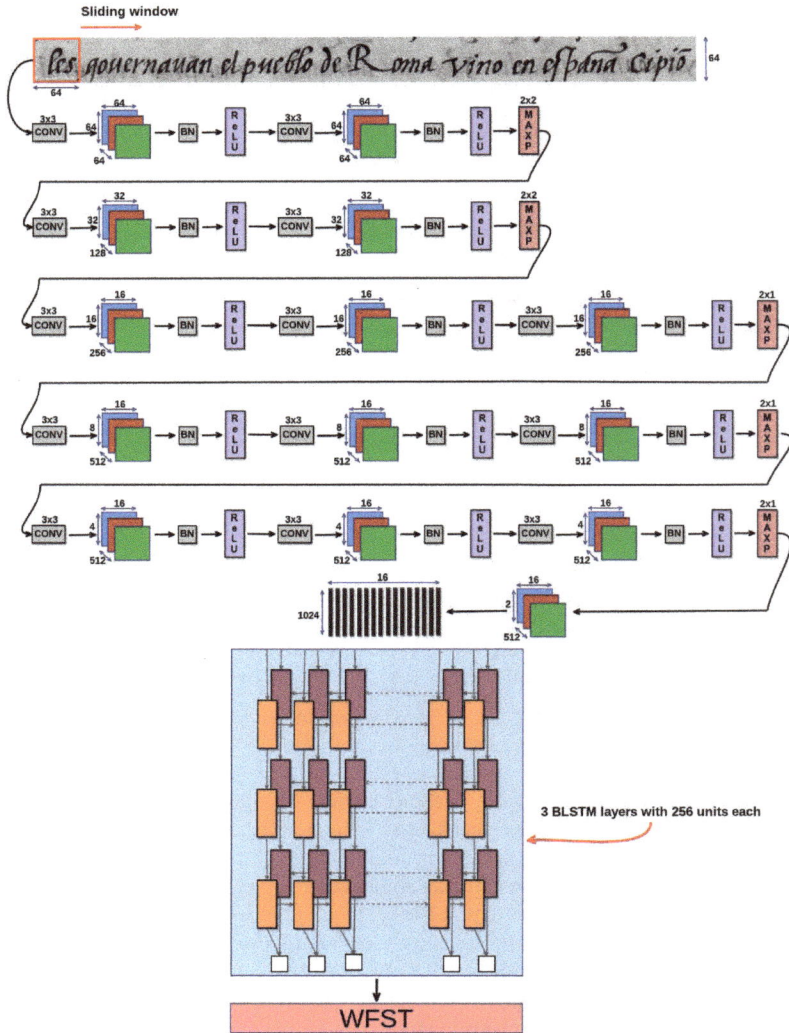

Figure 6. CRNN system architecture.

The Adam optimizer [38] was used to train the network with the initial learning rate of 0.001. This algorithm could be thought of as an upgrade for RMSProp [39], offering bias correction and momentum [40]. It provides adaptive learning rates for the stochastic gradient descent update computed from the first and second moments of the gradients. It also stores an exponentially decaying average of the past squared gradients (similar to Adadelta [41] and RMSprop) and the past gradients (similar to momentum). Batch normalization, as described in [42], was added after each convolutional layer in order to accelerate the training process. It basically works by normalizing each batch by both the mean and variance. The network was trained in an end-to-end fashion with the CTC loss function [35].

3.5. Decoding with Deep Optical Models

Decoding for both deep net systems was performed with Weighted Finite State Transducers (WFST). Our decoder is based on the CTC-specific implementation proposed by [43] for speech recognition. A "token" WFST was designed to handle all possible label sequences at the frame level, so as to allow for the occurrence of the blank label along with the repetition of non-blank labels. It can map a sequence of frame-level CTC labels to a single character. A search graph is built with three WFSTs (T, L and G) compiled independently and combined as follows:

$$S = T \circ \min(\det(L \circ G)) \tag{3}$$

T, L and G are the token, lexicon and grammar WFSTs respectively, whereas $\circ$, det and min denote composition, determination and minimization, respectively. The determination and minimization operations are needed to compress the search space, yielding a faster decoding.

3.6. Evaluation Metrics

The quality of the obtained transcriptions was assessed using the edit distance [44] with respect to the reference text, at the word and at the character level. The Word Error Rate (WER) is this edit distance at the word level and can be calculated as the minimum number of substitutions, deletions and insertions needed to transform the transcription into the reference, divided by the number of words of the reference:

$$\text{WER} = \frac{s + d + i}{n} \cdot 100 \tag{4}$$

where s is the number of substitutions, d the number of deletions, i the number of insertions and n the total number of words in the reference.

Similarly, this edit distance can be calculated at the character level, giving the Character Error Rate (CER). In this framework, the CER value is especially interesting, since transcription errors are usually corrected at the character level. The OOV Word Accuracy Rate (OOV WAR) was measured as the amount of recognized OOV words over the total amount of OOV words. The statistical significance of experimental results can be estimated by means of confidence intervals. Generally, when comparing two experimental results, it is always true that if the confidence intervals do not overlap, we can say that the difference is statistically significant [45]. In this work, confidence intervals of probability 95% ($\alpha = 0.025$) were calculated by using the bootstrapping method with 10,000 repetitions [46] for these rate measures.

Finally, as language models are probability distributions over entire sentences or texts, perplexity [47] can be used to evaluate their performance over a reference text. In this work, we use the perplexity presented by a character LM over the OOV words (as sequences of characters), to assess the differences between the recognized and unrecognized OOV words.

4. Experimental Results

In the test experiments, we compared the performance on the test partition of the *Rodrigo* corpus. Different systems were compared, the first one based on HMMs, the second one based on RNN and the third one on CRNN. For the three systems, experiments were performed at word, sub-word, and character levels. We first explore the influence of the size of the LM context (n-gram degree). Then, we develop an analysis of the difference between the structure of recognized and unrecognized OOV words. The last experiment compares the results obtained in three different cases: open vocabulary, closed vocabulary and when using the validation samples for training the LM.

We observed that in the training partition of *Rodrigo*, usually there are no spaces between words and punctuation marks, so we decided to remove those spaces from the hypotheses offered by the word-based systems. Therefore, in the word-based cases, the recognized OOV words correspond to words attached to punctuation marks, which were correctly recognized after removing the space between them (see Figure A2).

4.1. Study of the Context Size Influence

Figure 7 presents the results obtained for the word-based HMM system (in terms of WER and CER) by using n-gram LM with different context sizes $n = \{1, \ldots, 6\}$. As can be observed in this figure, the best result was obtained by using a three-gram LM; concretely, a WER equal to $43.3\% \pm 0.5$, a CER equal to $21.1\% \pm 0.3$ and an OOV WAR equal to $2.3\% \pm 0.4$.

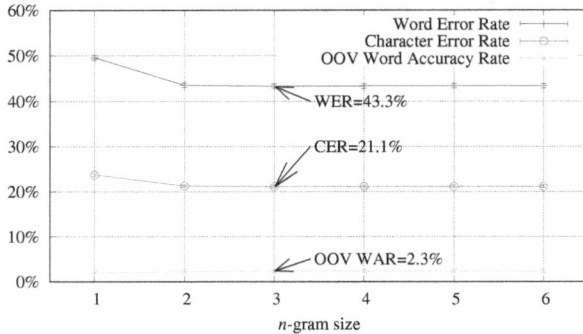

Figure 7. Results obtained by the HMM word-based system using n-gram language models with size $n = \{1, \ldots, 6\}$.

Then, the performance of the HMM system at the sub-word level was tested. Figure 8 presents the results obtained using sub-word n-gram LM with different sizes $n = \{1, \ldots, 6\}$ in terms of WER, CER and recognition accuracy of the OOV words. The best result was obtained with a sub-word language model of size $n = 4$ (a WER equal to $43.2\% \pm 0.5$ and a CER equal to $20.0\% \pm 0.3$). Regarding the recognition of OOV words, the sub-word approach was able to recognize correctly $9.3\% \pm 0.7$ of the OOV words.

Figure 9 presents the results obtained for the HMM system using character n-gram LM with different degrees $n = \{1, \ldots, 15\}$ in terms of WER, CER and recognition accuracy of the OOV words. Although similar results are obtained for $n \geq 6$, the overall best result was obtained with a character language model of degree $n = 10$ (a WER equal to $39.8\% \pm 0.5$ and a CER equal to $17.6\% \pm 0.3$). Regarding the recognition of OOV words, this character-based approach was able to recognize correctly $18.3\% \pm 0.9$ of the OOV words using no external resource or dictionary, but a character language model only.

Table 2 presents a summary of the obtained best results for the test experiments for the HMM system. As can be observed, the improvement offered by the sub-word approach is not statistically significant at the WER level compared to the results obtained from the word-based system. Nevertheless, the character-based approach offers 9.3% of statistically-significant relative improvement over the baseline in terms of WER and 17.0% of statistically-significant relative improvement over the baseline in terms of CER. Thus, using a dictionary and LM at the word level performs worse than using a single character-based n-gram LM, with n large enough. This demonstrates the interest in working at the character level for transcribing historical manuscripts. We study in the following the structure of the OOV words in comparison with the training words (Section 4.2). We also study the effect of reducing the OOV rate, either by using the validation set or by closing the vocabulary (Section 4.3).

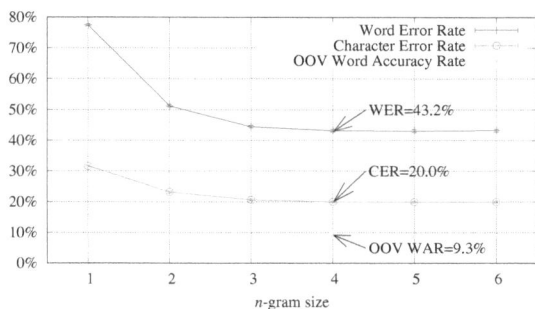

Figure 8. Results obtained by decoding at the HMM sub-word level by using *n*-gram language models with size $n = \{1, \ldots, 6\}$.

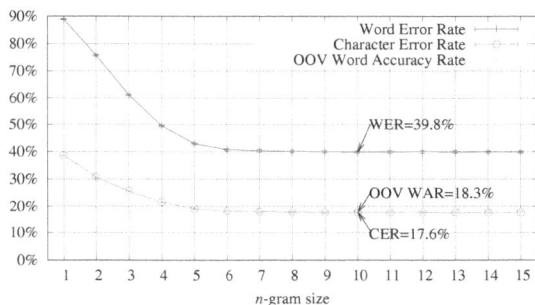

Figure 9. Results obtained by decoding at the HMM character level by using *n*-gram language models with size $n = \{1, \ldots, 15\}$.

Table 2. Overall best results on the *Rodrigo* test set in terms of WER, CER and OOV WAR for the HMM system.

Measure	Word 3-gram	Sub-Word 4-gram	Character 10-gram
WER	$43.9\% \pm 0.5$	$43.2\% \pm 0.5$	$39.8\% \pm 0.5$
CER	$21.2\% \pm 0.3$	$20.0\% \pm 0.3$	$17.6\% \pm 0.3$
OOV WAR	$2.3\% \pm 0.3$	$9.3\% \pm 0.7$	$18.3\% \pm 0.9$

4.2. Study of the Relation between the Structure of the OOV Words and the Training Words

The character-based approach is able to recognize some OOV words given that the character-based LM learns the structure of the words contained in the training set. In order to verify this hypothesis, we measured the perplexity presented by the best character-based LM (10-gram) for decoding each one of the 4918 OOV words as their corresponding character sequences. Figure 10 presents the obtained perplexity per OOV word separated into two distributions, recognized and unrecognized OOV words. Table 3 summarizes the main features of these distributions. As expected, the recognized OOV words present lower perplexity than the unrecognized OOV words. The overlap of both distributions makes us think that there is still room for improvement given that more OOV words could be recognized.

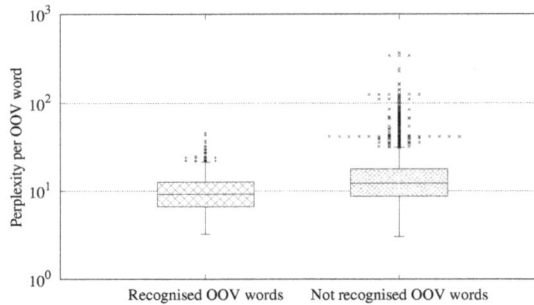

Figure 10. Distribution of the perplexity presented by the 10-gram character Language Model (LM) per recognized and unrecognized OOV words (decomposed into character sequences) by the HMM system.

Table 3. Features of the perplexity per OOV word recognized and unrecognized distributions for the HMM character-based 10-gram LM. Q_1, Q_2 and Q_3 are respectively the 1th, 2nd and 3rd quartile, IQR the interquartile range, Min. and Max. the minimum and maximum values and SD the standard deviation.

Distribution	Q_1	Q_2	Q_3	IQR	Min.	Max.	SD
Recognized	6.64	9.22	12.57	5.94	3.26	46.05	5.37
Unrecognized	8.70	12.21	17.75	9.05	3.06	367.07	16.25

4.3. Study of the Effect of Closing the Vocabulary and Adding the Transcription of the Validation Set for Training the LM

After the adjustment of the decoding parameters with the validation set, the transcription of the text lines contained in this partition can be used to train an improved LM that, hopefully, will reduce the amount of OOV words. Moreover, the OOV words can be included in the vocabulary as unigrams (closed vocabulary experiments) to verify their influence on the recognition. These conditions were experimented for the best language models at word and character levels (3-gram for the word based system and 10-gram for the character-based system). Given that the sub-word approach presented no significative difference in terms of WER, compared to the word-based system (see Table 2), this approach was not tested in this experiment.

Figures 11–13 allow comparing the obtained results for the word-based system and the character-based approach with open and closed vocabulary, with and without the use of the validation samples when training the LM (see Section 3.4). On the one hand, as can be seen in Figures 11 and 13, the use of the validation set does not significantly improve the word-based recognition in terms of WER or CER. However, this additional information is very useful in the character-based approach. As can be observed in Figure 11, a statistically-significant improvement in terms of CER is achieved ($16.9\% \pm 0.3$ instead of $17.6\% \pm 0.3$). This improvement allows increasing the OOV word recognition accuracy (see Figure 12). On the other side, although closing the vocabulary significantly improves the recognition performance, it is interesting to note the beneficial effect of the use of the validation samples in the character-based approach. It is also interesting to note in Figures 11 and 13 that the character-based system, even in the more difficult case ("open-vocabulary"), outperforms, in terms of CER, the word-based system in the best case ("closed-vocabulary"). In the closed vocabulary conditions, the word-based system recognizes more OOV words than the character-based system, $34.7\% \pm 1.2$ instead of $29.6\% \pm 1.1$ (see Figure 12). However, in the real-world case, i.e., the open-vocabulary conditions, the character-based system performs better.

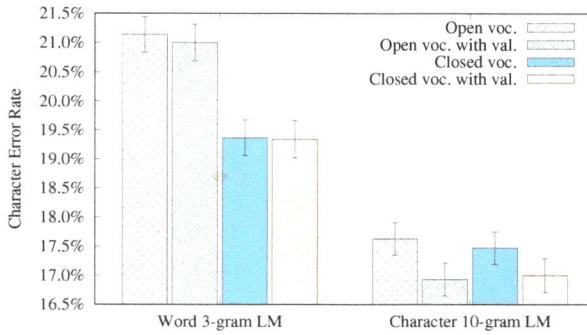

Figure 11. CER results obtained by the best word-based HMM system and the best character-based HMM system with open and closed vocabulary, with and without using the validation samples for training the LM.

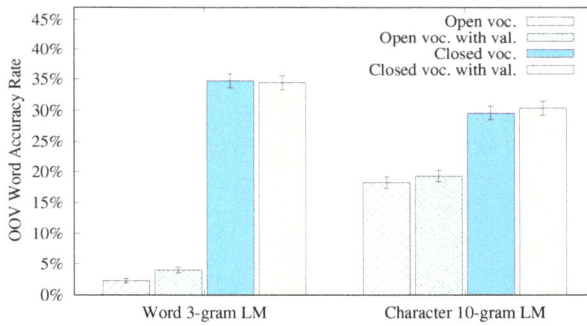

Figure 12. Recognition accuracy rate for OOV words by the best word-based HMM system and the best character-based HMM system with open and closed vocabulary, with and without using the validation samples for training the LM.

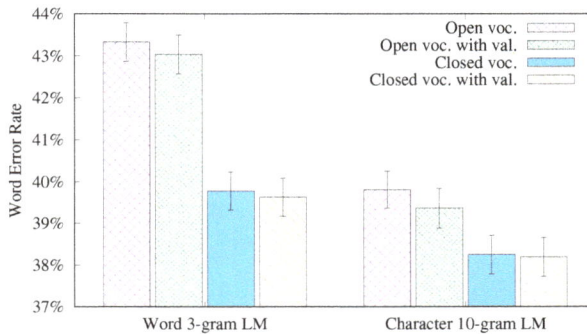

Figure 13. WER results obtained by the best word-based HMM system and the best character-based HMM system with open and closed vocabulary, with and without using the validation samples for training the LM.

4.4. Study of the Context Size Influence Using Deep Optical Models

This last part of the experimentation studies the influence of the different language units and the context size of the language model, on the HTR system based on deep neural networks (see Sections 3.4.2 and 3.4.3).

4.4.1. Results for Deep Models Based on Recurrent Neural Networks with BLSTMs

In Figure 14, the recognition results obtained for the word-based RNN system are presented. As explained before, in this case, the recognized OOV words correspond to words attached to punctuation marks, which were correctly recognized after removing the space between them (see the example presented in Figure A2). Compared with the word-based HMM system, the obtained results are significantly worse in terms of WER; however, in terms of CER and OOV word recognition accuracy, the obtained results are significantly better. Concretely, the best result was obtained by using a two-gram LM, and it presents a WER equal to $52.5\% \pm 0.8$, a CER equal to $17.2\% \pm 0.3$ and an OOV WAR equal to $16.3\% \pm 0.9$.

Figure 15 shows the results obtained using sub-word n-gram LM. As can be observed, the WFST approach has no context information about the separation between words when sub-word unigrams LM are used; therefore, it is unable to reconstruct words correctly in spite of obtaining a good CER. We will see this effect in the next experiments with the sub-word and character-based deep net systems. In this case, the best result was obtained with a five-gram language model (a WER equal to $38.6\% \pm 0.5$, a CER equal to $17.3\% \pm 0.3$ and an OOV WAR equal to $27.4\% \pm 1.1$).

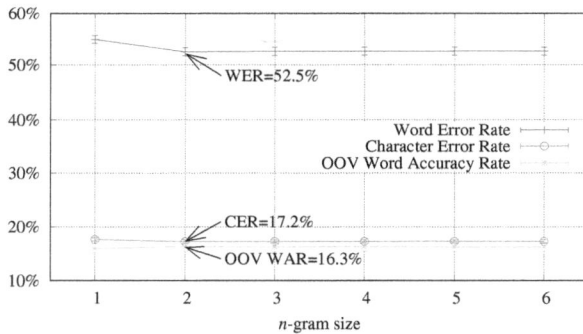

Figure 14. Results obtained by the RNN word-based system using n-gram language models.

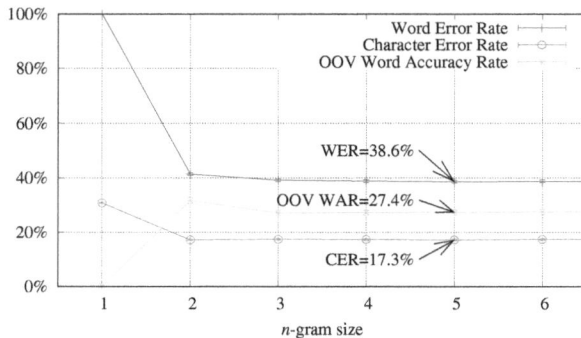

Figure 15. Results obtained by the RNN sub-word-based system using n-gram language models.

The results obtained with the RNN system using character *n*-gram LM are presented in Figure 16. As in the character-based HMM experiments, similar results are obtained for $n \geq 6$, and the overall best result was obtained with a 10-gram character language model: a WER equal to $37.7\% \pm 0.5$, a CER equal to $14.3\% \pm 0.3$ and an OOV WAR equal to $37.8\% \pm 1.1$.

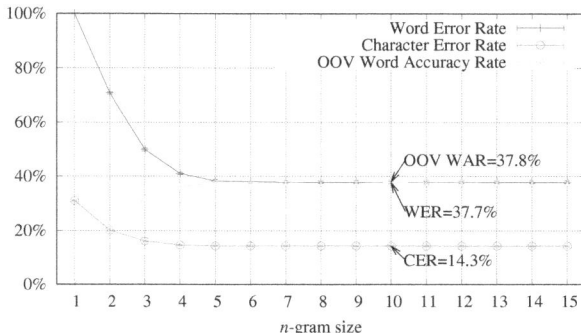

Figure 16. Results obtained by the RNN character-based system using *n*-gram language models.

A summary of the obtained best results for the test experiments for the RNN system is presented in Table 4. As can be observed, generally, the RNN approach performs better than the traditional HMM approach. Although the use of the word-based RNN system obtains a statistically-significant relative deterioration of 19.6% over the HMM system ($43.9\% \pm 0.5$) in terms of WER, 18.9% statistically-significant relative improvement in terms of CER ($21.2\% \pm 0.3$) can be considered. Moreover, 16.3% of OOV words, which correspond to words followed by punctuation marks, are well recognized.

Table 4. Summary of the best results in terms of WER, CER and OOV WAR for the RNN system.

Measure	Word 2-gram	Sub-Word 5-gram	Character 10-gram
WER	$52.5\% \pm 0.8$	$38.6\% \pm 0.5$	$37.7\% \pm 0.5$
CER	$17.2\% \pm 0.3$	$17.3\% \pm 0.3$	$14.3\% \pm 0.3$
OOV WAR	$16.3\% \pm 0.9$	$27.4\% \pm 1.1$	$37.8\% \pm 1.1$

The use of sub-word units offers better results than using words, allowing one to obtain significant improvements in terms of WER and CER over the HMM system. In this case, the use of a five-gram LM trained with hyphenated words allowed obtaining statistically-significant improvements at the WER level over the use of a two-gram LM of full words. However, as for the HMM system, the overall best results are obtained by using the character-based approach: a WER equal to $37.7\% \pm 0.5$, a CER equal to $14.3\% \pm 0.3$ and an OOV WAR equal to $37.8\% \pm 1.1$.

4.4.2. Results for Deep Models Based on Convolutional Recurrent Neural Networks

Figure 17 presents the recognition results obtained for the word-based CRNN system. As in the previous word-based systems, the recognized OOV words correspond to words attached to punctuation marks, which were correctly recognized after removing the space between them (see the example presented in Figure A2). The best result, obtained by using a three-gram LM, presents a WER equal to $17.9\% \pm 0.4$, a CER equal to $4.0\% \pm 0.1$ and an OOV WAR equal to $21.5\% \pm 1.0$.

The results obtained using sub-word *n*-gram LM are shown in Figure 18. The best result was obtained with a four-gram language model (a WER equal to $14.8\% \pm 0.3$ and a CER equal to $3.4\% \pm 0.1$).

Regarding the recognition of OOV words, the sub-word approach allowed correctly recognizing 42.4% ± 1.5 of the OOV words.

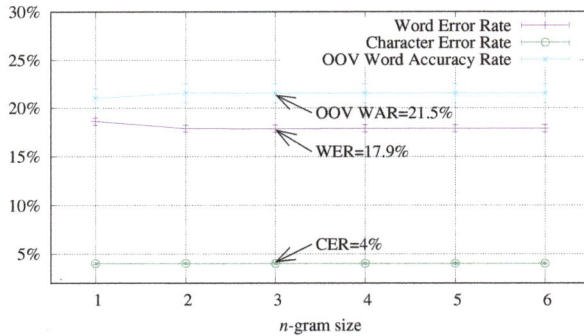

Figure 17. Results obtained by the CRNN word-based system using *n*-gram language models with size $n = \{1, \ldots, 6\}$.

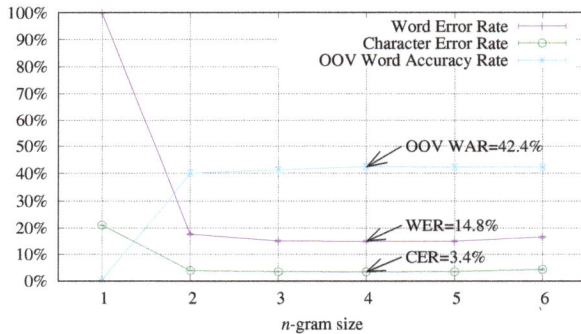

Figure 18. Results obtained by the CRNN sub-word-based system using *n*-gram language models with size $n = \{1, \ldots, 6\}$.

Figure 19 presents the results obtained with the CRNN system using character *n*-gram LM. As in the previous character-based experiments, similar results are obtained for $n \geq 6$, and the overall best result was obtained with a 10-gram character language model (a WER equal to 14.0% ± 0.3 and a CER equal to 3.0% ± 0.1). Regarding the recognition of OOV words, this approach was able to recognize correctly 69.2% ± 1.1 of the OOV words using no external resource or dictionary, but a character language model only.

Table 5 presents a summary of the obtained best results for the test experiments for the CRNN system. As can be observed, the use of deep optical models allows one to obtain a statistically-significant relative improvement of 59.2% over the HMM system (43.9% ± 0.5) in terms of WER and 81.1% statistically-significant relative improvement over the HMM system in terms of CER. Regarding OOV words, 21.5% of OOV words, which correspond to words followed by punctuation marks, are well recognized. It should be noted that these results are also significantly better than those obtained by the HMM system in the closed vocabulary experiments (Figures 11–13).

The use of sub-word units performs better than using words. In this case, the use of a four-gram LM trained with hyphenated words allowed obtaining statistically-significant improvements over the use of a three-gram LM of full words. However, the overall best results are obtained by using the character-based approach: a WER equal to 14.0% ± 0.3, a CER equal to 3.0% ± 0.1 and an OOV WAR equal to 69.2% ± 1.1. These results confirm the interest of working at the character level for transcribing historical manuscripts.

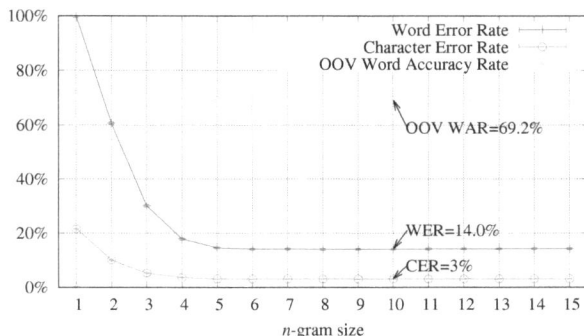

Figure 19. Results obtained by the CRNN character-based system using n-gram language models with size $n = \{1, \ldots, 15\}$.

Table 5. Overall best results on the *Rodrigo* test set in terms of WER, CER and OOV WAR for the CRNN system.

Measure	Word 3-gram	Sub-Word 4-gram	Character 10-gram
WER	17.9% ± 0.4	14.8% ± 0.3	14.0% ± 0.3
CER	4.0% ± 0.1	3.4% ± 0.1	3.0% ± 0.1
OOV WAR	21.5% ± 1.0	42.4% ± 1.5	69.2% ± 1.1

5. Conclusions

In this paper, we deal with the transcription of historical documents, for which no external linguistic resources are available. We have developed various HTR systems that model language at word and sub-lexical levels. We have shown that character-based language modeling performs best.

The strengths of the proposed work are:

- comparing several types of HTR systems (HMM-based, RNN-based).

- proposing a state-of-the-art HTR system for the transcription of ancient Spanish documents whose optical part is based on very deep nets (CRNNs).

- proposing to associate the optical HTR system with a dictionary and a language model based on sub-lexical units. These units are shown to be efficient in order to cope with OOV words.

- reaching with such optical and LM HTR components the best overall recognition results on a publicly available Spanish historical dataset of document images.

In future work, we would like to extend this work using other kinds of language models, such as models based on RNN.

Acknowledgments: Work partially supported by projects READ: Recognition and Enrichment of Archival Documents - 674943 (European Union's H2020) and CoMUN-HaT: Context, Multimodality and User Collaboration in Handwritten Text Processing - TIN2015-70924-C2-1-R (MINECO/FEDER), and a DGA-MRIS (Direction Générale de l'Armement - Mission pour la Recherche et l'Innovation Scientifique) scholarship.

Author Contributions: Emilio Granell and Edgard Chammas conceived and implemented the recognition systems (HMM, BLSTM, CRNN). All authors contributed in equal proportion to the design of the research and to the final manuscript.

Conflicts of Interest: The authors declare no conflict of interest.

Appendix A. Some Recognition Examples

This Appendix presents some recognition examples. Figures A1–A3 present the best hypothesis obtained for several lines of the *Rodrigo* corpus in the open vocabulary experiments, by using a 3-gram word-based LM, a 4-gram sub-word-based LM and a 10-gram character-based LM.

Text Image	
Text Reference	muerte e peor merescia el por quanto passara el mandami
Word-based 1-best	me & peor matara el por quanto pagana el manda
Sub-word-based 1-best	mun do <SPACE> & <SPACE> por <SPACE> ma ta ra <SPACE> el <SPACE> por <SPACE> quan to <SPACE> pa ga na <SPACE> el <SPACE> man da <SPACE>
	mundo & por matara el por quanto pagana el manda
Character-based 1-best	m u c h o <SPACE> & <SPACE> p o r <SPACE> m e r e s c i a <SPACE> e l <SPACE> p o r <SPACE> q u a n t o <SPACE> p a g a u a <SPACE> e l <SPACE> m a n d a m i
	mucho & por merescia el por quanto pagaua el mandami

Figure A1. Example of the best hypotheses obtained for the 12th line of page 500 of *Rodrigo*.

Text Image	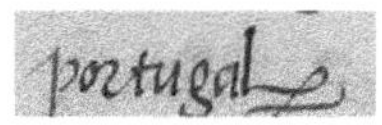
Text Reference	portugal.
Word-based 1-best	portugal.
	portugal.
Sub-word-based 1-best	pe tu gal zo
	petugalzo
Character-based 1-best	p o r t u g a z
	portugaz

Figure A2. Example of the best hypotheses obtained for the 9th line of page 619 of *Rodrigo*.

Text Image	
Text Reference	maron lo cauallero e seyendo Cauallero enfermo muy mal
Word-based 1-best	non lo Cauallero & seyendo Cauallero enfermo muy dia
Sub-word-based 1-best	na ron <SPACE> la <SPACE> Caua lle ro <SPACE> & <SPACE> se yen do <SPACE> Caua lle ro <SPACE> en fer mo <SPACE> muy <SPACE> dia <SPACE>
	naron la Cauallero & seyendo Cauallero enfermo muy dia
Character-based 1-best	m a r o n <SPACE> l a <SPACE> c a u a l l e r o <SPACE> & <SPACE> s e y e n d o <SPACE> C a u a l l e r o <SPACE> e n f e r m o <SPACE> m u y <SPACE> m a l
	maron la cauallero & seyendo Cauallero enfermo muy mal

Figure A3. Example of the best hypotheses obtained for the 4th line of page 514 of *Rodrigo*.

References

1. España-Boquera, S.; Castro-Bleda, M.J.; Gorbe-Moya, J.; Zamora-Martinez, F. Improving Offline Handwritten Text Recognition with Hybrid HMM/ANN Models. *IEEE Trans. Pattern Anal. Mach. Intell.* **2011**, *33*, 767–779.
2. Al-Hajj-Mohamad, R.; Likforman-Sulem, L.; Mokbel, C. Combining Slanted-Frame Classifiers for Improved HMM-Based Arabic Handwriting Recognition. *IEEE Trans. Pattern Anal. Mach. Intell.* **2009**, *31*, 1165–1177.
3. Vinciarelli, A. A survey on off-line cursive word recognition. *Pattern Recognit.* **2002**, *35*, 1433–1446.
4. Bianne-Bernard, A.L.; Menasri, F.; El-Hajj, R.; Mokbel, C.; Kermorvant, C.; Likforman-Sulem, L. Dynamic and Contextual Information in HMM modeling for Handwritten Word Recognition. *IEEE Trans. Pattern Anal. Mach. Intell.* **2011**, *99*, 2066–2080.
5. Graves, A. Supervised Sequence Labelling with Recurrent Neural Networks. Ph.D. Thesis, Technische Universität München, Munich, Germany, 2008.

6. Xie, Z.; Sun, Z.; Jin, L.; Feng, Z.; Zhang, S. Fully convolutional recurrent network for handwritten Chinese text recognition. In Proceedings of the 23rd International Conference on Pattern Recognition (ICPR), Cancun, Mexico, 4–8 December 2016; pp. 4011–4016.

7. Bluche, T.; Messina, R. Gated Convolutional Recurrent Neural Networks for Multilingual Handwriting Recognition. In Proceedings of the 13th International Conference on Document Analysis and Recognition (ICDAR), Kyoto, Japan, 13–15 November 2017.

8. Sudholt, S.; Fink, G.A. PHOCNet: A deep convolutional neural network for word spotting in handwritten documents. In Proceedings of the 15th International Conference on Frontiers in Handwriting Recognition (ICFHR), Shenzhen, China, 23–26 October 2016; pp. 277–282.

9. Brakensiek, A.; Rottland, J.; Kosmala, A.; Rigoll, G. Off-line handwriting recognition using various hybrid modeling techniques and character n-grams. In Proceedings of the 7th International Workshop on Frontiers in Handwritten Recognition, Amsterdam, The Netherlands, 11–13 September 2000; pp. 343–352.

10. Fischer, A.; Frinken, V.; Bunke, H.; Suen, C.Y. Improving hmm-based keyword spotting with character language models. In Proceedings of the 12th International Conference on Document Analysis and Recognition (ICDAR), Washington, DC, USA, 25–28 August 2013; pp. 506–510.

11. Santoro, A.; Parziale, A.; Marcelli, A. A Human in the Loop Approach to Historical Handwritten Documents Transcription. In Proceedings of the 15th International Conference on Frontiers in Handwriting Recognition (ICFHR), Shenzhen, China, 23–26 October 2016; pp. 222–227.

12. Stefano, C.D.; Marcelli, A.; Parziale, A.; Senatore, R. Reading Cursive Handwriting. In Proceedings of the 12th International Conference on Frontiers in Handwriting Recognition, Kolkata, India, 16–18 November 2010; pp. 95–100.

13. Oprean, C.; Likforman-Sulem, L.; Popescu, A.; Mokbel, C. Handwritten word recognition using Web resources and recurrent neural networks. *Int. J. Doc. Anal. Recognit. (IJDAR)* **2015**, *18*, 287–301.

14. Frinken, V.; Fischer, A.; Martínez-Hinarejos, C.D. Handwriting recognition in historical documents using very large vocabularies. In Proceedings of the 2nd International Workshop on Historical Document Imaging and Processing, Washington, DC, USA, 24 August 2013; pp. 67–72.

15. Swaileh, W.; Paquet, T. Handwriting Recognition with Multi-gram language models. In Proceedings of the 14h International Conference on Document Analysis and Recognition (ICDAR), Kyoto, Japan, 10–15 November 2017.

16. Kozielski, M.; Rybach, D.; Hahn, S.; Schlüter, R.; Ney, H. Open vocabulary handwriting recognition using combined word-level and character-level language models. In Proceedings of the 2013 International Conference on Acoustics, Speech and Signal Processing (ICASSP '13), Vancouver, BC, Canada, 26–31 May 2013; pp. 8257–8261.

17. Messina, R.; Kermorvant, C. Over-generative finite state transducer n-gram for out-of-vocabulary word recognition. In Proceedings of the 11th IAPR International Workshop on Document Analysis Systems (DAS), Tours, France, 7–10 April 2014; pp. 212–216.

18. Shi, B.; Bai, X.; Yao, C. An end-to-end trainable neural network for image-based sequence recognition and its application to scene text recognition. *IEEE Trans. Pattern Anal. Mach. Intell.* **2017**, *39*, 2298–2304.

19. Serrano, N.; Castro, F.; Juan, A. The RODRIGO Database. In Proceedings of the 7th International Conference on Language Resources and Evaluation (LREC), Valletta, Malta, 17–23 May 2010; pp. 2709–2712.

20. Pattern Recognition and Human Language Technology (PRHLT) Research Center. 2018. Available online: https://www.prhlt.upv.es (accessed on 5 January 2018).

21. Fischer, A. Handwriting Recognition in Historical Documents. Ph.D. Thesis, University of Bern, Bern, Switzerland, 2012.

22. Michel, J.B.; Shen, Y.K.; Aiden, A.P.; Veres, A.; Gray, M.K.; Brockman, W.; Team, T.G.B.; Pickett, J.P.; Hoiberg, D.; Clancy, D.; et al. Quantitative analysis of culture using millions of digitized books. *Science* **2010**, *331*, 176–182.

23. Pastor, M.; Toselli, A.H.; Vidal, E. Projection profile based algorithm for slant removal. In *Lecture Notes in Computer Science, Proceedings of the International Conference on Image Analysis and Recognition (ICIAR'04), Porto, Portugal, 29 September–1 October 2004*; Springer: Berlin, Germany, 2004; Volume 3212, pp. 183–190.

24. Toselli, A.H.; Juan, A.; González, J.; Salvador, I.; Vidal, E.; Casacuberta, F.; Keysers, D.; Ney, H. Integrated Handwriting Recognition and Interpretation using Finite-State Models. *Int. J. Pattern Recognit. Artif. Intell.* **2004**, *18*, 519–539.
25. Testhyphens – Testing hyphenation patterns. 2018. Available online: https://www.ctan.org/tex-archive/macros/latex/contrib/testhyphens (accessed on 5 January 2018)
26. Kneser, R.; Ney, H. Improved backing-off for M-gram language modeling. In Proceedings of the 1995 International Conference on Acoustics, Speech, and Signal Processing (ICASSP'95), Detroit, MI, USA, 9–12 May 1995; Volume 1, pp. 181–184.
27. Stolcke, A. SRILM—An extensible language modeling toolkit. In Proceedings of the 3rd Interspeech, Denver, CO, USA, 16–20 September 2002; pp. 901–904.
28. Young, S.; Evermann, G.; Gales, M.; Hain, T.; Kershaw, D.; Liu, X.; Moore, G.; Odell, J.; Ollason, D.; Povey, D.; et al. *The HTK Book (for HTK Version 3.4)*; Cambridge University Engineering Department: Cambridge, UK, 2006.
29. Luján-Mares, M.; Tamarit, V.; Alabau, V.; Martínez-Hinarejos, C.D.; Pastor, M.; Sanchis, A.; Toselli, A.H. iATROS: A Speech and Handwriting Recognition System. V Jornadas en Tecnologías del Habla, 2008; pp. 75–78. Available online: http://citeseerx.ist.psu.edu/viewdoc/download?doi=10.1.1.329.6708&rep=rep1&type=pdf (accessed on 5 January 2018)
30. Hermansky, H.; Ellis, D.P.W.; Sharma, S. Tandem connectionist feature extraction for conventional HMM systems. In Proceedings of the 2000 IEEE International Conference on Acoustics, Speech, and Signal Processing (ICASSP'00), Istanbul, Turkey, 5–9 June 2000; Volume 3, pp. 1635–1638.
31. Graves, A. RNNLIB: A Recurrent Neural Network Library for Sequence Learning Problems. 2016. Available online: http://sourceforge.net/projects/rnnl/ (accessed on 5 January 2018)
32. Chammas, E. Structuring Hidden Information in Markov Modeling with Application to Handwriting Recognition. Ph.D. Thesis, Telecom ParisTech, Paris, France, 2017.
33. Simonyan, K.; Zisserman, A. Very deep convolutional networks for large-scale image recognition. *arXiv* **2014**, arXiv:1409.1556.
34. Gu, J.; Wang, Z.; Kuen, J.; Ma, L.; Shahroudy, A.; Shuai, B.; Liu, T.; Wang, X.; Wang, G. Recent advances in convolutional neural networks. *arXiv* **2015**, arXiv:1512.07108.
35. Graves, A.; Fernández, S.; Gomez, F.; Schmidhuber, J. Connectionist temporal classification: labeling unsegmented sequence data with recurrent neural networks. In Proceedings of the 23rd international conference on Machine learning ACM, Pittsburgh, PA, USA, 25–29 June 2006; pp. 369–376.
36. Zeyer, A.; Schlüter, R.; Ney, H. Towards Online-Recognition with Deep Bidirectional LSTM Acoustic Models. In Proceedings of the 2016 INTERSPEECH, San Francisco, CA, USA, 8–12 September 2016; pp. 3424–3428.
37. Glorot, X.; Bengio, Y. Understanding the difficulty of training deep feedforward neural networks. In Proceedings of the Thirteenth International Conference on Artificial Intelligence and Statistics, Sardinia, Italy, 13–15 May 2010; pp. 249–256.
38. Kingma, D.; Ba, J. Adam: A method for stochastic optimization. *arXiv* **2014**, arXiv:1412.6980.
39. Tieleman, T.; Hinton, G. Lecture 6.5-rmsprop: Divide the gradient by a running average of its recent magnitude. *COURSERA Neural Netw. Mach. Learn.* **2012**, *4*, 26–31.
40. Qian, N. On the momentum term in gradient descent learning algorithms. *Neural Netw.* **1999**, *12*, 145–151.
41. Zeiler, M.D. ADADELTA: an adaptive learning rate method. *arXiv* **2012**, arXiv:1212.5701.
42. Ioffe, S.; Szegedy, C. Batch normalization: Accelerating deep network training by reducing internal covariate shift. In Proceedings of the International Conference on Machine Learning, Lille, France, 6–11 July 2015; pp. 448–456.
43. Miao, Y.; Gowayyed, M.; Metze, F. EESEN: End-to-end speech recognition using deep RNN models and WFST-based decoding. In Proceedings of the 2015 IEEE Workshop on Automatic Speech Recognition and Understanding (ASRU), Scottsdale, AZ, USA, 13–17 December 2015; pp. 167–174.
44. Levenshtein, V.I. Binary codes capable of correcting deletions, insertions, and reversals. *Sov. Phys. Dokl.* **1966**, *10*, 707–710.
45. Knezevic, A. *Overlapping Confidence Intervals and Statistical Significance*; StatNews; Cornell University Statistical Consulting Unit: Ithaca, NY, USA, 2008; Volume 73.

46. Bisani, M.; Ney, H. Bootstrap estimates for confidence intervals in ASR performance evaluation. In Proceedings of the IEEE International Conference on Acoustics, Speech, and Signal Processing ICASSP'04, Montreal, QC, Canada, 17–21 May 2004; Volume 1, pp. 409–412.
47. Brown, P.F.; Della Pietra, V.J.; Mercer, R.L.; Della Pietra, S.A.; Lai, J.C. An Estimate of an Upper Bound for the Entropy of English. *Comput. Linguist.* **1992**, *18*, 31–40.

Article

Journal of
Imaging

MDPI

A Study of Different Classifier Combination Approaches for Handwritten *Indic* Script Recognition

Anirban Mukhopadhyay *, Pawan Kumar Singh, Ram Sarkar * and Mita Nasipuri

Department of Computer Science and Engineering, Jadavpur University, Kolkata-700032, West Bengal, India; pawansingh.ju@gmail.com (P.K.S.); mitanasipuri@gmail.com (M.N.)
* Correspondence: anirbanmcse@gmail.com (A.M.); raamsarkar@gmail.com (R.S.)

Received: 15 December 2017; Accepted: 8 February 2018; Published: 13 February 2018

Abstract: Script identification is an essential step in document image processing especially when the environment is multi-script/multilingual. Till date researchers have developed several methods for the said problem. For this kind of complex pattern recognition problem, it is always difficult to decide which classifier would be the best choice. Moreover, it is also true that different classifiers offer complementary information about the patterns to be classified. Therefore, combining classifiers, in an intelligent way, can be beneficial compared to using any single classifier. Keeping these facts in mind, in this paper, information provided by one shape based and two texture based features are combined using classifier combination techniques for script recognition (word-level) purpose from the handwritten document images. *CMATERdb*8.4.1 contains 7200 handwritten word samples belonging to 12 *Indic* scripts (600 per script) and the database is made freely available at https://code.google.com/p/cmaterdb/. The word samples from the mentioned database are classified based on the confidence scores provided by Multi-Layer Perceptron (MLP) classifier. Major classifier combination techniques including majority voting, Borda count, sum rule, product rule, max rule, Dempster-Shafer (DS) rule of combination and secondary classifiers are evaluated for this pattern recognition problem. Maximum accuracy of 98.45% is achieved with an improvement of 7% over the best performing individual classifier being reported on the validation set.

Keywords: Classifier combination; Dempster-Shafer theory of evidence; *Indic* script identification; Histograms of Oriented Gradients; Modified Log-Gabor filter transform; Elliptical features

1. Introduction

In the domain of document images processing, Optical Character Recognition (OCR) systems are, in general, developed keeping a particular script in mind, which implies that such systems can read characters written in a specific script only. This is because the number of characters, shape of the characters or the writing style of using a particular character set is so different that designing a common feature set applicable for recognizing any character set is practically impossible. As an alternative, a pool of OCR systems that correspond to different scripts [1] can be used to solve this said problem. This statement infers that before the document images are fed to an OCR system, it is required to identify the script in which the document is written so that those document images can be suitably converted into a computer-editable format using that OCR system. This summarizes the problem of script identification. There are some important applications of script identification system such as automatic archiving as well as indexing of multi-script documents, searching required information from digitized archives of multi-scripts document images.

In this paper, script identification from handwritten document images written in different scripts is considered. In this regard, it is to be noted that hurdles are multi-fold when handwritten document images are considered compared to its printed counterpart. The main difficulty which researchers need to deal with is the non-uniformity of the shape and size of the characters written by different writers. Along with these, problems like skew, slant etc. are commonly seen in handwritten documents. Even the paper and ink qualities make things much difficult. Apart from the intrinsic complexities of handwritings, similarities among the characters belonging to different script augment the challenges of script recognition from the handwritten document images. It is worth mentioning that, usually, script recognition is performed at page, text-line or at word level. But in this paper, this is done at word-level because of two reasons: (a) feature extraction at word-level is less time consuming than at page or at text-line level and (b) sometimes, it is seen that a single document page or a single text line contains multiple scripts. In that case, word-level script identification is appropriate.

Script recognition articles for handwritten documents are relatively limited in comparison to its printed counterpart. Ubul et al. [2] comprehensively showed the state-of-the-art performance results for different identification, feature extraction and classification methodologies involved in the process. Recently, Singh et al. [1] provided a survey considering various feature extraction and classification techniques associated with the offline script identification of the *Indic* scripts. Spitz [3] proposed a method for distinguishing between *Asian* and *European* languages by analysing the connected components. Tan et al. [4] developed a method based on texture analysis for automatic script identification from document images using multiple channel (Gabor) filters and Gray level co-occurrence matrices(GLCM) for seven languages: *Chinese, English, Greek, Koreans, Malayalam, Persian* and *Russian*. Hochberg et al. [5,6] described an algorithm for script and language identification from handwritten document images using statistical features based on connected component analysis. Wood et al. [7] demonstrated a projection profile method to determine *Roman, Russian, Arabic, Korean* and *Chinese* characters. Chaudhuri et al. [8] discussed an OCR system to read two Indian languages *viz., Bangla* and *Devanagari* (*Hindi*). Pal et al. [9] proposed an algorithm for word-wise script identification from document containing *English, Devanagari* and *Telugu* text, based on conventional and water reservoir features. Chaudhury et al. [10] proposed a method for identification of Indian languages by combining Gabor filter based techniques and direction distance histogram classifier for *Hindi, English, Malayalam, Bengali, Telugu* and *Urdu*. Some analysis of the variability involved in the multi-script signature recognition problem as compared to the single-script scenario is discussed in [11,12].

Various classification algorithms are applied for different pattern recognition problems and the same fact also applies to the script recognition problem. Till date, for *Indic* script recognition purpose, different classifiers have been used such as *k*-Nearest Neighbours (*k*-NN) [13,14], Linear Discriminant Analysis (LDA) [15], Neural Networks (NN) [15,16], Support Vector Machine (SVM) [16,17], Tree based classifier [18,19], Simple Logistic [20] and MLP [21,22]. Though good results have already been achieved in this pattern recognition task but with a single classifier it is still hard to achieve acceptable accuracy. Studies expose that the fusion of multiple classifiers can be a viable solution to get better classification results as the error amassed by any single classifier is generally compensated using information from other classifiers. The reason for this is that different classifiers may offer complementary information about the patterns under consideration. Based on this fact, since long, a section of researchers has focused on devising different algorithms for combining classifiers in an intelligent way so that the combination can achieve better results than any of the individual classifier used for combining. The key idea is that instead of relying on a single decision maker, all the designs or their subsets are applied for the decision making by combining their individual beliefs in order to come up with a consensus decision. This fact motivates many researchers to apply the classifier combination methods to different pattern recognition problems. The popular methodologies for classifier combination include: Majority Voting [23,24], Subset-combining and re-ranking approach [25], Statistical model [26], Bayesian Belief Integration [27], Combination based on DS theory of evidence [27,28] and Neural Network combinator [29].

But till date, classifier combination approach for script recognition problem, either handwritten or printed, has not been tested much, though it has enormous potential. To bridge this research gap, this paper applies different classifier combination techniques in the field of *Indic* script recognition.

The main contribution of the present work is the comprehensive evaluation of the major classifier combination approaches which are either rule based or apply a secondary classifier for information fusion. The motivation is to improve the classification accuracy at the word-level handwritten script recognition by combining the results of the best performing classifier on three previously used feature sets. It is a multi-class classification problem and in the present case, 12 officially used *Indic*-scripts are considered which are: *Devanagari, Bangla, Odia, Gujarati, Gurumukhi, Tamil, Telugu, Kannada, Malayalam, Manipuri, Urdu* and *Roman*. Three different sets of feature vectors based on both shape and texture analysis have been estimated from each of the handwritten word images. Identification of the scripts in which the word images are written, is done with these feature values by feeding the same into different MLP classifiers. Soft-decisions provided by the individual classifiers are then combined using an array of classifier combination techniques. This kind of work is implemented for the first time assuming the number of *Indic* scripts undertaken and the range of combination techniques applied. The system developed for the script recognition task here, is a part of the general framework where different feature sets and classifier outputs can be modelled into a single system without much increase in the computation involved. Block diagram of the present work is shown in Figure 1.

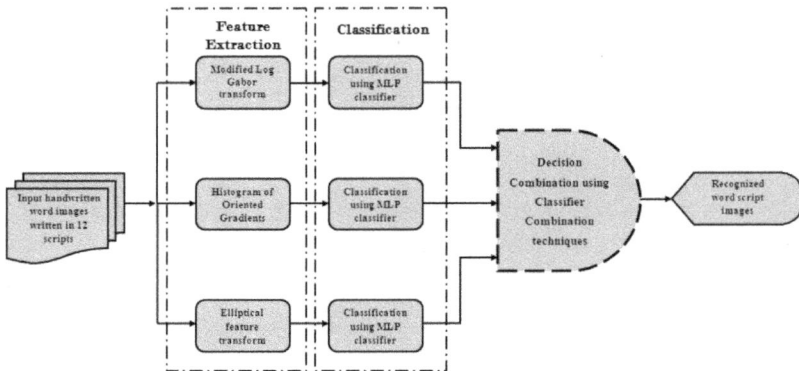

Figure 1. Schematic diagram of the proposed methodology.

2. Feature Extraction

In this paper, three popular feature extraction methodologies have been used for the combination *namely,* Elliptical Features [21], Histogram of Oriented Gradients (HOG) [30] and Modified log-Gabor filter transform [20]. The first feature set is applied to capture the overall structure present in the script word images whereas the rest two feature sets deal with the texture of the same. These features have already provided satisfactory results to this challenging task of handwritten script identification.

2.1. Elliptical Features

The word images are generally found to be elongated in nature which can better covered by an ellipse. That is why; elliptical features are extracted from the contour and the local regions of a word image so that it is easier to isolate a particular script. Two more important notations used in this subsection are: (a) Pixel ratio (P_r) and (b) Pixel count (P_c). Pr is defined as the ratio of the number of contour pixels (object) to the number of background pixels and the pixel count whereas P_c is defined as the number of contour pixels. The features are described in detail:

2.1.1. Maximum Inscribed Ellipse

The height and width of the bounding box are calculated for each word image. A representative ellipse is then inscribed (considering the orientation of the ellipse) inside this bounding box having major and minor axes equal to the width and the height of the bounding box and the centre of an ellipse is also the centre of the corresponding bounding box. This ellipse divides the word image into eight regions R_i, I = 1, 2 ... , 8. The bounding box along with the inscribed ellipse for a handwritten Bangla word image are shown in Figure 2b. Taking the values of P_r from these eight regions, as shown in Figure 2a, eight features (F1–F8) for each handwritten word image are estimated. Now, another type of feature, $P_c$along N (N = 8 for the present work) lines parallel to major/minor axis of the representative ellipse are computed. The mean and standard deviation of the values of $P_c$along major/minor axis are taken as four additional features (F9–F12).

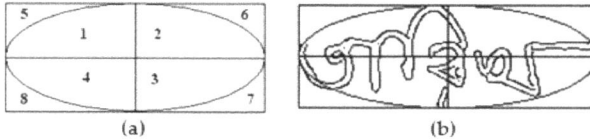

Figure 2. Illustration of fitting (**a**) an imaginary ellipse inside the minimum boundary box which divides a Bangla handwritten word image in 8 regions as shown in (**b**).

2.1.2. Sectional Inscribed Ellipse

Each of the word images surrounded by the minimum bounding box is again divided into four equal rectangles and a representative ellipse is fit into each of these rectangles using the same procedure as described in the previous subsection. As a result, every ellipse produces eight regions inside its rectangular area namely, R_{ij} where $1 \leq i \leq 4$ and $1 \leq j \leq 8$ which makes $8 \times 4 = 32$ regions in total. A total of 32 feature values (F13–F44) using the Pr values is computed from the 32 ellipses in similar fashion.

2.1.3. Concentric Ellipses

These feature values are computed by taking the entire topology of the word image. A primary ellipse is made circumscribing the word image with centre taken to be the midpoint of its minimum bounding box. The values of the major and minor axes of the ellipse are taken into consideration. After fitting the primary ellipse, three concentric ellipses are drawn inside the primary ellipse having the same centre point as the primary ellipse and major and minor axes equal to 1/4th, 2/4th and 3/4th of major and minor axes of the primary ellipse respectively. These four ellipses divide each of the word images into four regions- R_{e1}, R_{e2}, R_{e3} and R_{e4}. The partitioning of the four regions on a sample handwritten *Devanagari* word image is shown in Figure 3. From the four regions, four features values (F45–F48) considering the Pr's and four feature values (F49–F52) considering the Pc's of the regions R_{e1}, R_{e2}, R_{e3} and R_{e4} are estimated. The remaining six features (i.e., F53–F58) are taken as the corresponding differences of the Pr's and Pc's between the regions R_{e1} and R_{e2}, R_{e2} and R_{e3}, R_{e3} and R_{e4} respectively. The elliptical features (F1–F58) are suitably normalized by the height and width of the corresponding word image.

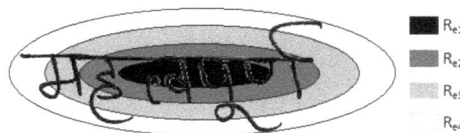

Figure 3. Figure showing the elliptical partition of four regions on a sample handwritten *Devanagari* word image.

2.2. Histogram of Oriented Gradients (HOG)

HOG descriptor [31] counts occurrences of gradient orientation in localized portions of an image which was first proposed for pedestrian detection in steady images. The essential thought behind the HOG descriptor is that local object appearance and shape within an image can be described by the distribution of intensity gradients or edge directions. At first, the values of the magnitude and direction of all the pixels for each of the word images are calculated. Next, each pixel is pigeonholed in certain category according to its direction which is known as orientation bins. Then, the word image is divided into n (here $n = 10$) connected regions, called cells and for each cell, a histogram of gradient directions or edge orientations is computed for the pixels within the cell. The combination of these histograms then represents the descriptor. Since the number of orientation bins is taken as 8 for the present work, an 80-D (i.e., 10×8) feature vector has been extracted using HOG descriptor [30]. The magnitude and direction of each pixel of a sample handwritten *Telugu* word image are also shown in Figure 4.

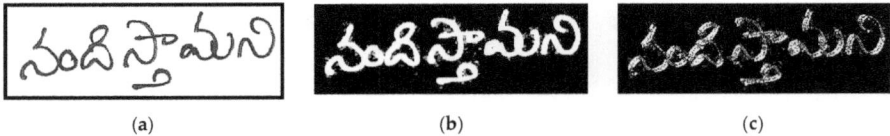

(a) (b) (c)

Figure 4. Illustration of: (**a**) handwritten *Telugu* word image, (**b**) its magnitude part and (**c**) its direction part.

2.3. Modified Log-Gabor Filter Transform (MLG Transform)

Modified log-Gabor filter transform-based features, proposed in Reference [20], had performed well in the script classification task and therefore are also chosen as one of the feature descriptors of our proposed methodology in order to identify the script of the word images. In order to preserve the spatial information, a Windowed Fourier Transform (WFT) is considered in the present work. WFT involves multiplication of the image by the window function and the resultant output is followed by applying the Fourier transform. WFT is basically a convolution of the image with the low-pass filter. Since for texture analysis, both spatial and frequency information are preferred, the present work tries to achieve a good trade-off between these two. Gabor transforms use a Gaussian function as the optimally concentrated function in the spatial as well as in the frequency domain [32]. Due to the convolution theorem, the filter interpretation of the Gabor transform allows the efficient computation of the Gabor coefficients by multiplication of the Fourier transformed image with the Fourier transform of the Gabor filter. The inverse Fourier transform is then applied on the resultant vector to get the output filtered images.

The images, after low pass filtering, are passed as input to a function that computes Gabor energy feature from them. The input image is then passed to a function to yield a Gabor array which is the array equivalent of the image after Gabor filtering. The function displays the image equivalent of the magnitude and the real part of the Gabor array pixels.

For the present work, both energy and entropy features [33] based on Modified log-Gabor filter transform have been extracted for 5 scales (1, 2, 3, 4 and 5) and 6 orientations ($0°$, $30°$, $60°$, $90°$, $120°$ and $150°$) to capture complementary information found in different script word images. Here, each filter is convolved with the input image to obtain 60 different representations (response matrices) for a given input image. Figure 5 shows output images formed after the application of Modified log-Gabor filter transform for a sample handwritten *Bangla* word image.

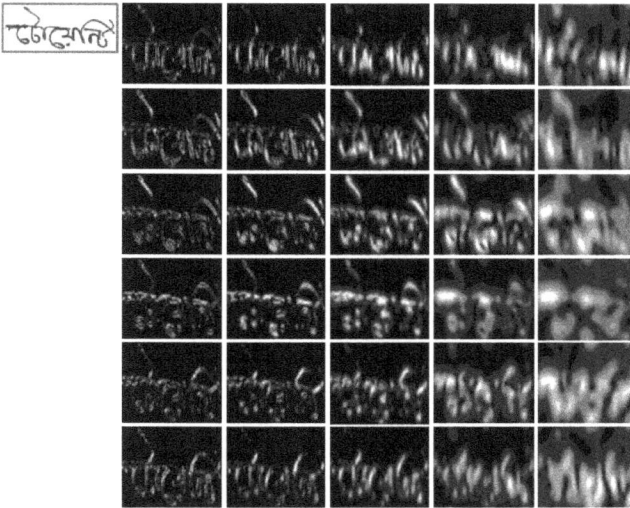

Figure 5. Output word images of Modified log-Gabor filter transform on a sample handwritten *Bangla* word image (shown on left-side) for 5 different scales and 6 different orientations (the first row shows the output for $n_o = 0^0$ and five scales, the second row shows the output for $n_o = 30^0$ and five scales and so on).

3. Classifier Combination

Classifier combination tries to improve on the task of pattern recognition performance through mathematical models. The outputs of classifiers can be represented as vectors of numbers where the dimension of vectors is equal to the number of classes. As a result, the combination problem can be defined as a problem of finding the combination function accepting N-dimensional score vectors from M classifiers and outputting N final classification scores (see Figure 6), where the function tries to minimize the misclassification cost.

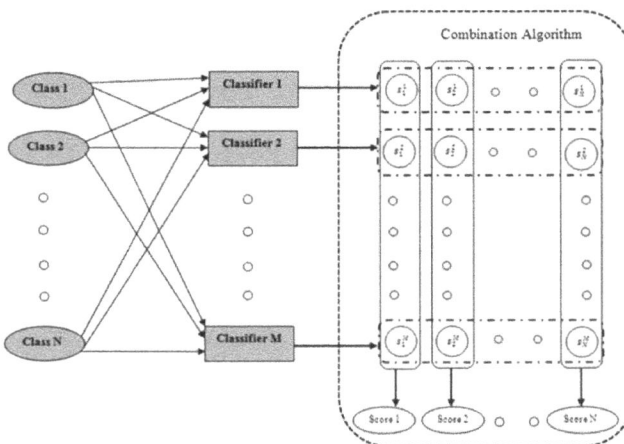

Figure 6. Classifier combination takes a set of s_i^j score for class i by classifier j and produces combination scores S_i for each class i [34].

The field of classifier combination can be grouped into different categories [35] based on the stage at which the process is applied, type of information (classifier output) being fused and the number and type of classifiers being combined.

Based on the operating level of the classifiers, classifier combination can be done at the feature level. Multiple features can be joined to provide a new feature set which provide more information about the classes. But with the increase in dimensionality of the data, training becomes expensive.

The classifier outputs after the extraction of the individual feature sets can be combined to provide better insights at the decision level. Decision level combination techniques are popular as it cannot need any understanding of the ideas behind the feature generation and classification algorithms.

Feature level combination is performed by concatenating the feature sets in all possible combinations and passing it through the base classifier, MLP in this case. Apart from that, all the other combination processes worked out operate at the decision level.

Classifier combination can also be classified by the outputs of the classifiers used in the combination. Three types of classifier outputs are usually considered [36]:

- Type I (Abstract level): This is the lowest level in a sense that the classifier provides the least amount of information on this level. Classifier output is a single class label informing the decision of the classifier.
- Type II (Rank level): Classifier output on the rank level is an ordered sequence of candidate classes, the so-called n-best list. The candidate classes are ordered from the most likely class at the front and the least likely class index featuring at the last of the list. There are no confidence scores attached to the class labels on rank level and the relative positioning provides the required information.
- Type III (Measurement level): In addition to the ordered n-best lists of candidate classes on the rank level, classifier output on the measurement level has confidence values assigned to each entry of the n-best list. These confidences, or scores, are generally real numbers generated using the internal algorithm for the classifier. This soft-decision information at the measurement level thus provides more information than the other levels.

In this paper, Type II (rank level) and Type III (measurement level) combination procedures are worked out because they allow the inculcation of a greater degree of soft-decision information from the classifiers and find use in most practical applications.

The focus of this paper is to explore the classifier combination techniques on a fixed set of classifiers. The purpose of the combination algorithm is to learn the behaviour of these classifiers and produce an efficient combination function based on the classifier outputs. Hence, we use non-ensemble classifier combinations which try to combine heterogeneous classifiers complementing each other. The advantage of complementary classifiers is that each classifier can concentrate on its own small sub problem and together the single larger problem is better understood and solved. The heterogeneous classifiers, here, are generated by training the same classifier with different feature sets and tuning them to optimal values of their parameters. This procedure does away with the need for normalization of the confidence scores provided by different classifiers which do not tend to follow a common standard and depend of the algorithm. For example, in the MLP classifier used here, the last layer has each node containing a final score for one class. These scores can then be used for the rank level and decision level combination along with the maximum being chosen for the individual classifier decision.

In the next sub-section, the set of major classification algorithms evaluated in this paper are categorized into two approaches based on how the combination process is implemented. In the first approach, rule based combination practices are demonstrated that apply a given function to combine the classifier confidences into a single set of output scores. The second approach employs another classifier, called the 'secondary' classifier that operates on the outputs of the base classifier and automatically account for the strengths of the participants. The classification algorithm is trained on these confidence values with output classes same as the original pattern recognition problem.

Essentially, both the approaches apply a function on the confidence score inputs, where the rule based functions are simpler operations like sum rule, max rule, etc. and classifiers like k-NN and MLP apply more complicated functions.

3.1. Rule Based Combination Techniques

Rules are applied on the abstract level, rank level and measurement level outputs from the classifiers to obtain a final set of confidence scores that can take into account the insights provided by the previous stage of classification. Elementary combination approaches like majority voting, Borda count, sum rule, product rule and the max rule come under this approach of classifier combination. DS theory of evidence is a relatively complex technique that is adopted for this purpose, utilising the rule of combination for information sources with the same frame of discernment.

3.1.1. Majority Voting

A straightforward voting technique is majority voting operating at the abstract level. It considers only the decision class provided by each classifier and chooses the most frequent class label among this set. In order to reduce the number of ties, the number of classifiers used for voting is usually odd.

3.1.2. Borda Count

Borda count is a voting technique on rank level [37]. For every class, Borda count adds the ranks in the n-best lists of each classifier so that for every output class the ranks across the classifier outputs get accumulated. The class with the most likely class label, contributes the highest rank number and the last entry has the lowest rank number. The final output label for a given test pattern X is the class with highest overall rank sum. In mathematical terms, this reads as follows: Let N be the number of classifiers and r_i^j the rank of class i in the n-best list of the j-th classifier. The overall rank r_i of class i is thus given by

$$r_i = \sum_{j=1}^{N} r_i^j \tag{1}$$

The test pattern X is assigned the class i with the maximum overall rank count r_i. Borda count is very simple to compute and requires no training. There is also a trainable variant that associates weights to the ranks of individual classifiers. The overall rank count for class i is then computed as given below

$$r_i = \sum_{j=1}^{N} w_j r_i^j \tag{2}$$

The weights can be the performance of each individual classifier measured on a training or validation set.

3.1.3. Elementary Combination Approaches on Measurement Level

Elementary combination schemes on measurement level apply simple rules for combination, such as sum rule, product rule and max rule. Sum rule simply adds the score provided by each classifier from a set of classifier for every class and assigns the class label with the maximum score to the given input pattern. Similarly, product rule multiplies the score for every class and then outputs the class with the maximum score. The max rule predicts the output by the selecting the class corresponding to the maximum confidence value among all the participating classifiers' output scores.

Interesting theoretical results, including error estimations, have been derived for these simple combination schemes. Kittler et al. showed that sum rule is less sensitive to noise than other rules [38]. Despite their simplicity, simple combination schemes have resulted in high recognition rates and shown comparable results to the more complex procedures.

3.1.4. Dempster-Shafer Theory of Evidence

The DS framework [39] is based on the view whereby propositions are represented as subsets of a given set W, referred to as a frame of discernment. Evidence can be associated to each proposition (subset) to express the uncertainty (belief) that has been observed or discerned. Evidence is usually computed based on a density function m called Basic Probability Assignment (BPA) and $m(p)$ represents the belief exactly committed to the proposition p.

DS theory has an operation called *Dempster's rule of combination* that aggregates two (or more) bodies of evidence defined within the same frame of discernment into one body of evidence. Let m_1 and m_2 be two $BPAs$ defined in W. The new body of evidence is defined by the $BPA m_{1,2}$ as:

$$m_{1,2}(A) = \begin{cases} 0 & if \ A = \varnothing \\ \frac{1}{1-K} \sum_{B \cap C = A} m_1(B) m_2(C) & if \ A \neq \varnothing \end{cases} \tag{3}$$

where, $K = \sum_{B \cap C = \varnothing} m_1(B) m_2(C)$ and A is the intersection of subsets B and C.

In other words, the Dempster's combination rule computes a measure of agreement between two bodies of evidence concerning various propositions determined from a common frame of discernment. The rule focuses only on those propositions that both bodies of evidence support.

The denominator is a normalization factor that ensures that m is a BPA, called the conflict. The Yagar's modification of the DS theory [40] has been implemented in the paper with the normalizing factor as 1. This reduces some of the issues regarding the conflict factor.

Earlier, DS theory based combination has been applied on different fields like handwritten digit recognition [41], skin detection [42], 3D palm print recognition [43] among other pattern recognition domains.

3.2. Secondary Classifier Based Combination Techniques

The confidence values provided by the classifiers act as the feature set for the secondary classifier which acts on the second stage of the framework. With the training from the classifier scores, it learns to predict the outcome for a set of new confidence scores from the same set of classifiers. The advantage of using such a generic combinator is that it can learn the combination algorithm and can automatically account for the strengths and score ranges of the individual classifiers. For example, Dar-Shyang Lee [29] used a neural network to operate on the outputs of the individual classifiers and to produce the combined matching score. Apart from the neural network, other classifiers like k-NN, SVM and Random Forest have been fitted and tested in this paper.

4. Results and Interpretation

4.1. Preparation of Database

At present, no standard benchmark database of handwritten *Indic* scripts is freely available in the public domain. Hence, we have created our own database of handwritten documents in the laboratory. The document pages for the database were collected from different sources on request. Participants of this data collection drive were asked to write few lines on A-4 size pages. No other restrictions were imposed regarding the content of the textual materials. The documents were written in 12 official scripts of India. The document pages are digitized at 300 dpi resolution and stored as grey tone images. The scanned images may contain noisy pixels which are removed by applying Gaussian filter [33]. The text words are automatically extracted from the handwritten documents by using a page-to-word segmentation algorithm described in [44]. A sample snapshot of word images written in 12 different scripts is shown in Figure 7. Finally, a total of 7200 handwritten word images are prepared, with exactly 600 text words per script.

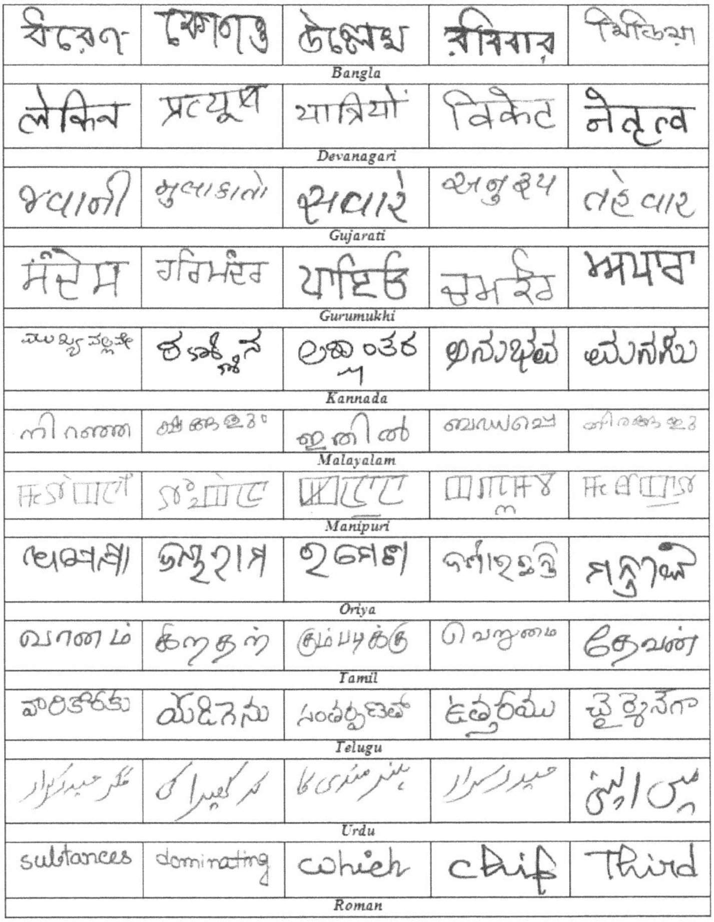

Figure 7. Sample word images written in 12 different Indian scripts.

Our developed database has been named as *CMATERdb*8.4.1, where *CMATER* stands for 'Centre for Microprocessor Applications for Training Education and Research,' a research laboratory at Computer Science and Engineering Department of Jadavpur University, India, where the current database is prepared. Here, *db* symbolizes database, the numeric value 8 represents handwritten multi-script *Indic* image database and the value 4 indicates word-level. In the present work, the first version of *CMATERdb*8.4 has been released as *CMATERdb*8.4.1. The database is made freely available at https://code.google.com/p/cmaterdb/.

4.2. Performance Analysis

The classifier combination approaches, described above, are applied on a dataset of 7200 words divided into 12 classes with equal number of instances in each of them. 12 classes refer to the 12 *Indic* scripts that have been studied before and for which the MLP classifier results can be obtained with high accuracy. The classes numbered from **A** to **L** are *Devanagari*, *Bangla*, *Oriya*, *Gujarati*, *Gurumukhi*, *Tamil*, *Telugu*, *Kannada*, *Malayalam*, *Manipuri*, *Urdu* and *Roman* in that particular order.

First, the confusion matrix that is obtained from the MLP based classifier on the dataset by using MLG feature along with the overall accuracy is presented. Then, the result generated by the same classifier on the HOG and Elliptical feature sets applied on the same dataset is also presented. Results have been cross-validated for the classifier parameter values to obtain the optimal results for the dataset and the values are provided in the result section.

The MLG feature set consisting of 60 feature values for every input image is fed into the MLP classifier with 30 hidden layer neurons and a learning rate of 0.8. Here, 500 iterations are allowed with an error tolerance of 0.1. The overall accuracy obtained is 91.42% and the confusion matrix generated in this case is given in Table 1. The **R** column in the table refers to the rejection of the input by the recognition module but the class confidences that are associated with them get accounted for during the combination process.

The HOG feature set, consisting of 80 feature values for every input data, is fed into the MLP classifier with 40 hidden layer neurons and a learning rate of 0.8. Same error tolerance and the number of iterations, as applied in case of MLG features, are allowed here. A maximum recognition accuracy of 78.04% has been noted. The confusion matrix is shown in Table 2.

The Elliptical feature set containing 58 feature values derived from each image data forms the training set for the MLP classifier with 30 hidden neurons with a learning rate of 0.7. The error tolerance and number of iterations remain the same as the previous cases. An accuracy of 79.2% is achieved and represented in the confusion matrix given in Table 3.

Table 1. Classification results for HOG feature set with MLP Classifier.

Class \ Class	A	B	C	D	E	F	G	H	I	J	K	L	R
A	345	9	6	22	13	21	64	42	27	0	44	7	27
B	27	548	0	7	9	0	1	0	1	0	7	0	0
C	0	0	557	0	6	13	1	19	2	1	0	1	38
D	38	4	0	516	3	3	4	0	9	0	20	3	10
E	10	6	1	12	449	26	5	2	0	0	13	76	30
F	30	0	23	3	46	417	33	36	6	1	4	1	27
G	27	2	15	10	12	16	446	34	12	1	24	1	10
H	10	0	27	17	16	41	8	420	28	11	14	8	38
I	38	2	4	16	0	10	34	33	455	0	8	0	0
J	0	0	17	0	7	0	0	16	0	553	1	6	38
K	38	6	5	35	22	14	42	31	0	2	404	1	2
L	2	2	14	6	15	24	1	9	0	13	5	509	0

Table 2. Classification results for MLG feature set with MLP Classifier.

Class \ Class	A	B	C	D	E	F	G	H	I	J	K	L	R
A	528	0	2	13	1	1	19	9	5	0	12	10	0
B	0	576	0	6	0	0	0	0	0	0	3	15	1
C	1	0	596	0	0	0	1	1	1	0	0	0	2
D	2	9	0	574	0	0	0	0	1	0	0	14	0
E	0	0	0	0	592	6	0	1	0	0	0	1	0
F	0	0	2	0	16	553	0	20	0	9	0	0	4
G	4	0	9	3	0	1	528	15	26	0	10	4	7
H	7	0	5	0	5	30	8	512	16	1	8	8	12
I	12	0	7	1	0	0	12	2	560	0	4	2	0
J	0	0	0	0	3	4	0	5	0	588	0	0	19
K	19	3	1	7	2	0	24	2	4	0	527	11	3
L	3	2	25	29	24	9	4	21	18	4	13	448	0

Table 3. Classification results for Elliptical feature set with MLP Classifier.

Class \ Class	A	B	C	D	E	F	G	H	I	J	K	L	R
A	**355**	29	49	48	0	25	3	4	42	10	6	29	2
B	2	**550**	0	7	0	1	32	0	0	0	0	8	27
C	27	0	**479**	8	1	19	2	11	31	1	8	13	32
D	32	9	0	**514**	0	13	23	0	3	2	4	0	66
E	66	1	2	1	**441**	42	4	7	10	20	4	2	96
F	96	3	6	15	6	**397**	16	4	19	12	14	12	55
G	55	13	7	54	6	17	**402**	1	3	19	22	1	25
H	25	0	2	3	0	26	0	**491**	28	10	4	11	7
I	7	0	23	3	33	8	7	3	**493**	10	4	9	0
J	0	0	1	0	16	5	2	2	9	**553**	6	6	2
K	2	0	16	7	1	7	12	0	2	7	**546**	0	8
L	8	22	1	0	6	6	20	13	6	9	1	**508**	0

Now, the confidence values provided to the classes for every input data by the classifiers on the three sets of features form the input for the classifier combination procedures. The confusion matrix resulting from the Majority voting procedure is presented in Table 4. An overall accuracy of 95.6% is achieved on this dataset containing 7200 samples divided equally among the 12 script classes. It is seen that *Devanagari* script has got the least accuracy and gets confused with *Telugu* whereas high accuracies are shown for *Manipuri* and *Odia* and *Bangla*.

Borda count algorithm gives an accuracy of 93.5% which is an increase of 2.1% over the best performing individual classifier. It provides the highest recognition rate for *Devanagari* among all the combination schemes and good accuracies for other popular scripts like *Bangla* and *Odia* and hence can be the preferred choice for wide usage. The trainable version of the algorithm with weights based on overall accuracy of the classifiers improves the results further. The increase is 2.9% with satisfactory results for scripts like *Telugu*, *Kannada* and *Urdu*. The accuracy for the *Gurumukhi* script remains low irrespective of the weights. The results are presented in Tables 5 and 6.

The simple rules at the measurement level to combine the decisions provide good results in the present work. The sum rule attains an accuracy of 97.76% with almost close to perfect recognition for *Urdu*, *Gurumukhi* and *Roman*. The product rule and max rule have accuracies of 95.73% and 94.60% respectively. Highest accuracy is found for *Odia* script whereas product rule suffers in case of *Gurumukhi* and max rule in case of *Devanagari*. The results for the elementary rules of combination are tabulated in Tables 7–9.

Sum rule outperforms all other rule based combination approaches in this work and testifies the results presented by Kittler et al. mentioned in [38] by being less prone to noise and unclean data. The DS theory results combine the results, two at a time and then all three together. The class-wise performance based BPA, which outperforms the global performance based *BPA*, has been implemented for the multi-classifier combination using the DS theory [45]. The rule applied for this process is quasi-associative and hence the results of combining two sources cannot be combined with the third. The rule has to be extended to include all the three sources together. Results for the combination of the classifier results on HOG and Elliptical features, MLG and Elliptical features, and, HOG and MLG features are presented in Tables 10, 11 and 12respectively. The combination result including all the three sources of information is given in Table 13.

There is no improvement shown by the combination of the results from MLG and HOG feature sets. But when the Elliptical feature set is involved in the combination process there is much improvement over the participating classifiers. Overall accuracies of 91.2% and 97.04% are achieved by combining sources having 78.1% and 79.4% accuracies and 91.4% and 79.4% accuracies respectively. So, improvements of 6% and 10% are found by applying the DS theory of evidence. Combining all three, an accuracy of 95.64%, more than 4% over the better performing classifier is seen. In both the schemes all the script classes have accuracies over 90% and with almost 100% accuracy for certain scripts like *Manipuri, Gujarati* and *Urdu,* thus proving to be the model to be used where these scripts are widely used.

In order to understand why the results from the Elliptical feature set combine so well with the two other feature sets, correlation analysis is performed on the confidence score outputs. Spearman rank correlation is done on the rank level information provided by the classifiers to arrive at mean values for the measure of the correlation. HOG and MLG show an index of 0.619 which is almost the double of the scores obtained by comparing the Elliptical features with these two. With values of 0.32 and 0.27, the low correlation index is an indication of better possibilities for the combination processes. Thus, complementary information is provided by the output of Elliptical feature set which helps in the improvement the overall combined accuracy.

Secondary classifiers are applied to learn the patterns from the primary classifier outputs and develop a way to combine them. The confidence scores from the three sources are concatenated to form a larger training set with its correct label. This set is the new feature set which undergoes classification using well-known algorithms. Classifiers like *k*-NN, Logistic Regression, MLP and Random Forest are applied to report final results which are tabulated in Tables 14–17 respectively. The results are reported after 3-fold cross validation and tuning of the parameters involved. This process is computationally costly and takes a processing step along with much higher complexity but is compensated by the high accuracy results that are obtained. 3-NN provides an accuracy of 98.30%, Random Forest classified 98.33% of the 7200 samples correctly and Logistic Regression attained 98.48% accuracy. Using MLP again as the secondary classifier, 98.36% accuracy is obtained. *Devanagari* is the most confused script in all the cases but still has accuracy over 95%. The other scripts are predicted to almost certainty.

Table 4. Classification results after combination using Majority voting procedure.

Class \ Class	A	B	C	D	E	F	G	H	I	J	K	L
A	534	2	3	10	3	2	16	7	13	0	5	5
B	1	590	0	3	0	0	0	0	0	0	0	6
C	0	0	597	0	0	2	0	0	1	0	0	0
D	2	3	0	590	0	0	0	0	0	0	2	3
E	0	1	0	0	591	2	0	0	0	1	0	5
F	12	0	5	0	13	561	1	7	1	0	0	0
G	2	0	6	4	5	1	554	14	7	3	4	0
H	4	0	4	3	1	6	0	567	7	2	5	1
I	9	1	1	2	0	1	7	2	572	0	5	0
J	0	0	1	0	2	0	0	3	0	594	0	0
K	4	0	2	5	4	1	10	4	1	0	567	2
L	0	2	10	3	6	2	1	1	1	5	3	566

Table 5. Classification results after combination using Borda count procedure without weight.

Class \ Class	A	B	C	D	E	F	G	H	I	J	K	L
A	567	0	5	7	0	0	5	4	7	0	3	2
B	16	580	0	4	0	0	0	0	0	0	0	0
C	1	0	586	0	0	5	0	4	3	0	0	1
D	25	1	0	572	0	0	0	0	0	0	1	1
E	6	0	0	0	466	108	0	11	1	0	0	8
F	16	0	2	0	7	571	0	1	1	1	0	1
G	25	0	4	2	0	2	548	3	1	0	15	0
H	30	0	5	0	0	21	0	533	6	0	1	4
I	39	0	2	1	0	2	6	6	540	0	4	0
J	0	0	2	0	2	0	0	7	0	589	0	0
K	5	0	0	3	0	0	12	0	1	0	579	0
L	4	0	10	2	2	4	0	1	3	3	3	568

Table 6. Classification results after combination using Borda count procedure using weight.

Class \ Class	A	B	C	D	E	F	G	H	I	J	K	L
A	**563**	0	4	7	0	0	8	6	7	0	3	2
B	14	**582**	0	4	0	0	0	0	0	0	0	0
C	0	0	**586**	0	0	5	0	5	3	0	0	1
D	15	2	0	**580**	0	0	0	0	0	0	2	1
E	5	0	0	0	**466**	102	0	17	2	0	0	8
F	13	0	0	0	6	**576**	0	2	1	1	0	1
G	14	0	3	2	0	1	**558**	3	1	0	18	0
H	22	0	5	0	0	14	0	**546**	6	0	1	6
I	30	0	2	1	0	1	6	5	**551**	0	4	0
J	0	0	2	0	2	0	0	6	0	**590**	0	0
K	2	0	0	2	0	0	9	0	0	0	**587**	0
L	4	0	10	1	2	4	0	1	2	3	3	**570**

Table 7. Classification result after combination using Sum rule.

Class \ Class	A	B	C	D	E	F	G	H	I	J	K	L
A	**549**	1	4	6	0	1	13	10	11	0	3	2
B	3	**589**	0	2	0	0	0	0	0	0	1	5
C	0	0	**597**	0	0	1	0	1	1	0	0	0
D	1	3	0	**593**	0	0	0	0	0	0	0	3
E	0	0	0	0	**595**	2	0	0	0	0	0	3
F	6	0	2	0	8	**576**	0	7	0	0	0	1
G	3	0	5	3	1	0	**568**	10	3	2	5	0
H	2	0	2	0	0	5	0	**582**	4	1	3	1
I	16	0	1	1	0	1	6	3	**569**	0	3	0
J	0	0	0	0	0	0	0	1	0	**599**	0	0
K	0	0	0	2	2	0	5	3	0	0	**588**	0
L	1	0	6	1	3	0	0	0	1	3	2	**583**

Table 8. Classification result after combination using Product rule.

Class \ Class	A	B	C	D	E	F	G	H	I	J	K	L
A	**566**	0	4	6	0	0	6	6	9	0	1	2
B	7	**584**	0	8	0	0	0	0	1	0	0	0
C	0	0	**597**	0	0	2	0	0	1	0	0	0
D	5	2	0	**591**	0	0	0	0	0	0	1	1
E	4	0	0	0	**467**	97	0	10	2	0	7	13
F	8	0	0	0	5	**582**	0	3	0	1	0	1
G	5	0	2	2	0	1	**578**	3	1	0	8	0
H	12	0	4	0	0	15	0	**562**	4	0	0	3
I	31	0	0	0	0	1	8	2	**556**	0	2	0
J	0	0	1	0	1	0	0	4	0	**594**	0	0
K	2	0	0	1	0	0	4	0	0	0	**593**	0
L	2	0	2	0	0	1	0	1	0	2	3	**589**

Table 9. Classification result after combination using Max rule.

Class＼Class	A	B	C	D	E	F	G	H	I	J	K	L
A	512	6	4	13	4	2	17	13	17	0	8	4
B	1	587	0	3	0	0	2	0	0	0	0	7
C	0	0	599	0	1	0	0	0	0	0	0	0
D	2	5	0	580	0	0	2	0	3	0	3	5
E	1	1	0	4	566	3	0	1	0	1	0	23
F	7	0	6	0	10	556	2	12	2	1	1	3
G	3	1	6	5	4	1	553	8	7	4	8	0
H	3	0	4	4	1	5	0	566	11	2	3	1
I	8	1	1	1	2	1	6	5	569	0	5	1
J	0	0	1	0	1	0	0	4	0	594	0	0
K	3	1	4	5	6	4	5	6	1	0	564	1
L	0	3	3	9	5	2	2	2	2	5	2	565

Table 10. Classification results after combination using DS theory for HOG and Elliptical features.

Class＼Class	A	B	C	D	E	F	G	H	I	J	K	L
A	512	6	4	13	4	2	17	13	17	0	8	4
B	1	587	0	3	0	0	2	0	0	0	0	7
C	0	0	599	0	1	0	0	0	0	0	0	0
D	2	5	0	580	0	0	2	0	3	0	3	5
E	1	1	0	4	566	3	0	1	0	1	0	23
F	7	0	6	0	10	556	2	12	2	1	1	3
G	3	1	6	5	4	1	553	8	7	4	8	0
H	3	0	4	4	1	5	0	566	11	2	3	1
I	8	1	1	1	2	1	6	5	569	0	5	1
J	0	0	1	0	1	0	0	4	0	594	0	0
K	3	1	4	5	6	4	5	6	1	0	564	1
L	0	3	3	9	5	2	2	2	2	5	2	565

Table 11. Classification results after combination using DS theory for Elliptical and MLG features.

Class＼Class	A	B	C	D	E	F	G	H	I	J	K	L
A	564	0	3	8	0	1	7	4	8	0	0	5
B	0	593	0	1	0	0	0	0	0	0	0	6
C	0	0	600	0	0	0	0	0	0	0	0	0
D	0	7	0	590	0	0	1	0	0	0	0	2
E	1	0	0	0	589	8	0	1	0	1	0	0
F	0	0	0	0	5	580	1	7	0	4	1	2
G	5	0	4	3	0	0	563	3	10	3	8	1
H	3	0	1	0	1	16	0	567	7	2	1	2
I	6	0	1	0	2	0	2	1	583	2	1	2
J	0	0	0	0	0	0	0	0	0	600	0	0
K	1	0	1	0	0	0	5	0	1	0	592	0
L	3	6	3	3	7	0	4	2	2	2	2	566

Table 12. Classification results after combination using DS theory for HOG and MLG features.

Class \ Class	A	B	C	D	E	F	G	H	I	J	K	L
A	511	2	4	12	5	1	19	19	12	0	8	7
B	6	573	0	5	2	0	0	0	0	0	5	9
C	1	0	590	0	1	1	0	5	2	0	0	0
D	4	4	0	580	0	0	1	0	2	0	4	5
E	3	1	1	6	540	6	0	1	0	0	3	39
F	7	0	5	0	18	549	1	18	0	2	0	0
G	7	1	7	5	2	0	530	21	11	1	13	2
H	10	0	10	6	8	13	4	524	11	5	7	2
I	26	1	2	4	0	0	17	11	535	0	4	0
J	0	0	3	0	5	0	0	9	0	580	0	3
K	22	3	4	11	3	1	15	11	3	2	522	3
L	3	1	19	8	12	4	1	2	2	6	3	539

Table 13. Classification results after combination using DS theory for Elliptical, HOG and MLG features.

Class \ Class	A	B	C	D	E	F	G	H	I	J	K	L
A	548	1	5	7	3	1	12	5	12	0	2	4
B	4	584	0	5	0	0	0	0	0	0	1	6
C	0	0	598	0	0	0	0	1	1	0	0	0
D	5	2	0	592	0	0	0	0	0	0	0	1
E	0	1	0	5	548	22	0	3	1	2	3	15
F	10	1	2	0	5	572	1	6	0	1	0	2
G	10	0	3	3	3	0	556	9	7	1	8	0
H	8	0	4	0	2	8	0	568	4	1	4	1
I	17	1	1	2	0	0	12	3	561	0	3	0
J	0	0	0	0	1	0	0	2	0	597	0	0
K	6	0	1	2	0	2	7	0	0	0	582	0
L	2	1	5	4	2	0	2	0	0	3	1	580

Table 14. Classification results using 3-NN secondary classifier.

Class \ Class	A	B	C	D	E	F	G	H	I	J	K	L
A	573	1	3	5	0	0	8	5	3	0	1	1
B	1	597	0	2	0	0	0	0	0	0	0	0
C	1	0	598	0	0	0	0	0	1	0	0	0
D	1	2	0	595	0	0	0	0	0	0	0	2
E	0	0	0	0	599	1	0	0	0	0	0	0
F	0	0	0	0	4	590	0	5	0	0	0	1
G	7	1	5	3	0	0	571	1	5	1	6	0
H	2	0	0	0	0	9	1	585	1	0	2	0
I	11	0	2	0	0	0	0	0	586	0	1	0
J	0	0	0	0	0	0	0	2	0	598	0	0
K	1	0	0	0	1	0	4	0	0	0	594	0
L	1	0	3	0	1	1	0	1	0	1	0	592

Table 15. Classification results using Logistic Regression secondary classifier.

Class \ Class	A	B	C	D	E	F	G	H	I	J	K	L
A	**583**	0	1	2	0	0	5	4	2	0	1	2
B	0	**598**	0	2	0	0	0	0	0	0	0	0
C	1	0	**598**	0	0	0	0	0	1	0	0	0
D	1	2	0	**594**	0	0	0	1	0	0	0	2
E	1	0	0	0	**595**	3	0	0	0	0	0	1
F	0	0	0	0	4	**589**	0	5	1	0	0	1
G	4	0	3	2	0	0	**578**	2	3	0	8	0
H	3	0	0	0	0	6	1	**586**	1	0	1	2
I	6	0	1	0	0	0	2	3	**587**	0	1	0
J	0	0	0	0	0	2	0	1	0	**596**	0	1
K	1	0	0	0	0	0	5	0	0	0	**594**	0
L	1	0	1	0	2	0	2	1	0	0	0	**593**

Table 16. Classification results using MLP secondary classifier.

Class \ Class	A	B	C	D	E	F	G	H	I	J	K	L
A	**575**	1	2	4	0	0	4	4	4	0	2	4
B	1	**597**	0	2	0	0	0	0	0	0	0	0
C	1	0	**598**	0	0	0	0	0	1	0	0	0
D	3	2	0	**594**	0	0	0	1	0	0	0	0
E	0	0	0	0	**599**	1	0	0	0	0	0	0
F	0	0	0	0	2	**594**	0	3	0	0	0	1
G	7	1	2	1	3	0	**577**	0	2	1	6	0
H	4	0	0	0	1	7	1	**583**	1	1	0	2
I	7	0	2	1	0	0	3	0	**586**	0	1	0
J	0	0	1	0	0	0	0	0	0	**599**	0	0
K	0	0	0	0	0	0	7	0	0	0	**593**	0
L	1	0	0	0	4	0	1	0	0	0	0	**594**

Table 17. Classification results using Random Forest secondary classifier.

Class \ Class	A	B	C	D	E	F	G	H	I	J	K	L
A	**581**	0	3	2	0	0	8	1	4	0	0	1
B	0	**597**	0	2	0	0	1	0	0	0	0	0
C	1	0	**595**	0	0	0	2	0	1	0	0	1
D	2	2	0	**593**	0	0	0	0	0	0	0	3
E	0	0	0	0	**598**	2	0	0	0	0	0	0
F	0	0	0	0	7	**585**	0	6	0	1	0	1
G	3	0	4	1	0	0	**585**	1	2	0	4	0
H	4	0	2	0	0	8	0	**582**	1	0	1	2
I	9	0	1	0	0	0	3	2	**584**	0	1	0
J	0	0	0	0	0	2	0	2	0	**595**	0	1
K	0	0	0	0	0	0	6	0	0	0	**593**	1
L	1	0	4	0	1	0	0	1	0	1	0	**592**

Script recognition is a difficult task given the variation in the words for a particular script. But the results are really encouraging for building a model that can identify the script with certainty. The results obtained after the combination exceeds the reported accuracies for this certain task and hence set the new benchmark. Table 18 provides the class wise accuracy along with the overall accuracy achieved by each procedure used in the paper. It shows that the Logistic Regression classifier acting on the MLP classifier outputs provide the best result where 98.45% accuracy is obtained with an improvement of 7.05%. Results are also obtained from the feature level combination. Natural combination or concatenation of two features at a time and all three together are done. The new feature set formed in each case undergoes the same process of classification through the MLP classifier. The comprehensive results for comparison are given in the following Table 19.

Table 18. Class-wise percentage accuracy for the classifier combination methods.

Class	Abstract Level	Rank Level	Measurement Level Combination Rules			DS Theory of Evidence		3-NN	Secondary Classifier		
	Majority Voting	Borda Count	Sum Rule	Product Rule	Max Rule	DS with 2 Sources	DS with 3 Sources		Random Forest	MLP	Logistic Regression
A	89.0	93.8	91.5	94.3	85.3	94.0	91.3	95.5	96.8	96.5	97.1
B	98.3	97.0	98.2	97.3	97.8	98.8	97.3	99.5	99.5	99.8	99.6
C	99.5	97.7	99.5	99.5	99.8	100.0	99.6	99.6	99.1	99.5	99.6
D	98.3	96.7	98.8	98.5	96.6	98.3	98.6	99.1	98.8	98.8	99.0
E	98.5	77.7	99.1	77.8	94.3	98.1	91.3	99.8	99.6	99.5	99.1
F	93.5	96.0	96.0	97.0	92.6	96.6	95.3	98.3	97.5	98.1	98.1
G	92.3	93.0	94.6	96.3	92.1	93.8	92.6	95.1	97.5	95.5	96.3
H	94.5	91.0	97.0	93.6	94.3	94.5	94.6	97.5	97.0	97.1	97.6
I	95.3	91.8	94.8	92.6	94.8	97.1	93.5	97.6	97.3	97.6	97.8
J	99.0	98.3	99.8	99.0	99.0	100.0	99.5	99.6	99.1	99.5	99.3
K	94.5	97.8	98.0	98.8	94.0	98.6	97.0	99.0	98.8	99.5	99.0
L	94.3	95.0	97.2	98.1	94.1	94.3	96.6	98.6	98.6	98.6	98.8
Overall	95.6	94.3	97.8	95.7	94.6	97.0	95.6	98.3	98.3	98.3	98.5

Table 19. Final combination results at feature level.

Feature/Methodology	Recognition Accuracy (%)
MLG	91.42
HOG	78.04
Elliptical	79.57
MLG + HOG	86.03
HOG + Elliptical	86.57
MLG + Elliptical	93.44
MLG + HOG + Elliptical	91.03
Best result after classifier combination	98.45

5. Conclusions

This is the first application of classifier combination approaches in the domain of script recognition considering the number of scripts being undertaken and the range of classifier combination procedures that are evaluated. Combination is performed at the feature level as well as decision level using abstract level, rank level and measurement level information provided by the classifiers. Encouraging results are obtained from the experiments. High accuracies in the range of 95–98% have been achieved by using combination techniques as shown in the previous Result section. There is an increase of over 7% with the best performing MLP classifier when Logistic Regression is used as the secondary classifier for 7200 samples from 12 different scripts. So, this model proves to be useful for this complex pattern recognition problem and makes a better decision based on the information provided by the base classifier.

Though, in the present work, three sources of information with different feature sets have been combined using their respective classifier results but this process can be extended to include more input sources along with different classifier. With the increase in the number of sources, an intelligent and dynamic selection procedure needs to be employed in order to facilitate combination in a more meaningful way. The combination being an overhead to the classification task, it is important to develop methods that can indicate if the combination would work or not qualitatively. In future, the work can be extended for a larger dataset so that the robustness of the procedures can be established. The script recognition system here is a general framework which can be applied to other similar pattern recognition tasks like block and line level recognition of scripts to establish its usefulness in document analysis research.

Acknowledgments: The authors are thankful to the Center for Microprocessor Application for Training Education and Research (*CMATER*) and Project on Storage Retrieval and Understanding of Video for Multimedia (*SRUVM*) of Computer Science and Engineering Department, Jadavpur University, for providing infrastructure facilities during progress of the work. The authors of this paper are also thankful to all those individuals who willingly contributed in developing the handwritten *Indic* script database used in the current research.

Author Contributions: Anirban Mukhopadhyay and Pawan Kumar Singh conceived and designed the experiments; Anirban Mukhopadhyay performed the experiments; Anirban Mukhopadhyay and Pawan Kumar Singh analyzed the data; Ram Sarkar amd Mita Nasipuri contributed reagents/materials/analysis tools; Anirban Mukhopadhyay and Pawan Kumar Singh wrote the paper.

Conflicts of Interest: The authors declare no conflict of interest. The founding sponsors had no role in the design of the study; in the collection, analyses, or interpretation of data; in the writing of the manuscript and in the decision to publish the results.

References

1. Singh, P.K.; Sarkar, R.; Nasipuri, M. Offline Script Identification from Multilingual Indic-script Documents: A state-of-the-art. *Comput. Sci. Rev.* **2015**, *15–16*, 1–28. [CrossRef]
2. Ubul, K.; Tursun, G.; Aysa, A.; Impedovo, D.; Pirlo, G.; Yibulayin, T. Script Identification of Multi-Script Documents: A Survey. *IEEE Access* **2017**, *5*, 6546–6559. [CrossRef]
3. Spitz, A.L. Determination of the script and language content of document images. *IEEE Trans. Pattern Anal. Mach. Intell.* **1997**, *19*, 234–245. [CrossRef]
4. Tan, T.N. Rotation Invariant Texture Features and their use in Automatic Script Identification. *IEEE Tran. Pattern Anal. Mach. Intell.* **1998**, *20*, 751–756. [CrossRef]
5. Hochberg, J.; Kelly, P.; Thomas, T.; Kerns, L. Automatic script identification from document images using cluster-based templates. *IEEE Trans. Pattern Anal. Mach. Intell.* **1997**, *19*, 176–181. [CrossRef]
6. Hochberg, J.; Bowers, K.; Cannon, M.; Keely, P. Script and language identification for hand-written document images. *IJDAR* **1999**, *2*, 45–52. [CrossRef]
7. Wood, S.; Yao, X.; Krishnamurthi, K.; Dang, L. Language identification for printed text independent of segmentation. In Proceedings of the International Conference on Image Processing, Washington, DC, USA, 23–26 October 1995; pp. 428–431.

8. Chaudhuri, B.B.; Pal, U. An OCR system to read two Indian language scripts: Bangla and Devnagari (Hindi). In Proceedings of the 4th IEEE International Conference on Document Analysis and Recognition (ICDAR), Ulm, Germany, 18–20 August 1997; pp. 1011–1015.
9. Pal, U.; Sinha, S.; Chaudhuri, B.B. Word-wise Script identification from a document containing English, Devnagari and Telgu Text. In Proceedings of the -Second National Conference on Document Analysis and Recognition, PES, Mandya, Karnataka, India, 11–12 July 2003; pp. 213–220.
10. Chaudhury, S.; Harit, G.; Madnani, S.; Shet, R.B. Identification of scripts of Indian languages by Combining trainable classifiers. In Proceedings of the Indian Conference on Computer Vision, Graphics and Image Processing, Bangalore, India, 20–22 December 2000; pp. 20–22.
11. Das, A.; Ferrer, M.; Pal, U.; Pal, S.; Diaz, M.; Blumenstein, M. Multi-script vs. single-script scenarios in automatic off-line signature verification. *IET Biom.* **2016**, *5*, 305–313. [CrossRef]
12. Diaz, M.; Ferrer, M.A.; Sabourin, R. Approaching the Intra-Class Variability in Multi-Script Static Signature Evaluation. In Proceedings of the 23rd International Conference on Pattern Recognition (ICPR), Cancun, Mexico, 4–8 December 2016; pp. 1147–1152.
13. Padma, M.C.; Vijaya, P.A. Global Approach for Script Identification using Wavelet Packet Based Features. *Int. J. Signal Process. Image Process. Pattern Recognit.* **2010**, *3*, 29–40.
14. Hiremath, P.S.; Shivshankar, S.; Pujari, J.D.; Mouneswara, V. Script identification in a handwritten document image using texture features. In Proceedings of the 2nd IEEE International Conference on Advance Computing, Patiala, India, 19–20 February 2010; pp. 110–114.
15. Pati, P.B.; Ramakrishnan, A.G. Word level multi-script identification. *Pattern Recognit. Lett.* **2008**, *29*, 1218–1229. [CrossRef]
16. Dhanya, D.; Ramakrishnan, A.G.; Pati, P.B. Script identification in printed bilingual documents. *Sadhana* **2002**, *27*, 73–82. [CrossRef]
17. Chanda, S.; Pal, S.; Franke, K.; Pal, U. Two-stage Approach for Word-wise Script Identification. In Proceedings of the 10th IEEE International Conference on Document Analysis and Recognition (ICDAR), Barcelona, Spain, 26–29 July 2009; pp. 926–930.
18. Pal, U.; Chaudhuri, B.B. Identification of different script lines from multi-script documents. *Image Vis. Comput.* **2002**, *20*, 945–954. [CrossRef]
19. Pal, U.; Sinha, S.; Chaudhuri, B.B. Multi-Script Line identification from Indian Documents. In Proceedings of the 7th IEEE International Conference on Document Analysis and Recognition (ICDAR), Edinburgh, UK, 6 August 2003; pp. 880–884.
20. Singh, P.K.; Chatterjee, I.; Sarkar, R. Page-level Handwritten Script Identification using Modified log-Gabor filter based features. In Proceedings of the 2nd IEEE International Conference on Recent Trends in Information Systems (ReTIS), Kolkata, India, 9–11 July 2015; pp. 225–230.
21. Singh, P.K.; Sarkar, R.; Nasipuri, M.; Doermann, D. Word-level Script Identification for Handwritten Indic scripts. In Proceedings of the 13th IEEE International Conference on Document Analysis and Recognition (ICDAR), Tunis, Tunisia, 23–26 August 2015; pp. 1106–1110.
22. Singh, P.K.; Das, S.; Sarkar, R.; Nasipuri, M. Line Parameter based Word-Level Indic Script Identification System. *Int. J. Comput. Vis. Image Process.* **2016**, *6*, 18–41. [CrossRef]
23. Nadal, C.; Legault, R.; Suen, C.Y. Complementary algorithms for the recognition of totally uncontrained handwritten numerals. In Proceedings of the 10th International Conference on Pattern Recognition, Atlantic City, NJ, USA, 16–21 June 1990; Volume A, pp. 434–449.
24. Suen, C.Y.; Nadal, C.; Mai, T.; Legault, R.; Lam, L. Recognition of totally unconstrained handwritten numerals based on the concept of multiple experts. In Proceedings of the International Workshop on Frontiers in Handwriting Recognition, Montreal, QC, Canada, 2–3 April 1990; pp. 131–143.
25. Ho, T.K. A Theory of Multiple Classifier Systems and Its Application to Visual Word Recognition. Ph.D. Thesis, State University of New York, Buffalo, NY, USA, 1992.
26. Ho, T.K.; Hull, J.J.; Srihari, S.N. Decision combination in multiple classifier systems. *IEEE Trans. Pattern Anal. Mach. Intell.* **1994**, *16*, 66–75.
27. Xu, L.; Krzyzak, A.; Suen, C. Methods of combining multiple classifiers and their applications to handwritten recognition. *IEEE Trans. Syst. Man Cybern.* **1992**, *22*, 418–435. [CrossRef]
28. Mandler, E.; Schuerman, J. *Pattern Recognition and Artificial Intelligence*; North-Holland: Amsterdam, The Netherlands, 1988.

29. Lee, D. A Theory of Classifier Combination: The Neural Network Approach. Ph.D. Thesis, State University of New York, Buffalo, NY, USA, 1995.

30. Singh, P.K.; Mondal, A.; Bhowmik, S.; Sarkar, R.; Nasipuri, M. Word-level Script Identification from Multi-script Handwritten Documents. In *Proceedings of the 3rd International Conference on Frontiers in Intelligent Computing Theory and Applications (FICTA)*; Springer: Cham, Switzerland, 2014; AISC Volume 1, pp. 551–558.

31. Dalal, N.; Triggs, B. Histograms of Oriented Gradients for Human Detection. In Proceedings of the IEEE Computer Society Conference on Computer Vision and Pattern Recognition (CVPR), San Diego, CA, USA, 20–25 June 2005; pp. 886–893.

32. Daugman, J.G. Uncertainty relation for resolution in space, spatial-frequency and orientation optimized by two-dimensional visual cortical filters. *J. Opt. Soc. Am.* **1985**, *2*, 1160–1169. [CrossRef]

33. Gonzalez, R.C.; Woods, R.E. *Digital Image Processing*; Prentice-Hall: India, 1992; Volume I.

34. Singh, P.K.; Sarkar, R.; Nasipuri, M. Correlation Based Classifier Combination in the field of Pattern Recognition. *Comput. Intell.* **2017**. [CrossRef]

35. Tulyakov, S.; Jaeger, S.; Govindaraju, V.; Doermann, D. Review of classifier combination methods. In *Machine Learning in Document Analysis and Recognition*; Springer: Berlin/Heidelberg, Germany, 2008; pp. 361–386.

36. Kittler, J. Combining Classifiers: A Theoretical Framework. *Pattern Anal. Appl.* **1998**, *1*, 18–27. [CrossRef]

37. Van Erp, M.; Vuurpijl, L.G.; Schomaker, L. An Overview and Comparison of Voting Methods for Pattern Recognition. In Proceedings of the 8th International Workshop on Frontiers in Handwriting Recognition (IWFHR-8), Niagara-on-the-Lake, ON, Canada; 2002; pp. 195–200.

38. Kittler, J.; Hatef, M.; Duin, R.; Matas, J. On combining classifiers. *IEEE Trans. Pattern Anal. Mach. Intell.* **1998**, *20*, 226–239. [CrossRef]

39. Shafer, G. *A Mathematical Theory of Evidence*; Princeton University Press: Princeton, NJ, USA, 1976.

40. Yager, R.R. On the Dempster-Shafer framework and new combination rules. *Inf. Sci.* **1987**, *41*, 93–137. [CrossRef]

41. Basu, S.; Sarkar, R.; Das, N.; Kundu, M.; Nasipuri, M.; Basu, D.K. Handwritten Bangla Digit Recognition Using Classifier Combination through DS Technique. In Proceedings of the 1st International Conference on Pattern Recognition and Machine Intelligence (PReMI), Kolkata, India, 20–22 December 2005; pp. 236–241.

42. Shoyaib, M.; Abdullah-Al-Wadud, M.; Chae, O. A Skin detection approach based on the Dempster-Shafer theory of evidence. *Int. J. Approx. Reason.* **2012**, *53*, 636–659. [CrossRef]

43. Ni, J.; Luo, J.; Liu, W. 3D Palmprint Recognition Using Dempster-Shafer Fusion Theory. *J. Sens.* **2015**, *2015*, 7. [CrossRef]

44. Singh, P.K.; Chowdhury, S.P.; Sinha, S.; Eum, S.; Sarkar, R. Page-to-Word Extraction from Unconstrained Handwritten Document Images. In *Proceedings of the 1st International Conference on Intelligent Computing and Communication (ICIC2)*; Springer: Singapore, 2016; AISC Volume 458, pp. 517–524.

45. Zhang, B.; Srihari, S.N. Class-wise multi-classifier combination based on Dempster-Shafer theory. In Proceedings of the 7th IEEE International Conference on Control, Automation, Robotics and Vision (ICARCV 2002), Singapore, 2–5 December 2002; Volume 2.

Journal of
Imaging

MDPI

Article

DocCreator: A New Software for Creating Synthetic Ground-Truthed Document Images

Nicholas Journet [1,*,†], Muriel Visani [2,†], Boris Mansencal [1,†], Kieu Van-Cuong [3] and Antoine Billy
[1]

[1] Laboratoire Bordelais de Recherche en Informatique UMR 5800, Université de Bordeaux, CNRS,
    Bordeaux INP, 33400 Talence, France; boris.mansencal@labri.fr (B.M.); antoine.billy@labri.fr (A.B.)
[2] Laboratoire Informatique, Image et Interaction (L3i), Université de La Rochelle, 17000 La Rochelle, France;
    muriel.visani@univ-lr.fr
[3] LIPADE Laboratory, Paris Descartes University, 45, rue des Saints-Pères, 75270 Paris, CEDEX 6, France;
    van-cuong.kieu@parisdescartes.fr
*   Correspondence: journet@labri.fr
†   These authors contributed equally to this work. Other authors: Kieu Van-Cuong worked on degradation
    models, Antoine Billy worked on synthetic document reconstruction.

Received: 30 October 2017; Accepted: 5 December 2017; Published: 11 December 2017

Abstract: Most digital libraries that provide user-friendly interfaces, enabling quick and intuitive access to their resources, are based on Document Image Analysis and Recognition (DIAR) methods. Such DIAR methods need ground-truthed document images to be evaluated/compared and, in some cases, trained. Especially with the advent of deep learning-based approaches, the required size of annotated document datasets seems to be ever-growing. Manually annotating real documents has many drawbacks, which often leads to small reliably annotated datasets. In order to circumvent those drawbacks and enable the generation of massive ground-truthed data with high variability, we present DocCreator, a multi-platform and open-source software able to create many synthetic image documents with controlled ground truth. DocCreator has been used in various experiments, showing the interest of using such synthetic images to enrich the training stage of DIAR tools.

Keywords: synthetic image generation; document degradation models; performance evaluation; data augmentation for retraining and fine-tuning; DIAR

1. Introduction

Almost every researcher in the field of Document Image Analysis and Recognition (DIAR) had to face the problem of obtaining a ground-truthed document image dataset. Indeed, many DIAR tools (image restoration, layout analysis, text-graphic separation, binarization, OCR, etc.) rely on a preliminary stage of supervised training. Moreover, ground-truthed document image datasets are needed to evaluate these DIAR tools. Digital curators are the first users of these tools, e.g., for announcing expected OCR recognition rates together with automatic transcriptions of books [1]. One common solution is to use ground-truthed training and benchmarking datasets publicly available on the internet. For document images, the following databases are the most commonly used. For printed documents: Washington UW3 [2], LRDE [3], RETAS-OCR [4], PaRADIIT [5], etc.; for handwritten documents IAM database [6], RIMES [7], GERMANA [8], etc.; for graphical documents: chemical symbol database [9], logo databases [10,11], architectural symbol database [12] or musical symbol database CVC-MUSICMA [13]; camera-based document image analysis [14,15]. The International Association for Pattern Recognition, for instance, gathered some interesting datasets [16] mostly used for different conference competitions over the last two decades. The main international conference in document image analysis, ICDAR, references on its websites many contest datasets. However, very few of them are reliably annotated, copyright-free, up-to-date or easily available to download. An alternative for researchers and digital curators is to create their own ground truth by manually annotating document images. In order to assist them in the tedious task of ground truth creation, multiple software have been proposed during the last two decades.

As detailed in Table 1, some are fully manual stand-alone software (Pink Panther (1998) [17], trueViz (2003) [18]), while others provide semi-automatic annotation modules (GEDI (2010) [19], Aletheia (2011) [20,21]). Some of the most recent solutions are based on an online collaborative platform (Transcriptorium (2014) [22], DIVADIAWI [23] (2015), [24] (2016), Recital manuscript platform [25] (2017)). Among non open-source solutions, some have an academic licence: [20,26]. These software assist the user in creating the ground truth associated with real documents, intrinsically limited in number because of acquisition procedures and copyright issues. Moreover, despite the use of such software, manual annotation remains a costly task that cannot always be performed by a non-specialist.

Another solution is available for getting (quickly and with lower human cost) large ground-truthed document image datasets. This solution, investigated since the beginning of the nineties [27], is to generate synthetic images with controlled ground truth. The authors of [28,29] propose two similar systems. They consist of using a text editor (e.g., Word-office, Latex, etc.) to automatically create multiple documents with varied contents (in terms of font, background, layout). Alternative approaches consist of re-arranging, in a new way, elements extracted from real images so as to generate (manually, semi-automatically or automatically) multiple semi-synthetic document images [12,30]. Recently, in particular with the advent of deep learning techniques which require huge masses of training data, the need for synthetic data generation seems to be ever-growing. In [31], among the 60,000 character patches that were used to train a convolutional network for text recognition, only 3000 were real.

In this paper we present DocCreator, an open-source and multi-platform software that is able to create virtually unlimited amounts of different ground-truthed synthetic document images based on a small number of real images.

Table 1. Technical and functional characteristics of existing annotation software. Six features are presented: export format, source availability, desktop/online software, groundtruthing assistance (whether the software provides features that help the user to quickly create the groundtruth), collaborative/crowd-sourcing software, and year of distribution.

	Export	Open-Source	Desktop/Online	Groundtruthing Assistance	Collaborative	Year
Software for manual ground truth creation						
Pink Panther [17]	ASCII	n/a	desktop	no	no	1998
TrueViz [18]	XML	yes	desktop	no	no	2003
PerfectDoc [32]	XML	yes	desktop	?	no	2005
PixLabeler [33]	XML	no	desktop	no	no	2009
GEDI [19]	XML	yes	desktop	yes	no	2010
DAE [34]	no	yes	online	yes	yes	2011
Aletheia [20,26]	XML	no	online/desktop	yes	no	2011
Transcriptorium [22]	TEI-XML	no	online	yes	yes	2014
DIVADIAWI [23]	XML	n/a	online	yes	n/a	2015
Recital [25]	no	yes	online	yes	yes	2017
Algorithms for synthetic data augmentation						
Baird et al. [27]	no	n/a	n/a	n/a	no	1990
Zhao et al. [28]	no	n/a	n/a	n/a	no	2005
Delalandre et al. [12]	no	n/a	n/a	n/a	no	2010
Yin et al. [30]	no	n/a	n/a	n/a	no	2013
Mas et al. [24]	no	n/a	n/a	n/a	yes	2016
Seuret et al. [35]	no	n/a	n/a	n/a	no	2015
Software for semi-automatic ground truth creation and data augmentation capabilities						
DocCreator	XML	yes	online/desktop	yes	no	2017

As illustrated in Figure 1, DocCreator can handle the creation of ground-truthed synthetic images from a limited set of real images. Various realistic degradation models can be applied on original document images, the resulting images being called semi-synthetic images in the rest of the paper. If there is no ground truth associated to the real images, DocCreator can create, with a given text, synthetic images that look like the real ones and their associated ground truth. Depending on the needs and expertise of the user, DocCreator can be used in a fully automatic mode, or in a semi-automatic mode where the user can interact with the system and tune its parameters. Visual feedback of the results is returned by the system. Degradations available in DocCreator can be applied on any type of document images. The DocCreator ability to create synthetic documents that mimic real ones is effective for typewritten and handwritten characters (as long as the characters are apart from one another). Images created with DocCreator have already been used in many DIAR contexts: text/background/image pixel classification [36]; staff removal [13,37,38]; and handwritten character recognition [39]. In this article we present how DocCreator can be useful to enhance a binarization algorithm and for OCR performance prediction. DocCreator could also be used, for example, for camera-based document image analysis and word spotting.

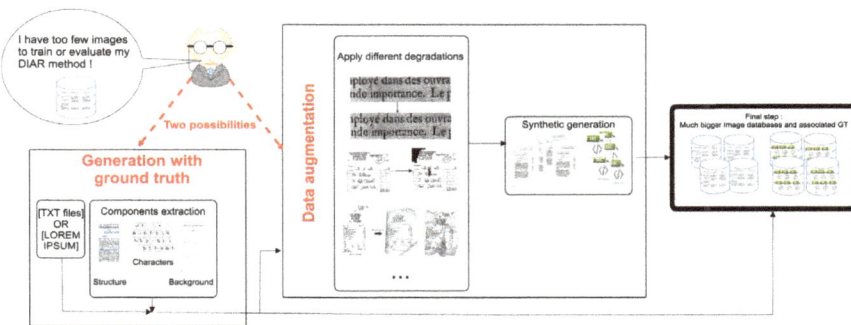

Figure 1. According to the needs of the DIAR researcher, it is possible to generate synthetic document images (and their ground truth) in different ways. First possibility: if a researcher has real document images but without any ground truth, DocCreator can generate synthetic images that look like the real ones, and of course, with the associated ground truth. Second possibility: a researcher has a ground-truthed database but it is too small or not heterogeneous enough. DocCreator provides several degradation algorithms to augment the dataset. By degrading text ink, paper shape or background colours it is possible to create a representative document image database where many defects are present. This complete database is finally useful for very precise performance evaluation or to provide multiple cases for retraining processes (in algorithms embedding a learning step).

DocCreator features compared to existing software are highlighted in Table 1. First of all, DocCreator is the only one that can create synthetic documents that mimic real ones. Besides, as it includes several degradation models, it provides an integrated solution to carry out data augmentation. DocCreator thus makes quickly available ground-truthed databases. It makes DocCreator a unique software that can be seen as a complementary tool to those mentioned in Table 1.

This paper is organized as follows. In Section 2, we present the methods used to extract document characteristics and to generate synthetic documents, while in Section 3 document degradation models are discussed. Section 4 highlights the advantages of DocCreator on various DIAR tasks, both for benchmarking and for retraining DIAR tools using data augmentation.

2. How to Create a Synthetic Document (with Ground Truth) That Looks Like a Real One?

The left part of Figure 1 illustrates the pipeline used in order to generate synthetic documents that look realistic. Given an original image, we extract the three main required components: (1) the font; (2) the background; and (3) the layout of the document. The system can then write any text with the extracted font, onto the reconstructed background, with a layout similar to one of the original documents (see Figure 2).

The font is extracted using a semi-automatic method. First, an Optical Character Recognition system (OCR) automatically associates a label (Unicode value) to each character on the image. Tesseract OCR engine [40] is used. Then the user can modify the character properties (label, baseline, margins, etc.). Our software allows us to pair a given symbol with several character images in the extracted font. Thus, when the font is used to write a text, the software will randomly choose an image for the required symbol. This will make the final output look more realistic by reducing the strict uniformity between similar letters.

In order to correctly write text with this font, the baseline of each character has to be computed. The baseline is the imaginary straight line on which semi cursive or cursive text are aligned and upon which most letters "sit" and, below which, descenders extend. In order to extract each character baseline and deal with various documents, we propose a different approach from classical ones [40,41]. The main originality of our baseline extraction method is that the baseline is computed for each character individually instead of finding a baseline for the whole line. We evaluated this method on more than 5000 manually annotated baselines considered as the ground truth. The baseline extraction error rate is relatively close the one obtained using [40] (the same baseline extraction error rate). However, our method has the main advantage of being robust to skew orientation, the hand-written wavy pattern and unaligned columns in a same page. Besides, the inter words distance is automatically computed as the average of characters width. For the inter characters distance, none is specified by default. However, the user can interactively specify the left and right margin to better position the character relative to others. The GUI of DocCreator also allows the user to modify several parameters to improve the extracted characters. The user can change the baseline or the letter assigned to a character and smooth the border of a character. Via this semi-automatic font extraction method, the user is able to correct mistakes made by the OCR (frequent on old documents). From several testing sessions, we evaluate the time needed to correctly extract a font between 30 seconds (when the OCR works accurately) and 60 min (when the OCR fails and the user has to manually extract the characters).

Once the font is extracted, the background of the document can be computed. This background extraction is performed completely automatically. For that purpose, we apply an inpainting method to remove all the characters. We use the OpenCV implementation of [42].

To construct a realistic document, the layout of the document image is also extracted. Document image physical layout analysis algorithms can be categorized into three classes: top-down approaches [43], bottom-up approaches [44] and hybrid approaches [45,46]. As word segmentation is already available via Tesseract OCR (but not the complete document layout), we use a hybrid approach proposed by [45]. With only one parameter to adjust the number of extracted blocks, this method ensures a good layout segmentation of many different classes of typewritten documents. DocCreator, as an interactive software, leaves once more the possibility to adapt to the wished segmentation results. This method has the advantage of a very low computational cost, without any preprocessing training required.

At this point, the three characteristics used in the synthetic image generation process have been extracted (background, font and layout). The next step is to assemble these elements with a given text in order to build the final output, which is the created synthetic image and the associated XML ground truth. Figure 2 illustrates a synthetic image (right) created automatically from a given original document image (left). As this example illustrates, a complete automatic generation may still produce perfectible results. In particular, if the original image suffers from local deformations (as the original image in Figure 2), the characters extracted to build the font may have different forms or sizes, and, when assembled to compose the final document, may locally look too different and thus not realistic.

Obviously, one can combine fonts, background images, layout from different images and various texts, to generate many of synthetic document images.

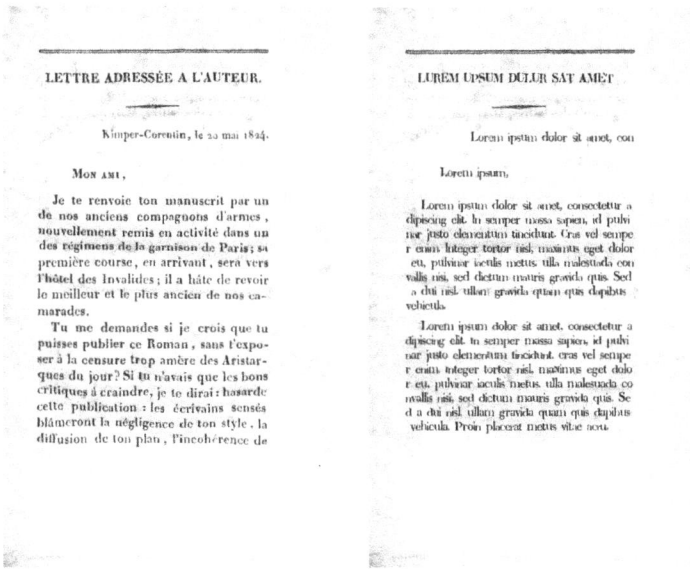

Figure 2. Synthetic document image generation. (**Left**) original document image. (**Right**) synthetic document image generated automatically with the random text "Lorem ipsum". The automatically generated image looks similar to the original one. The result is still perfectible. Here, as the original image suffers from local deformations, the characters extracted to build the font are quite different and may look too random when assembled on the synthetic document. A better font extraction or composition using the context to choose new characters may alleviate this problem.

3. Document Degradation Models

Physical degradation due to ageing, storage conditions or poor quality of printing materials may be present on documents.

DocCreator currently proposes seven degradation models.

As detailed in Figure 1 (right part), all these models can be applied on real images to extend any document image database. The user can interact with DocCreator in order to set the quantity of defects to generate.

In the following sections, we describe the main ideas of these seven degradation models. As DocCreator is an open source software, readers can consult the source code to get more details about the implementation of these models.

3.1. Ink Degradation

DocCreator provides a grayscale ink degradation model (detailed in [41]) able to simulate the most common character degradations due to the age of the document itself and printing/writing process, such as ink splotches, white specks or streaks. This model locally degrades the image in the neighbourhood of the characters boundaries. Noise is then generated to create some small ink spots near characters or to erase some characters ink area. Contrary to the well known Kanungo noise model [47] that works only on black and white images, this degradation method can process grayscale images. See Figure 3 for an ink degradation example.

Figure 3. Ink degradation on an old document. (**Left**) original image. (**Right**) degraded image.

3.2. Phantom Character

The phantom character is a typical defect that appears on documents that have been manually printed (using a wooden or metal character). After many uses, a printing character can be eroded. It is thus possible that ink reaches the borders of the piece; borders are then printed on the sheet of paper. DocCreator provides an algorithm that reproduces such ink apparition around the characters. To be as realistic as possible, we have manually extracted more than 30 phantom defects from real images. These defects are then automatically put between characters following a patch-based algorithm.

The degradation algorithm works as follow: (1) the user provides an image and the percent of character to degrade; (2) characters are extracted using a connected component algorithm; (3) a list of characters is randomly set; (4) for each selected character; (4.1) a phantom defect is randomly selected from the manually extracted available defects; (4.2) the phantom defect is resized to fit with the character size; (4.3) to be realistic, the phantom defect is used only as a pattern; the pixels within the pattern are transformed using a patch algorithm inspired from [48] where a zone from another part of the document image is selected and copied within the patch.

See Figure 4 for an example.

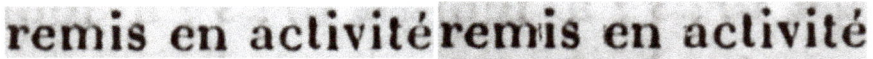

Figure 4. Phantom character apparition. (**Left**) original image. (**Right**) degraded image.

3.3. Paper Holes

Many old or recent document images contain holes. These holes have different shapes, sizes and locations. DocCreator provides an algorithm that creates different kinds of holes in a document image. This algorithm simply randomly applies holes extracted from real document images on a given document image. See Figure 5 for examples.

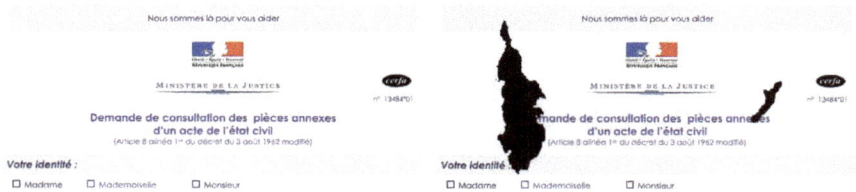

Figure 5. *Cont.*

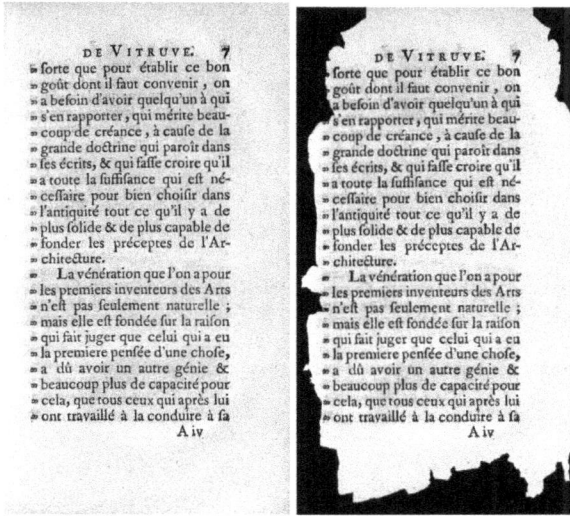

Figure 5. Hole degradation. (**Left**) original images. (**Right**) degraded images.

3.4. Bleed-Through

With DocCreator it is possible to add bleed-through defects. This algorithm is directly inspired from [49] that initially proposes an algorithm for erasing the bleed-through from a document image. By just giving an input recto image, an input verso image and the amount of wished degradation, a physical model is applied. This model mimics the verso ink that seeps through the recto side. This model simulates an anisotropic diffusion and each pixel at time $t + 1$ is modified according to the pixels values at time t with the following equation:

$$I_{i,j}^{t+1} = I_{i,j}^t + \lambda * (c_{N_{i,j}}^t \cdot \nabla_N I_{i,j}^t + c_{S_{i,j}}^t \cdot \nabla_S I_{i,j}^t + c_{E_{i,j}}^t \cdot \nabla_E I_{i,j}^t + c_{W_{i,j}}^t \cdot \nabla_W I_{i,j}^t)$$

where I is the recto image, V is the verso image, lambda a constant value in $[0; 0.25]$ and N, S, E, W are the mnemonic subscripts for North, South, East, West. $\nabla_N I_{i,j}^t$ and $c_{N_{i,j}}^t$ are defined as: $\nabla_N I_{i,j}^t = V_{i-1,j}^t - I_{i,j}^t$ and $c_{N_{i,j}}^t = \frac{1}{1+\frac{(V_{i-1,j}^t - I_{i,j}^0)^2}{\sigma^2}}$. The user sets the number of iterations and thus the quantity of generated bleed-through.

See Figure 6 for a bleed-through example.

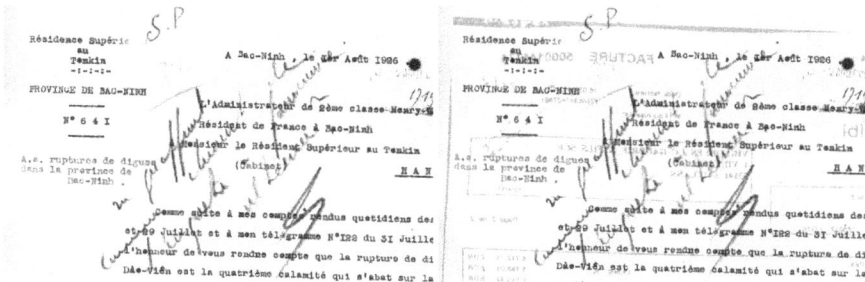

Figure 6. Bleed-through defect. (**Left**) original image. (**Right**) degraded image.

3.5. Adaptive Blur

The blur defect is a very common defect encountered during digitization campaigns. The difficulty here is to create a realistic blur defect that mimics the very slight blur that appears when the scanner is incorrectly set (a large blur is easily detected by scanners). To do so, we propose a method inspired by the blur detection from [50]. First, the user chooses a blur image to mimic among a real blur example available in DocCreator. Then, using a dichotomic algorithm, we compute the size of the kernel of a Gaussian blur that, once applied on the input image, produces a blur similar to the chosen real blur image. In this method, the Fourier Transform of the image is first computed. Then the module of the Fourier Transform is binarized according to its mean. The resulting binarized image produces a disc for images with only text. As the high frequencies decrease when the blur increases, the disc radius in the binarized image also decreases when the blur increases. This radius is used to characterize the images. The dichotomic algorithm is used to search the kernel size that produces a radius similar to the one found on the selected example image. See Figure 7 for an example.

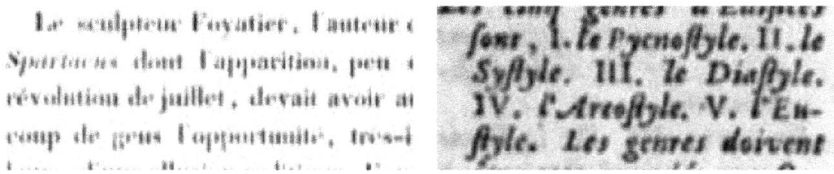

Figure 7. Adaptive blur defect. (**Left**) image with real blur. (**Right**) image with synthetic blur that mimics the real one

3.6. 3D Paper Deformation

The paper on which a book is printed may have several types of deformation (along curvature, rotation, fold, hole, etc.). We propose a 3D deformation model that generate realistic small or large paper deformations.

The full process is detailed in [51]. The main idea is: first, a 3D scanner is used to acquire a 3D mesh from a real document. This mesh preserves all representative distortions. Then, the mesh is unfolded into a 2D plan. Therefore, each vertex in the mesh has a corresponding 2D point. The coordinates of such a point are considered as texture coordinates. Finally, the mesh can be rendered with any 2D image mapped as a texture. For the rendering, we use the Phong reflection model as the illumination model. Changing light properties and position allows to accentuate or minimize distortion effects. DocCreator currently provides 17 parameterized meshes, enabling one to produce numerous distorted images. Figure 8 shows such a deformation.

This 3D paper deformation model can be used to simulate mobile document capture. The user can add a background plane with texture, on top of which the document stands. By changing viewpoint and light positions, the user can generate many images. These images can be used for camera-based document image analysis. Figure 9 shows examples of two points of view generated with the same document image.

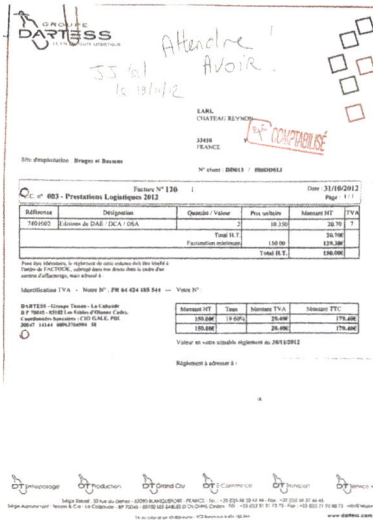

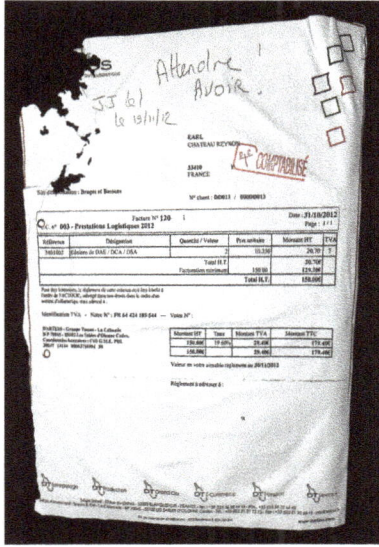

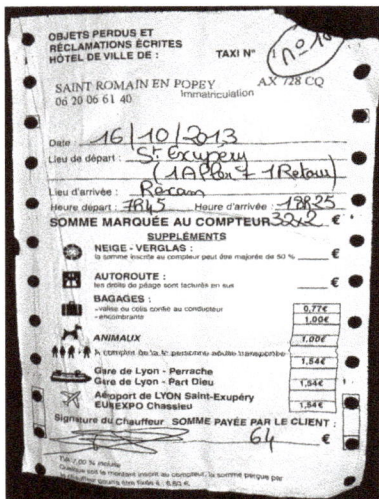

Figure 8. Examples of 3D deformations of a 2D receipt images. (**Left**) original images. (**Right**) degraded images.

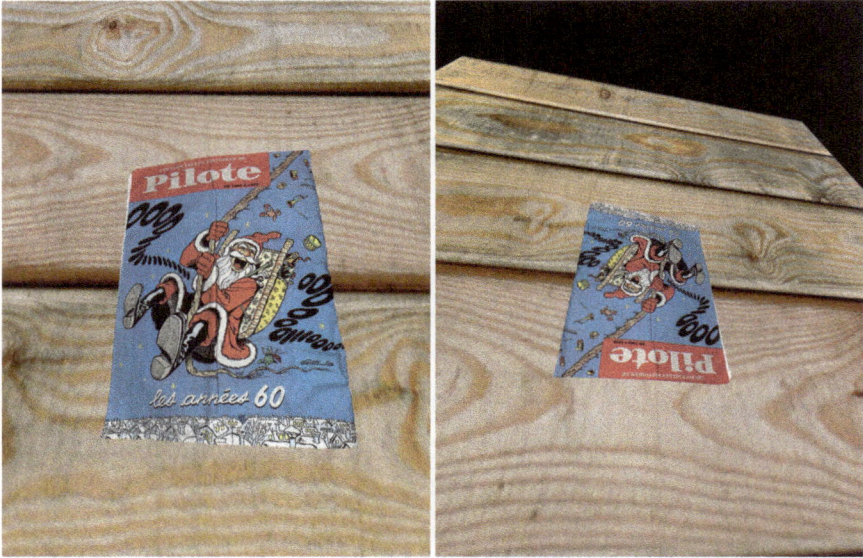

Figure 9. Examples of two viewpoints of the same document image, that could be used in the context of camera-based document image analysis.

3.7. Nonlinear Illumination Model

DocCreator provides an implementation of the nonlinear illumination model proposed in [52]. When scanning thick documents, the page to be photocopied may not be flat on the document glass and thus the illumination is not constant on the whole document. This model consider that a border of the document is bend in one direction by a radius ρ. The illumination at a point P' on the document pages is inversely proportional to the distance of point P' from the light source L. The illumination at point P' is computed with the following equation:

$$I_{P'} = I_0 \left(\frac{l_0}{(l_0 + \rho(1 - cos\phi))} \right)^2$$

where I_0 is the original intensity, l_0 the distance between the document glass and the light source L, and ϕ the angle between the normal at P' (where the page is curved) and the normal to the glass document. Figure 10 shows an example of this illumination defect. It is noteworthy that this illumination defect simulates just a particular case of what our 3D paper deformation model presented in the previous section can portray.

Except for the ink degradation model, the other degradation models work both on grayscale and colour document images.

DocCreator aims at providing other degradation models. In particular, we are currently working on the integration of a colour ink spot generation model described in [35].

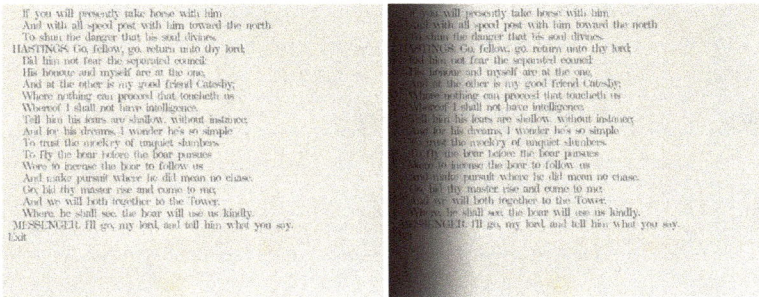

Figure 10. Examples of nonlinear illumination model defect. (**Left**) original image. (**Right**) degraded image with illumination defect applied on the left border.

4. Use of DocCreator for Performance Evaluation Tasks or Retraining

Here, we describe rapidly how DocCreator was used by other researchers and the conclusions they drew.

4.1. Published Results Using DocCreator

4.1.1. Document Image Generation for Performance Evaluation

The segmentation system proposed by [36] is based on a texture feature extraction without any a priori knowledge on the physical and logical document layout. To assess the noise robustness of their system, they used DocCreator and applied the character degradation model. From 25 simplified real document images, they generated a semi-synthetic database of 150 document images. This database is made up of several subsets where the degradation levels are different. The performance evaluations presented in [36] highlight that the texture descriptors are slightly perturbed by the degradations. When characters are highly disconnected (our algorithm has erased important character ink areas), a drop of the segmentation performances was observed.

DocCreator was also used during the ICDAR contest: staff-line removal from musical scores. The 3D distortion and the character degradation models were used in order to generate an extended database from the 1000 images of the MUSCIMA database [13]. As a result, the extended database contains 6000 semi-synthetic grayscale images and 6000 semi-synthetic binary images. This database has been used in the second edition of the music score competition ICDAR 2013 [37]. Five participants submitted eight methods. Participants were given a training set of 4000 semi-synthetic images and then 2000 semi-synthetic images to test their methods on. Regarding the results on the 3D distortion set, the submitted methods seem less robust to global distortion than to the presence of small curves and folds. For more details about the participants, the methods and the contest protocol, refer to [37]. This database has already become a benchmark database for musical document images analysis and recognition, as stated in [53]. So far, the database has indeed been used for benchmarking in multiple scientific publications about musical document processing and recognition [38,53–56] and even in the more general field of machine learning [57].

4.1.2. Document Image Generation for Retraining Task

The IAM-HistDB [58] database contains 127 handwritten historical manuscript images together with their ground truth. This database consists of three sets: the Saint Gall set containing 60 images (1.410 text lines) in Latin, the Parzival set containing 47 images (4.477 text lines) in Medieval German, and the Washington set containing 20 images in English. The authors of [39] used the character degradation model to create two extended databases of the IAM-HistDB. The first one is composed of 17.661 images degraded with the ink model. The 1.524 images from the second dataset have been created using the 127 original images and transformed using our 3D distortion model. The tests presented in [39,59] confirm the conclusion of [60] about the impact of the degradation level on re-training, either for a task of character recognition or layout extraction.

4.2. New Results on Performance Prediction Using DocCreator

Here, we show whether DocCreator can be useful for performance prediction of existing methods.

4.2.1. Increase the Prediction Rate of Predictive Binarization Algorithm

In [61], we have presented an algorithm to predict error rates of 11 binarization methods on given document images so that the best binarization method is automatically chosen for any image depending on its quality. This method requires ground-truthed data as input of the training step. The DIBCO database [62] was used. However, the DIBCO database contains only 36 images.

We propose here to extend the original DIBCO database by using the ink degradation model. Since the DIBCO database contains 36 images, we extend it with the same number of semi-synthetic document images. This extended dataset is then used to train the prediction model of [61].

Our retraining tests show that the use of this extended dataset allows one to increase the performance of the prediction model of [61]. More precisely, the error rate of the prediction model decreases (until it levels off) when the number of semi-synthetic images in the training set increases. On average, the error rate drops of about 15% compared with using only real images in the training set. The error rate converges when the proportion of semi-synthetic images is around 50% of the training set.

4.2.2. Predict OCR Recognition Rate Using Synthetic Images

Many on-line digital libraries propose a text search engine. To this end, the text within the document images has to be transcribed. Depending on the OCR recognition rate quality, three options are available: (1) directly use the OCR result when the recognition rate is close to 100%; (2) manually correct the OCR result when the automatic transcription gives "acceptable quality", or (3) do a complete manual transcription (often quite expansive). As a consequence, it is very important to be aware of the OCR recognition rate before deciding between one of these three solutions. The amount of recent publications on this subject ([63–66]) reflects the scientific interest in predicting OCRs recognition rate.

We propose here to use synthetic images to predict the OCR rate of a digitized book as follows: (1) font, background and layout are extracted from original images (with methods described in II). It is noteworthy to mention that the fonts were extracted thoroughly, in particular to include even characters not recognized by the OCR, or even to adjust margins of correctly labeled characters. (2) An adapted Lorem ipsum text is randomly generated and used to create synthetic images with the font and background previously extracted. This adapted Lorem ipsum is generated with accentuated characters (é, à, ù, etc.) and old characters (ff, fi, ſ, fl, ffi) if the original text contains such characters. Generating such characters is important to have a representative dataset for fair OCR testing. As a result, images like the one presented in Figure 2 are generated with the associated XML ground truth. (3) An OCR (Tesseract) is finally used to recognize the text on these synthesized images. This text is compared with the Lorem ipsum ground truth text, giving an OCR recognition rate. We consider that this recognition rate is a prediction of the OCR rate if the OCR software was applied on original images. Table 2 Column 1 provides the average OCR recognition rate obtained on the original images, Table 2 Column 3 refers to the average OCR rates computed on the synthetic "Lorem ipsum" images versions.

We also propose to evaluate the capacity of our method to correctly predict the OCR recognition rate by comparing original images with their synthetic version generated with exactly the same text (see Figure 11 to compare the original images and their synthetic versions). These images are generated following this protocol: (1) pages from three books (2 typewritten and 1 manuscript book) have been manually transcribed; (2) font, background and layout are automatically extracted from original images; and (3) font, background and layout are finally used with the transcribed texts to automatically generate synthetic versions of the original ones. The whole database is composed of 93 document images containing 18.240 words and 115.622 characters.

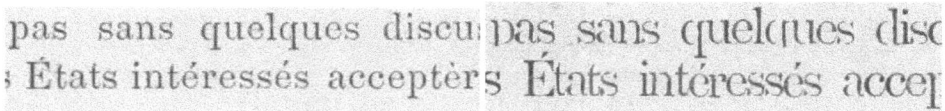

(**a**) (DB1) Contemporary french typewritten document.

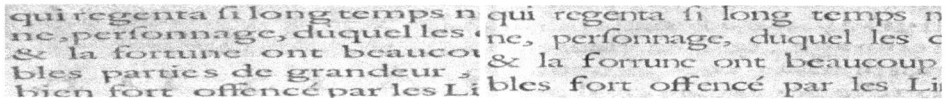

(**b**) (DB2) Old french typewritten document.

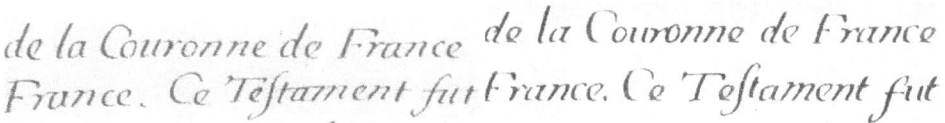

(**c**) (DB3) Old french Manuscript document.

Figure 11. Images extracted from the database used for testing our prediction algorithm. (**Left**) original images. (**Right**) synthetic generated images.

To evaluate the OCR text recognition rate, we use the Levenshtein distance (a metric measuring the difference between two strings) between the whole original transcribed text and the whole recognized text. We compute the mean of the Levenshtein distances for the N documents of each database. Using this Levenshtein distance, the difference between the OCR text recognition rate computed on real images and the one computed on "Lorem ipsum" version (Table 2 Column 1 and Column 2) is, on average, only overestimated by 0.04. The difference between the real OCR rate and the one computed on the synthetic versions (Table 2 Column 1 and Column 3) is, on average, only overestimated by 0.03. Most of the success of different existing OCR prediction methods ([63–66]) are related to the quality and quantity of the needed ground truth. Our prediction method presented here provides comparable results with the ones form the state of the art.

Table 2. Comparison between OCR recognition rates obtained on three different books original images and their synthetic versions. Column 1: OCR recognition rate on original images, Column 2: OCR recognition rate on synthetic images generated with both the text and the font from the original images, Column 3: OCR recognition rate on synthetic images generated with lorem ipsum random text and the font from the original images.

	Original Image	Font From	Same Text	Lorem Text
DB1	0.95	DB1	0.94	0.88
DB2	0.80	DB2	0.85	0.84
DB3	0.24	DB3	0.21	0.23

5. Conclusions

DocCreator gives to DIAR researchers a simple and rapid way to extend existing document image databases or to create new ones avoiding the tedious task of manual ground truth generation. DocCreator embeds many fonts, backgrounds, meshes and realistic degradation models which, when combined, result in an interesting combination of ground-truthed databases. The experiments detailed in this paper show semi-synthetic and synthetic documents created with DocCreator are useful for performance evaluation, retraining tasks or performance prediction. In future work, we plan to improve the synthetic document creation to avoid to have too different characters in the composed document. For example, we should investigate if adding some constraints on the font extraction phase or taking into account the context when adding new characters to the synthetic document may lead to more realistic synthetic documents. We also consider to set up a cognitive experiment to evaluate the perceived realness of the degraded documents or even the created synthetic documents. We are also planning to investigate how the generation of highly diversified data can improve the results of tasks based on deep learning methods.

DocCreator (source, Linux, Mac, Windows packaged versions), all the databases used for the tests, a video and an extra database (31.000 synthetic images generated with William Shakespeare sonnet text files) are available at [http://doc-creator.labri.fr/].

Author Contributions: Nicholas Journet, Muriel Visani and Boris Mansencal contributed in equal proportion to the creation and tests of DocCreator (model degradation and synthetic document reconstruction) ; they also wrote this article. Kieu Van-Cuong contributed to the degradation models, Antoine Billy contributed to the synthetic document reconstruction.

Conflicts of Interest: The authors declare no conflict of interest.

References

1. L'Affaire Alexis. Available online: http://gallica.bnf.fr/ark:/12148/bpt6k8630878m/f1.item.texteImage (accessed on 9 December 2017).
2. Shahab, A.; Shafait, F.; Kieninger, T.; Dengel, A. An Open Approach Towards the Benchmarking of Table Structure Recognition Systems. In Proceedings of the 9th IAPR International Workshop on Document Analysis Systems, Boston, MA, USA, 9–11 June 2010; ACM: New York, NY, USA, 2010; pp. 113–120.
3. Lazzara, G.; Levillain, R.; Géraud, T.; Jacquelet, Y.; Marquegnies, J.; Crépin-Leblond, A. The SCRIBO Module of the Olena Platform: a Free Software Framework for Document Image Analysis. In Proceedings of the 2011 International Conference on Document Analysis and Recognition (ICDAR), Beijing, China, 18–21 September 2011.
4. Yalniz, I.; Manmatha, R. A Fast Alignment Scheme for Automatic OCR Evaluation of Books. In Proceedings of the 2011 International Conference on Document Analysis and Recognition (ICDAR), Beijing, China, 18–21 September 2011; pp. 754–758.
5. Roy, P.; Ramel, J.; Ragot, N. Word Retrieval in Historical Document Using Character-Primitives. In Proceedings of the 2011 International Conference on Document Analysis and Recognition (ICDAR), Beijing, China, 18–21 September 2011; pp. 678–682.
6. IAM Handwriting Database. Available online: http://www.iam.unibe.ch/fki/databases/iam-handwriting-database (accessed on 9 December 2017).
7. Grosicki, E.; Carré, M.; Brodin, J.M.; Geoffrois, E. Results of the second RIMES evaluation campaign for handwritten mail processing. In Proceedings of the 2009 10th International Conference on Document Analysis and Recognition (ICDAR), Barcelona, Spain, 26–29 July 2009.
8. Perez, D.; Tarazon, L.; Serrano, N.; Castro, F.; Terrades, O.R.; Juan, A. The GERMANA Database. In Proceedings of the 2009 10th International Conference on Document Analysis and Recognition (ICDAR), Barcelona, Spain, 26–29 July 2009; IEEE Computer Society: Washington, DC, USA, 2009; pp. 301–305.
9. Nakagawa, K.; Fujiyoshi, A.; Suzuki, M. Ground-truthed Dataset of Chemical Structure Images in Japanese Published Patent Applications. In Proceedings of the 9th IAPR International Workshop on Document Analysis Systems, Boston, MA, USA, 9–11 June 2010; ACM: New York, NY, USA, 2010; pp. 455–462.
10. Eurecom. Available online: http://www.eurecom.fr/huet/work.html (accessed on 9 December 2017).
11. University of California, San Francisco. *The Legacy Tobacco Document Library (LTDL)*; University of California: San Francisco, CA, USA, 2007.
12. Delalandre, M.; Valveny, E.; Pridmore, T.; Karatzas, D. Generation of Synthetic Documents for Performance Evaluation of Symbol Recognition & Spotting Systems. *Int. J. Doc. Anal. Recognit.* **2010**, *13*, 187–207.

13. Fornés, A.; Dutta, A.; Gordo, A.; Lladós, J. CVC-MUSCIMA: A ground truth of handwritten music score images for writer identification and staff removal. *Int. J. Doc. Anal. Recognit.* **2012**, *15*, 243–251.

14. Burie, J.C.; Chazalon, J.; Coustaty, M.; Eskenazi, S.; Luqman, M.M.; Mehri, M.; Nayef, N.; Ogier, J.M.; Prum, S.; Rusiñol, M. ICDAR2015 competition on smartphone document capture and OCR (SmartDoc). In Proceedings of the 2015 13th International Conference on Document Analysis and Recognition (ICDAR), Nancy, France, 23–26 August 2015; pp. 1161–1165.

15. Nayef, N.; Luqman, M.M.; Prum, S.; Eskenazi, S.; Chazalon, J.; Ogier, J.M. SmartDoc-QA: A dataset for quality assessment of smartphone captured document images-single and multiple distortions. In Proceedings of the 2015 13th International Conference on Document Analysis and Recognition (ICDAR), Nancy, France, 23–26 August 2015; pp. 1231–1235.

16. TC11 Online Resources. Available online: http://tc11.cvc.uab.es/datasets/ (accessed on 9 December 2017).

17. Yanikoglu, B.; Vincent, L. Pink Panther: A complete environment for ground-truthing and benchmarking document page segmentation. *Pattern Recognit.* **1998**, *31*, 1191–1204.

18. Ha Lee, C.; Kanungo, T. The architecture of TRUEVIZ: A groundTRUth/metadata Editing and VIsualiZing toolkit. *Pattern Recognit.* **2003**, *36*, 811–825.

19. Doermann, D.; Zotkina, E.; Li, H. GEDI—A Groundtruthing Environment for Document Images. In Proceedings of the 9th IAPR International Workshop on Document Analysis Systems (DAS 2010), Boston, MA, USA, 9–11 June 2010.

20. Clausner, C.; Pletschacher, S.; Antonacopoulos, A. Efficient OCR Training Data Generation with Aletheia. In Proceedings of the International Association for Pattern Recognition (IAPR), Tours, France, 7–10 April 2014.

21. Garz, A.; Seuret, M.; Simistira, F.; Fischer, A.; Ingold, R. Creating ground truth for historical manuscripts with document graphs and scribbling interaction. In Proceedings of the 2016 12th IAPR Workshop on Document Analysis Systems (DAS), Santorini, Greece, 11–14 April 2016; pp. 126–131.

22. Gatos, B.; Louloudis, G.; Causer, T.; Grint, K.; Romero, V.; Sánchez, J.A.; Toselli, A.H.; Vidal, E. Ground-truth production in the tranScriptorium project. In Proceedings of the 2014 11th IAPR International Workshop on Document Analysis Systems Document Analysis Systems (DAS), Tours, France, 7–10 April 2014; pp. 237–241.

23. Wei, H.; Chen, K.; Seuret, M.; Würsch, M.; Liwicki, M.; Ingold, R. *DIVADIAWI— A Web-Based Interface for Semi-Automatic Labeling of Historical Document Images*; Digital Humanities: Sydney, Australia, 2015.

24. Mas, J.; Fornés, A.; Lladós, J. An Interactive Transcription System of Census Records using Word-Spotting based Information Transfer. In Proceedings of the 12th IAPR International Workshop on Document Analysis Systems (DAS 2016), Santorini, Greece, 11–14 April 2016.

25. Recital Manuscript Platform. Available online: http://recital.univ-nantes.fr/ (accessed on 9 December 2017).

26. Clausner, C.; Pletschacher, S.; Antonacopoulos, A. Aletheia—An Advanced Document Layout and Text Ground-Truthing System for Production Environments. In Proceedings of the International Conference on document Analysis and Recognition, Beijing, China, 18–21 September 2011; pp. 48–52.

27. Baird, H.S. Document Image Defect Models. In Proceedings of the IAPR workshop on Syntatic and Structural Pattern Recognition, Murray Hill, NJ, USA, 13–15 June 1990; pp. 13–15.

28. Jiuzhou, Z. *Creation of Synthetic Chart Image Database with Ground Truth*; Technical Report; National University of Singapore: Singapore, 2005.

29. Ishidera, E.; Nishiwaki, D. A Study on Top-down Word Image Generation for Handwritten Word Recognition. In Proceedings of the 2003 7th International Conference on Document Analysis and Recognition (ICDAR), Edinburgh, UK, 3–6 August 2003; IEEE Computer Society: Washington, DC, USA, 2003.

30. Yin, F.; Wang, Q.F.; Liu, C.L. Transcript Mapping for Handwritten Chinese Documents by Integrating Character Recognition Model and Geometric Context. *Pattern Recognit.* **2013**, *46*, 2807–2818.

31. Opitz, M.; Diem, M.; Fiel, S.; Kleber, F.; Sablatnig, R. End-to-End Text Recognition Using Local Ternary Patterns, MSER and Deep Convolutional Nets. In Proceedings of the 2014 11th IAPR International Workshop on Document Analysis Systems (DAS), Tours, France, 7–10 April 2014; pp. 186–190.

32. Yacoub, S.; Saxena, V.; Sami, S. Perfectdoc: A ground truthing environment for complex documents. In Proceedings of the Eighth International Conference on Document Analysis and Recognition, Seoul, Korea, 31 August–1 September 2005; pp. 452–456.

33. Saund, E.; Lin, J.; Sarkar, P. Pixlabeler: User interface for pixel-level labeling of elements in document images. In Proceedings of the International Conference on Document Analysis and Recognition, Barcelona, Spain, 26–29 July 2009; pp. 646–650.

34. Lamiroy, B.; Lopresti, D. An Open Architecture for End-to-End Document Analysis Benchmarking. In Proceedings of the 2011 International Conference on Document Analysis and Recognition (ICDAR), Beijing, China, 18–21 September 2011; pp. 42–47.

35. Seuret, M.; Chen, K.; Eichenbergery, N.; Liwicki, M.; Ingold, R. Gradient-domain degradations for improving historical documents images layout analysis. In Proceedings of the 2015 13th International Conference on Document Analysis and Recognition (ICDAR), Tunis, Tunisia, 23–26 August 2015; pp. 1006–1010.

36. Mehri, M.; Gomez-Krämer, P.; Héroux, P.; Mullot, R. Old Document Image Segmentation Using the Autocorrelation Function and Multiresolution Analysis. In Proceedings of the IS & T/SPIE Electronic Imaging, Burlingame, CA, USA, 3–7 February 2013.

37. Visani, M.; Kieu, V.; Fornés, A.; Journet, N. The ICDAR 2013 Music Scores Competition: Staff Removal. In Proceedings of the 2013 12th International Conference on Document Analysis and Recognition (ICDAR), Washington, DC, USA, 25–28 August 2013; pp. 1407–1411.

38. Montagner, I.d.S.; Hirata, R., Jr.; Hirata, N.S.T. A Machine Learning based method for Staff Removal. In Proceedings of the 2014 22nd International Conference on Pattern Recognition (ICPR), Stockholm, Sweden, 24–28 August 2014; pp. 3162–3167.

39. Fischer, A.; Visani, M.; Kieu, V.C.; Suen, C.Y. Generation of Learning Samples for Historical Handwriting Recognition Using Image Degradation. In Proceedings of the the 2nd International Workshop on Historical Document Imaging and Processing, Washington, DC, USA, 24 August 2013; pp. 73–79.

40. Smith, R. An Overview of the Tesseract OCR Engine. In Proceedings of the 2007 9th International Conference on Document Analysis and Recognition (ICDAR), Parana, Brazil, 23–26 September 2007; pp. 629–633.

41. Bahaghighat, M.K.; Mohammadi, J. Novel approach for baseline detection and Text line segmentation. *Int. J. Comput. Appl.* **2012**, *51*, doi:10.5120/8013-1039.

42. Telea, A. An image inpainting technique based on the fast marching method. *J. Graph. Tools* **2004**, *9*, 23–34.

43. Mehri, M.; Héroux, P.; Lerouge, J.; Gomez-Krämer, P.; Mullot, R. A structural signature based on texture for digitized historical book page categorization. In Proceedings of the 2015 13th International Conference on Document Analysis and Recognition (ICDAR), Tunis, Tunisia, 23–26 August 2015; pp. 116–120.

44. Breuel, T.M. Two geometric algorithms for layout analysis. In *International Workshop on Document Analysis Systems*; Springer: Berlin, Germany, 2002; pp. 188–199.

45. Ramel, J.Y.; Leriche, S.; Demonet, M.; Busson, S. User-driven page layout analysis of historical printed books. *Int. J. Doc. Anal. Recognit.* **2007**, *9*, 243–261.

46. Garz, A.; Seuret, M.; Fischer, A.; Ingold, R. A User-Centered Segmentation Method for Complex Historical Manuscripts Based on Document Graphs. *IEEE Trans. Hum.-Mach. Syst.* **2017**, *47*, 181–193.

47. Kanungo, T.; Haralick, R. Automatic generation of character groundtruth for scanned documents: A closed-loop approach. In Proceedings of the 13th International Conference on Pattern Recognition, Vienna, Austria, 25–29 August 1996; Volume 3, pp. 669–675.

48. Shakhnarovich, G. Learning Task-Specific Similarity. Ph.D. Thesis, Massachusetts Institute of Technology, Cambridge, MA, USA, 2005.

49. Moghaddam, R.F.; Cheriet, M. Low Quality Document Image Modeling and Enhancement. *Int. J. Doc. Anal. Recognit.* **2009**, *11*, 183–201.

50. Lelégard, L.; Bredif, M.; Vallet, B.; Boldo, D. Motion blur detection in aerial images shot with channel-dependent exposure time. In Proceedings of the ISPRS-Technical-Commission III Symposium on Photogrammetric Computer Vision and Image Analysis (PCV), Saint-Mandé, France, 1–3 September 2010; pp. 180–185.

51. Kieu, V.; Journet, N.; Visani, M.; Mullot, R.; Domenger, J. Semi-synthetic Document Image Generation Using Texture Mapping on Scanned 3D Document Shapes. In Proceedings of the 2013 12th International Conference on Document Analysis and Recognition (ICDAR), Washington, DC, USA, 25–28 August 2013; pp. 489–493.

52. Kanungo, T.; Haralick, R.M.; Phillips, I. Global and Local Document Degradation Models. In Proceedings of the 1993 2nd International Conference on Document Analysis and Recognition (ICDAR), Tsukuba City, Japan, 20–22 October 1993; pp. 730–734.

53. Calvo-Zaragoza, J.; Micó, L.; Oncina, J. Music staff removal with supervised pixel classification. *Int. J. Doc. Anal. Recognit.* **2016**, *19*, 1–9.

54. Bui, H.N.; Na, I.S.; Kim, S.H. Staff Line Removal Using Line Adjacency Graph and Staff Line Skeleton for Camera Based Printed Music Scores. In Proceedings of the 2014 22nd International Conference on Pattern Recognition (ICPR), Stockholm, Sweden, 24–28 August 2014.

55. Géraud, T. A morphological method for music score staff removal. In Proceedings of the 2014 IEEE International Conference on Image Processing (ICIP), Paris, France, 27–30 October 2014; pp. 2599–2603.

56. Zaragoza, J.C. Pattern Recognition for Music Notation. Ph.D. Thesis, Universidad de Alicante, Alicante, Spain, 2016.

57. Montagner, I.S.; Hirata, N.S.; Hirata, R., Jr.; Canu, S. Kernel approximations for W-operator learning. In Proceedings of the International Conference on Graphics, Patterns and Images (SIBGRAPI), Sao Paulo, Brazil, 4–7 October 2016.

58. IAM-HistDB Database. Available online: https://diuf.unifr.ch/main/hisdoc/iam-histdb (accessed on 9 December 2017).

59. Wei, H.; Baechler, M.; Slimane, F.; Ingold, R. Evaluation of SVM, MLP, and GMM Classifiers for Layout Analysis of Historical Documents. In Proceedings of the 2013 12th International Conference on Document Analysis and Recognition (ICDAR), Washington, DC, USA, 25–28 August 2013; pp. 1220–1224.

60. Varga, T.; Bunke, H. Effects of training set expansion in handwriting recognition using synthetic data. In Proceedings of the 11th Conference of the International Graphonomics Society, Scottsdale, AZ, USA, 2–5 November 2003; pp. 200–203.

61. Rabeux, V.; Journet, N.; Vialard, A.; Domenger, J.P. Quality Evaluation of Degraded Document Images for Binarization Result Prediction. *Int. J. Doc. Anal. Recognit.* **2013**, 1–13.

62. Pratikakis, I.; Gatos, B.; Ntirogiannis, K. ICDAR 2011 Document Image Binarization Contest (DIBCO 2011). In Proceedings of the 2011 11th International Conference on Document Analysis and Recognition (ICDAR), Washington, DC, USA, 25–28 August 2011; pp. 1506–1510.

63. Bhowmik, T.K.; Paquet, T.; Ragot, N. OCR performance prediction using a bag of allographs and support vector regression. In Proceedings of the 2014 11th IAPR International Workshop on Document Analysis Systems (DAS), Tours, France, 7–10 April 2014; pp. 202–206.

64. Peng, X.; Cao, H.; Natarajan, P. Document image OCR accuracy prediction via latent Dirichlet allocation. In Proceedings of the 2015 13th International Conference on Document Analysis and Recognition (ICDAR), Tunis, Tunisia, 23–26 August 2015; pp. 771–775.

65. Ye, P.; Doermann, D. Document image quality assessment: A brief survey. In Proceedings of the 2013 12th International Conference on Document Analysis and Recognition (ICDAR), Washington, DC, USA, 25–28 August 2013; pp. 723–727.

66. Clausner, C.; Pletschacher, S.; Antonacopoulos, A. Quality prediction system for large-scale digitisation workflows. In Proceedings of the 2016 12th IAPR Workshop on Document Analysis Systems (DAS), Santorini, Greece, 11–14 April 2016; pp. 138–143.

Journal of
Imaging

MDPI

Article

Open Datasets and Tools for Arabic Text Detection and Recognition in News Video Frames

Oussama Zayene [1,2,*], Sameh Masmoudi Touj [1], Jean Hennebert [3], Rolf Ingold [2]
and Najoua Essoukri Ben Amara [1]

[1] LATIS Lab, National Engineering School of Sousse (Eniso), University of Sousse, Sousse 4054, Tunisia;
     samehmasmouditouj@yahoo.fr (S.M.T.); najoua.benamara@eniso.rnu.tn (N.E.B.A.)
[2] DIVA Group, Department of Informatics, University of Fribourg (Unifr), Fribourg 1700, Switzerland;
     rolf.ingold@unifr.ch
[3] ICoSys Institute, HES-SO, University of Applied Sciences, Fribourg 1705, Switzerland;
     jean.hennebert@hefr.ch
*   Correspondence: oussama.zayene@unifr.ch

Received: 26 November 2017; Accepted: 26 January 2018; Published: 31 January 2018

Abstract: Recognizing texts in video is more complex than in other environments such as scanned documents. Video texts appear in various colors, unknown fonts and sizes, often affected by compression artifacts and low quality. In contrast to Latin texts, there are no publicly available datasets which cover all aspects of the Arabic Video OCR domain. This paper describes a new well-defined and annotated Arabic-Text-in-Video dataset called AcTiV 2.0. The dataset is dedicated especially to building and evaluating Arabic video text detection and recognition systems. AcTiV 2.0 contains 189 video clips serving as a raw material for creating 4063 key frames for the detection task and 10,415 cropped text images for the recognition task. AcTiV 2.0 is also distributed with its annotation and evaluation tools that are made open-source for standardization and validation purposes. This paper also reports on the evaluation of several systems tested under the proposed detection and recognition protocols.

Keywords: video text detection; video text recognition; AcTiV dataset; Arabic Video OCR

1. Introduction

Broadcast news and public-affairs programs are a prominent source of information that provides daily updates on national and world news. Nowadays, TV newscasters archive a tremendous number of news video clips thanks to the rapid progress in mass storage technology. As the archive size grows rapidly, the manual annotation of all video clips becomes impractical.

Since the 80s, research in OCR techniques has been an attractive field in the document analysis and recognition community. Prior work has addressed specific research problems that have bordered on printed and handwritten texts in scanned documents. Recently, embedded text in videos has received increasing attention as it often gives crucial information about the media content [1–3]. News videos generally contain two types of texts [2]: scene text and artificial text (Figure 1). The first type is naturally recorded as part of scene during video capturing, such as traffic and shop signs. The second type of text is artificially superimposed on the video during the editing process. Compared with scene text, the artificial one usually provides brief and direct description of video content, which is important for automatic broadcast annotation. Typically, artificial text in news video indicates speaker's name, location, event information, scores of a match, etc. Therefore, in this context, we particularly focus on this category of text.

Recognizing text in videos, often called Video OCR [4], is an essential task in many applications such as news indexing and retrieval [5], video categorization, large archive managing and speaker identification [6]. A Video OCR system is generally composed of four stages: detection, tracking, extraction and recognition. The two first steps consist in locating text regions in video frames and generating the bounding boxes of text lines as an output. Text extraction aims at extracting text pixels and removing background ones. The recognition task converts image regions into text strings. In this work, we focus especially on the detection and recognition steps.

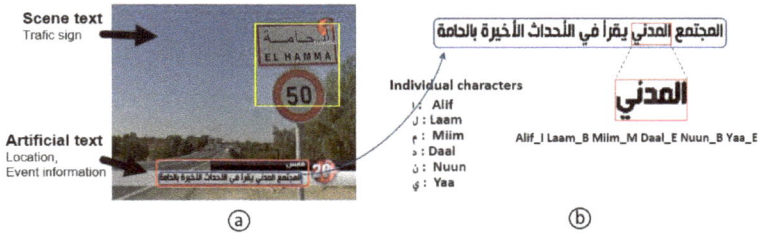

Figure 1. Example of an Arabic video frame including scene and artificial texts (**a**). Decomposition of an Arabic word into characters (**b**).

Compared to scanned documents, text detection and recognition in video frames is more challenging. The major challenges are:

- Text patterns variability: unknown font-size and font-family, different colors and alignment (even in the same TV channel).
- Background complexity: text-like objects in video frames, such as fences, bricks and signs, can be confused with text characters.
- Video quality: acquisition conditions, compression artifacts and low resolution.

All these challenges may give rise to failures in video text detection. The present study focuses on the Arabic video OCR problem. This introduces many additional challenges related to Arabic script [7]. Compared to Latin, the Arabic text has special characteristics such as presence of diacritics, non-uniform inter/intra-word distance and cursiveness of the script, i.e., characters may have up to four shapes depending on their position in the word (for examples, see Figure 1b).

Several techniques have been proposed in the conventional field of Arabic OCR in scanned documents [7–10]. However, few attempts have been made on the development of detection and recognition systems for overlaid text in Arabic news video [11–13]. These systems were tested on private datasets with different evaluation protocols and metrics that make direct comparison and objective benchmarking rather impractical. For instance, in [11], the proposed text detector was evaluated on a private set of 150 video images. In [13], Yousfi et al. evaluated their text detection system on two private test sets of 164 and 201 video frames. Therefore, the availability of an annotated and public dataset is of key importance for the Arabic video text analysis community.

In this paper, we present AcTiV 2.0 as an open Arabic-Text-in-Video dataset dedicated to benchmarking and comparison of systems for Arabic text detection, tracking and recognition. AcTiV 2.0 is an important extension of the one published in ICDAR 2015 [14]. It includes 189 video clips with an average length of 10 min per sequence for a global duration of about 31 h. These video sequences have been collected from four different Arabic news channels during the period between October 2013 and March 2016. In the present work, three video resolutions were chosen: HD (High Definition, 1920 × 1080), SD (Standard Definition, 720 × 576) and SD (480 × 360). The latter resolution concerns video clips that have been downloaded from the official YouTube channel of TunisiaNat1 TV.

The paper is organized as follows: In Section 2, we present related work on datasets for text detection/recognition problems. Then, we present in Section 3 the AcTiV 2.0 dataset in terms of features, statistics and annotations. We detail the evaluation protocols in Section 4 and present the experimental results in Section 5. In Section 6, we draw the conclusions and discuss future work.

2. Literature Review

Recently, several approaches have been proposed to detect and recognize texts in videos and natural scene images [1,2,15,16].

All mentioned work so far are dedicated to Latin or Chinese text detection and recognition methods. Much of the progress that has been made in this field of research is attributed to the availability of standard datasets. The most popular of these is the dataset of ICDAR 2003 Robust Reading Competitions (RRC) [17], prepared for scene text localization, character segmentation (removing background pixels) and word recognition. This dataset includes 509 text images in real environments captured with hand-held devices. 258 images from the database are used for training and the remaining 251 images constitute the test set. Some examples are depicted in Figure 2a. This dataset was also used in the ICDAR 2005 Text Locating Competition [18]. Figure 3 shows the evolution of the Latin text detection research between 2003 and 2013 [18–20] taking as a benchmark the ICDAR 2003 dataset. As can be observed, the method of Huang et al. [19] outperforms other approaches by a large margin. This method enhances the Stroke Width Transform (SWT) algorithm using color information and introduces Text Covariance Descriptors (TCDs). For the word-recognition task, the best accuracy of 93.1%, was achieved by Jaderberg et al. [21] using their proposed Convolutional Neural Networks (CNN) model. The dataset in ICDAR 2011 RRC [22] was inherited from the benchmark used in the previous ICDAR competitions (i.e., 2003 and 2005) but have undergone extension and modification, since there are some missing ground truth information and imprecise word bounding boxes. The final datasets consisted of 485 full images and 1564 cropped word images for localization and word-recognition tasks, respectively. On this dataset, the text detection method of Liao et al. [23] obtains state-of-the-art performance with an F-score of 82%. This algorithm is based on a fully convolutional network (FCN) followed by a standard non-maximum suppression process.

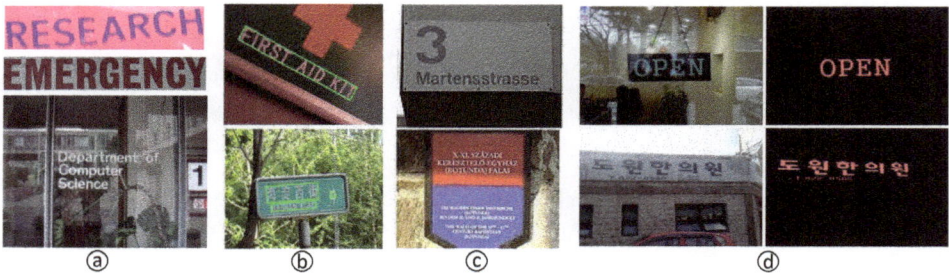

Figure 2. Typical samples from ICDAR2003 (**a**), MSRA-TD500 (**b**), NEOCR (**c**) and KAIST (**d**) datasets.

In the 2013 edition of ICDAR RRC [24], a new database was proposed for video text detection, tracking and recognition. It contains 28 short video sequences. An updated version of this dataset was provided in ICDAR 2015 [25] including a training set of 25 videos and a test set of 24 videos.

The MSRA-TD500 dataset [26] works on multi-oriented scene texts detection. This dataset includes 500 images (300 for training and 200 for testing) with horizontal and slant/skewed texts in complex natural scenes (see Figure 2b for examples). The method of Liu et al. [27] achieves state-of-the-art performance on this database with an F-score of 75%. This method makes use of the Maximally Stable Extremal Regions (MSER) technique as text candidates extractor as well as a set of heuristic rules and an AdaBoost classifier as a two-stages filtering process.

The Street View Text (SVT) dataset [28] is used for scene text detection, segmentation and recognition in outdoor images. It includes 350 full images with 904 word-level annotated bounding boxes. The method of Shi et al. [29] shows superiority over existing techniques with 80.8% as a recognition accuracy. This method is based on Convolutional Recurrent Neural Network (CRNN), which integrates the advantages of both CNN and Recurrent Neural Networks (RNN). For the segmentation task, the best F-score, 90%, was obtained by Mishra et al. [30]. The algorithm is mainly based on two steps: a GMM refinement using stroke and color features and a graph cut procedure.

The KAIST dataset [31] consists of 3000 images taken in indoor and outdoor scenes (see Figure 2d for examples). This is a multilingual dataset, which includes English and Korean texts. KAIST can be used for both detection and segmentation tasks, as it provides binary masks for each character in the image. The text segmentation algorithm of Zhu and Zhang [32] outperforms existing methods on this dataset with an F-score of 88%. The method is based on superpixel clustering. First, an adaptive SLIC text superpixel generation procedure is performed. Next, a DBSCAN-based superpixel clustering is used to fuse stroke superpixels. Finally, a stroke superpixel verification process is applied.

The NEOCR dataset [33] contains 659 natural scene images with multi-oriented texts of high variability (see Figure 2c for examples). This database is intended for scene text recognition and provided multilingual evaluation environments, as it includes texts in eight European languages.

In 2016, Veit et al. [34] proposed a dataset for English scene text detection and recognition called COCO-Text. The dataset is based on the Microsoft COCO dataset, which contains images of complex everyday scenes. The best result on this dataset (67.16%) was obtained by the winner of the COCO-Text ICDAR2017 competition [35]. Note that the participating methods on this competition were ranked based on their Average precision (AP) with an Intersection over Union (IoU) of 0.5.

Recently, Chng and Chan [36] introduced a new dataset, namely Total-text, for curved scene text detection and recognition problems. It contains 1555 scene images and 9330 annotated words with three different text orientations.

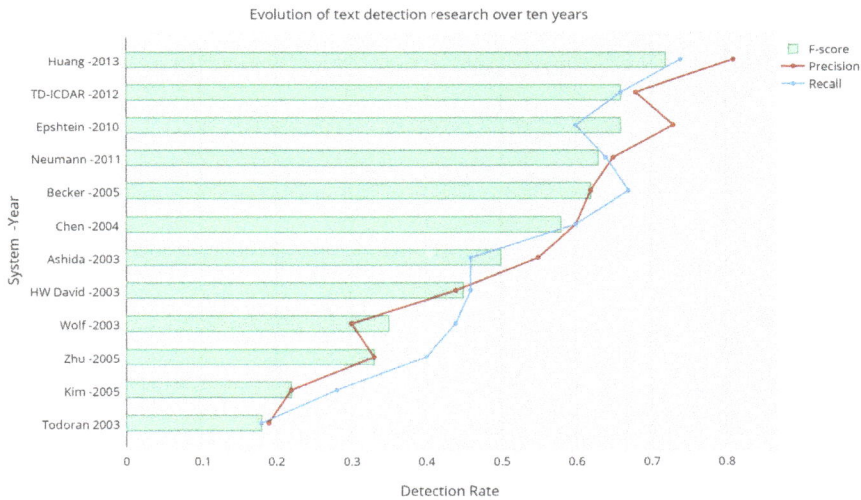

Figure 3. Some examples of text detection systems [18–20] showing the evolution of this area of research over ten years.

As for Arabic language, major contributions have already been made in the conventional field of printed and handwritten OCR systems [7,10]. Much progress of such systems has been triggered thanks to the availability of public datasets. Examples include the IFN/ENIT [37] and KHATT [38] datasets for offline handwriting recognition and writer identification; the APTI database [39] for printed word recognition; and the ADAB dataset [40] that works on online handwriting recognition.

However, handling Arabic text detection and recognition for multimedia documents is limited to very few studies [41–43].

Table 1 presents commonly used datasets for text processing in images and videos, and summarizes their features in terms of text categories, sources, tasks, script, information of training/test samples and best achieved result. As depicted by this table, publicly available datasets for Arabic Video OCR systems are limited to one work for the recognition task and are even non-existent for detection and tracking problems. Yousfi et al. [44] put forward a dataset for superimposed text recognition, called Alif. The dataset was composed of 6532 static cropped text images extracted from diverse Arabic TV channels and with about 12% extracted from web sources. This dataset offered only one image resolution.

Table 1. Most important existing datasets for text processing in videos and scene images. "D", "S" and "R" respectively denote "Detection", "Segmentation" and "Recognition".

Dataset (Year)	Category	Source	Task	# of Images (Train/Test)	# of Text (Train/Test)	Script	Best Scores
ICDAR'03 [18] (2003)	Scene text	Camera	D/R	509 (258/251)	2276 (1110/1156)	English	93.1% (R)
KAIST [31] (2010)	Scene text	Camera, mobile phone	D/S	3000	>5000	English, Korean	88% (S)
SVT [28] (2010)	Scene text	Google Street View	D/S/R	350 (100/250)	904 (257/647)	English	80.8% (R) 90% (S)
NEOCR [33] (2011)	Scene text	Camera	D/R	659	5238	Eight languages	
ICDAR'11 [22] (2011)	Scene text	Camera	D/R	485	1564	English	82% (D)
MSRA-TD500 [26] (2012)	scene text	Camera	D	500 (300/200)	–	English, Chinese	75%
ICDAR'13 [24] (2013)	Scene text Artificial text Video scene	Camera Web Camera	D/S/R D/S/R D/T/R	229/233 410/141 28 videos	848/1095 3564/1439 –	Spanish, French, English	
ALIF [44] (2015)	Artificial text	Video frames	R		6532 (4152/2199)	Arabic	55.03%
COCO-Text [34] (2016)	Scene text	MS COCO dataset	D/R	63,686 (43.6k/10k)	173,000	English	67.16% (D)
Total-Text [36] (2017)	Curved scene text	web	D/R	1555 (1255/300)	9330 (words)	English	

3. Proposed Datasets

In this section, we describe the AcTiV 2.0 dataset in terms of characteristics, statistics and annotation guidelines.

3.1. Data Characteristics and Statistics

As mentioned in the introduction, AcTiV 1.0 (http://tc11.cvc.uab.es/datasets/AcTiV_1) was presented in the ICDAR'15 conference [14] as the first publicly accessible annotated dataset designed to assess the performance of different Arabic Video OCR systems. This database is currently used by several research groups around the world. It was partially used as a benchmark in the first edition of the "AcTiVComp" contest in conjunction with the ICPR'16 conference [45]. The two main challenges addressed by this dataset are text pattern variability and presence of complex backgrounds with various text-like objects. AcTiV 1.0 consists of 80 video clips recorded from four different Arabic news channels: TunisiaNat1, France24, Russia Today and AljazeeraHD. AcTiV 1.0 is composed of video clips and their corresponding XML files (detailed in Section 3.2). We selected from these video clips 1843 frames dedicated to the detection task. In [14,46], the first results using AcTiV 1.0 were presented.

Based on the obtained results under different evaluation protocols and considering the AcTiV 1.0 users' feed-backs, it was necessary to extend the content in terms of video clips and resolutions offering more training samples, especially for deep learning-based methods.

The new dataset AcTiV 2.0 includes 189 video sequences, 4063 key frames, 10,415 text images and three video-stream resolutions, i.e., the new one is SD (480 × 360). A brief comparison in terms of content between the initial and new version of the proposed dataset is presented in Table 2. The architecture of the new dataset is completely different from the old one. In addition to the videos and their annotation XML files, AcTiV 2.0 includes two appropriate datasets for detection and recognition tasks, (see Figure 4).

Table 2. Statistics of AcTiV 1.0 and AcTiV 2.0.

	#Resolution	#Videos	#Frames	#Cropped Images
AcTiV 1.0	2	80	1843	-
AcTiV 2.0	3	189	4063	10,415

D

Resolution	1920 x 1080		720 x 576		460 x 380
TV Channel	AlJazeeraHD	France24	RussiaToday	TunisiaNat1	TunisiaNat1+
# Frames	909	874	882	1099	299

R

Resolution	1920 x 1080		720 x 576		460 x 380
TV Channel	AlJazeeraHD	France24	RussiaToday	TunisiaNat1	TunisiaNat1+
#Lines	2367	2276	2633	2411	631
#Words	9958	7084	16543	10998	2635
#Characters	57189	40520	96990	64493	15371

189

Figure 4. Architecture of AcTiV 2.0 and statistics of the detection (D) and recognition (R) datasets.

- *AcTiV-D* represents a dataset of non-redundant frames used to build and evaluate methods for detecting text regions in HD/SD frames. A total of 4063 frames have been hand-selected with a particular attention to achieve a high diversity in depicted text regions. Figure 5 provides examples from AcTiV-D for typical problems in video text detection. To test the systems' ability to locate texts under different situations, the proposed dataset includes some frames which contain the same text region but with different backgrounds and some others without any text component.
- *AcTiV-R* is a dataset of textline images that can be utilized to build and evaluate Arabic text recognition systems. Different fonts (more than 6), sizes, backgrounds, colors, contrasts and occlusions are represented in the dataset. Figure 6 illustrates typical examples from AcTiV-R. The collected text images cover a broad range of characteristics that distinguish video frames from scanned documents. AcTiV-R consists of 10,415 textline images, 44,583 words and 259,192 characters. To have an easily accessible representation of Arabic text, it is transformed into a set of Latin labels with a suffix that refers to the letter's position in the word, _B: Begin, _M: Middle; _E: End; and _I: Isolate. An example is shown in Figure 1. During the annotation process, we have considered 164 Arabic character forms:

  - 125 letters, i.e., taking into account this "positioning" variability;
  - 15 additional characters, i.e., combined with the diacritic sign "Chadda";
  - 10 digits; and
  - 14 punctuation marks including the *white space*.

The different character labels can be observed in Table 3. The same table gives for each character its frequency in the dataset.

More details about the statistics of the detection and recognition datasets are in Figure 4.

Table 3. Distribution of letters in the AcTiV-R dataset.

Character Label	# of Occurrence	In Arabic				Character Label	# of Occurrence	In Arabic			
		I	B	M	E			I	B	M	E
Alif	28,433	ا	-	-	ا	HamzaAboveAlif	1653	أ	-	-	أ
Baa	7417	ب	ب	ب	ب	HamzaUnderAlif	1049	إ	-	-	إ
Taaa	8948	ت	ت	ت	ت	TildAboveAlif	87	آ	-	-	آ
Thaa	851	ث	ث	ث	ث	HamzaAboveAlifBroken	1022	ئ	-	-	ئ
Jiim	3270	ج	ج	ج	ج	HamzaAboveWaaw	268	ؤ	-	-	ؤ
Haaa	3976	ح	ح	ح	ح	LaamHamzaAboveAlif	925	لأ	-	-	لأ
Xaa	1345	خ	خ	خ	خ	LaamHamzaUnderAlif	563	لإ	-	-	لإ
Daal	6656	د	-	-	د	LaamTildAboveAlif	63	لآ	-	-	لآ
Thaal	459	ذ	-	-	ذ	Space	31,458				
Raa	11,460	ر	-	-	ر	Digit_0	606			0	
Zaay	1478	ز	-	-	ز	Digit_1	655			1	
Siin	6348	س	س	س	س	Digit_2	475			2	
Shiin	2353	ش	ش	ش	ش	Digit_3	276			3	
Saad	2123	ص	ص	ص	ص	Digit_4	306			4	
Daad	1085	ض	ض	ض	ض	Digit_5	248			5	
Thaaa	2184	ط	ط	ط	ط	Digit_6	203			6	
Taa	481	ظ	ظ	ظ	ظ	Digit_7	138			7	
Ayn	5989	ع	ع	ع	ع	Digit_8	148			8	
Ghayn	890	غ	غ	غ	غ	Digit_9	116			9	
Faa	4942	ف	ف	ف	ف	Point	313			.	
Gaaf	4443	ق	ق	ق	ق	Colon	424			:	
Kaaf	2999	ك	ك	ك	ك	Comma	118			،	
Laam	18,868	ل	ل	ل	ل	Slash	76			/	
Miim	11,907	م	م	م	م	Percent	101			%	
Nuun	10,027	ن	ن	ن	ن	QuestionMark	8			?	
Haa	2608	ه	ه	ه	ه	ExclamationMark	12			!	
Waaw	10,614	و	-	-	و	Quote	445			""	
Yaa	18,153	ي	ي	ي	ي	Hyphen	457			–	
AlifBroken	1211	ى	-	-	ى	ParenthesisO	30			)	
Hamza	696			ء		ParenthesisC	29			(	
TaaaClosed	7239	ة	-	-	ة	Bar	13			I	
LaamAlif	1916	لأ	-	-	لا	Overall	259,192				

Figure 5. Typical video frames from AcTiV-D dataset. From left to right: Examples of RussiaToday Arabic, France24 Arabe, TunisiaNat1 (El Wataniya 1) and AljazeeraHD frames.

Figure 6. Example of text images from AcTiV-R depicting typical characteristics of video text images.

3.2. Annotation Guidelines

We utilized the AcTiV-GT tool [47] to annotate our collection of data. Figure 7 illustrates the user interface of this tool. In the annotation process, we collect the following information for each text rectangle.

- *position*: x, y, width and height.
- *content*: text strings, text color, background color, background type (transparent, opaque).
- *Interval*: apparition interval of the textline (Frame_S (Start), Frame_E (End)).

Note that a text rectangle can include multiple lines if they share the same font, color and size, and if they are not far from each other.

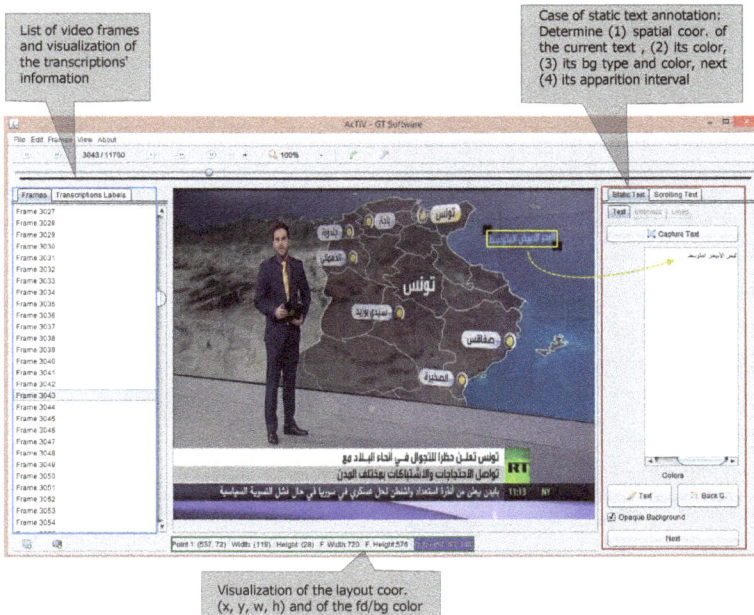

Figure 7. AcTiV-GT open-source tool displaying a labeled frame.

This set of information is saved in a meta file called global XML file (an extract is illustrated in Figure 8). This file can be used for tracking and end-to-end tasks. In AcTiV 2.0, two additional types of XML files have been generated, based on the information contained in the global XML file, one for the detection dataset and the other for the recognition dataset. The detection XML file is provided at the line level for each frame. Figure 9a depicts a part of the detection XML file of France24 TV channel. One bounding box is described by the element *Rectangle* which contains the rectangle attributes: (x, y) coordinates, width and height. The recognition ground-truth files are provided at the line level for each text image. The XML file is composed of two markup sections: *ArabicTranscription* and *LatinTranscription*. Figure 9b depicts an example of a ground-truth XML file and its textline image.


```
<?xml version="1.0" encoding="UTF-8"?>
- <video id="1" channel="AljazeeraHD" resolution="1080p" duration="00:09:17" fps="25"
  nbOfFrames="13925">
  - <staticText font="aljazeeraFont" nbOftextBox="17">
    - <textBox id="1" nbOfaInterval="1">
        <aInterval id="1" frame_S="1338" frame_E="1405"/>
        <position x="1161" y="909" width="268" height="63"/>
      - <content nbTextLines="1" textColor="51,65,90" bgColor="240,242,237" bgType="opaque">
          <textLine id="1" ArabicTranscription="حسن جنول" LatinTranscription="Haaa_B Siin_M Nuun_E
            Space Jiim_B MiimChadda_M Waaw_I Laam_I"/>
        </content>
      </textBox>
    - <textBox id="2" nbOfaInterval="3">
        <aInterval id="1" frame_S="3371" frame_E="6691"/>
        <aInterval id="2" frame_S="6960" frame_E="7916"/>
        <aInterval id="3" frame_S="8329" frame_E="9833"/>
        <position x="1352" y="514" width="251" height="50"/>
      - <content nbTextLines="1" textColor="240,250,255" bgColor="11,93,93" bgType="opaque">
          <textLine id="1" ArabicTranscription="ريتشارد فولك" LatinTranscription="Raa_I Yaa_B Taaa_M
            Shiin_M Alif_E Raa_I Daal_I Space Faa_B Waaw_E Laam_B Kaaf_E"/>
        </content>
      </textBox>
    - <textBox id="3" nbOfaInterval="3">
        <aInterval id="1" frame_S="3372" frame_E="6691"/>
        <aInterval id="2" frame_S="6960" frame_E="7916"/>
        <aInterval id="3" frame_S="8329" frame_E="9833"/>
        <position x="1129" y="562" width="475" height="43"/>
      - <content nbTextLines="1" textColor="240,250,255" bgColor="0,24,66" bgType="opaque">
          <textLine id="1" ArabicTranscription="المقرر الخاص لمجلس الأمم المتحدة لحقوق الإنسان السابق"
            LatinTranscription="Alif_I Laam_B Miim_M Gaaf_M Raa_E Raa_I Space Alif_I
            Laam_B Xaa_M Alif_E Saad_I Space Laam_B Miim_M Jiim_M Laam_M Siin_E
            Space Alif_I Laam_EHamzaAboveAlif_E Miim_B Miim_E Space Alif_I Laam_B
            Miim_M Taaa_M Haaa_M Daal_E TaaaClosed_I Space Laam_B Haaa_M Gaaf_M
            Waaw_E Gaaf_I Space Alif_I Laam_EHamzaUnderAlif_E Nuun_B Siin_M Alif_E
            Nuun_I Space Alif_I Laam_B Siin_M Alif_E Baa_B Gaaf_E"/>
        </content>
      </textBox>
```


Figure 8. A part of a global XML annotating a video sequence of Aljazeera TV. This figure contains ground-truth information about three text-boxes from a total of 17.


```
<?xml version="1.0" encoding="UTF-8"?>
- <Protocol4 channel="France24">
  - <frame id="3" source="vd03">
      <rectangle id="1" x="621" y="437" width="33" height="20"/>
      <rectangle id="2" x="400" y="461" width="253" height="30"/>
    </frame>
  - <frame id="15" source="vd03">
      <rectangle id="1" x="615" y="437" width="39" height="20"/>
      <rectangle id="2" x="385" y="460" width="268" height="31"/>
    </frame>
  - <frame id="29" source="vd03">
      <rectangle id="1" x="582" y="437" width="69" height="24"/>
    </frame>
    <frame id="72" source="vd03"> </frame>
    <frame id="101" source="vd03"> </frame>
  - <frame id="107" source="vd03">
      <rectangle id="1" x="555" y="437" width="101" height="26"/>
      <rectangle id="2" x="467" y="465" width="189" height="24"/>
    </frame>
```


(a)

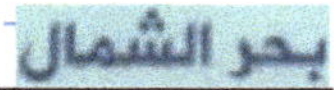


```
<?xml version="1.0" encoding="UTF-8" standalone="true"?>
<Image id="AljazeeraHD_vd02_frame_208-5">
  <ArabicTranscription>بحر الشمال</ArabicTranscription>
  <LatinTranscription> Baa_B Haaa_M Raa_E Space Alif_I Laam_B Shiin_M
    Miim_M Alif_E Laam_I </LatinTranscription>
</Image>
```


(b)

Figure 9. Example of AcTiV 2.0 specific XML files: (**a**) a part of the detection XML file of France24 TV; and (**b**) a recognition ground-truth file and its corresponding textline image.

4. Evaluation Protocols and Metrics

As mentioned before, the proposed AcTiV datasets are mainly dedicated to train and evaluate the existing systems for Arabic text detection and recognition in news video. To objectively compare and measure the performance of these systems, we proposed to partition each of the AcTiV-D and AcTiV-R datasets into train, test and closed test subsets taking advantage of the variability in data content. It is to note that the latter subset contains private data (quite similar to the test set) that are used in the context of competitions only. In addition, we suggested a set of evaluation protocols such that different techniques could be directly compared. In other words, the proposed protocols allow us to closely analyze the system behavior towards a given resolution (HD/SD) and/or quality (DBS/Web).

4.1. Detection Protocols and Metrics

Table 4 depicts the detection protocols.

- **Protocol 1** aims to measure the performance of single-frame based methods to detect texts in HD frames.
- **Protocol 4** is similar to Protocol 1, differing only by the channel resolution. All SD (720 × 576) channels in our database can be targeted by this protocol which is split in four sub-protocols: three channel-dependent (Protocols 4.1, 4.2 and 4.3) and one channel-free (Protocol 4.4).
- **Protocol 4bis** is dedicated to the new added resolution (480 × 360) for the Tunisia Nat1 TV channel. The main idea of this protocol is to train a given system with SD (720 × 576) data i.e., Protocol 4.3 and test it with different data resolution and quality.
- **Protocol 7** is the generic version of the previous protocols where text detection is evaluated regardless of data quality.

Table 4. Detection Evaluation Protocols.

Protocol	TV Channel	Training-Set 1 # Frames	Training-Set 2 # Frames	Test-Set 1 # Frames	Test-Set 2 # Frames	Closed-Set # Frames
1	AlJazeeraHD	337	610	87	196	103
4	France24	331	600	80	170	104
	Russia Today	323	611	79	171	100
	TunisiaNat1	492	788	116	205	106
	All SD	1146	1999	275	546	310
4bis	TunisiaNat1+	-	-	-	149	150
7	All	1483	2609	362	891	563

Metrics: The performance of a text detector is evaluated based on precision, recall and F-measure metrics that are defined as:

$$Precision = \frac{\sum_{i=1}^{|D|} matchD(D_i)}{|D|} \tag{1}$$

$$Recall = \frac{\sum_{i=1}^{|G|} matchG(G_i)}{|G|} \tag{2}$$

$$Fmeasure = 2 * \frac{Precision * Recall}{Precision + Recall} \tag{3}$$

where D is the list of detected rectangles, G is the list of ground-truth rectangles and $matchD/matchG$ are the matching functions, respectively. These measures are calculated using our evaluation tool [48] which takes into account all types of matching cases between G bounding boxes and D ones, i.e., one-to-one, one-to-many and many-to-one matching. In the matching procedure, two quality constraints, namely, t_p and t_r are utilized. $t_p \in [0, 1]$ is the constraint on area precision and $t_r \in [0, 1]$ is the recall constraint. Figure 10 depicts the user interface of our evaluation tool as well as the precision and recall curves, where x-axis denotes t_r values and y-axis denotes t_p ones. The proposed performance metrics and their underlying constraints are similar to those used in ICDAR 2013 [24] and ICDAR 2015 [25] RRCs. It is worth noting that our annotation and evaluation tools are fully implemented in Java and are made open-source for standardization and validation purposes.

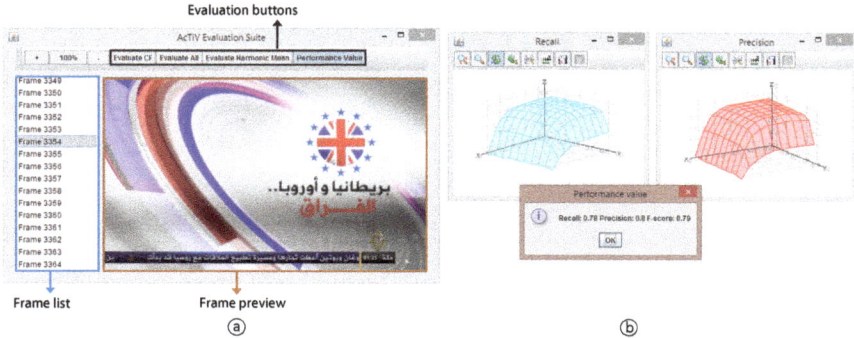

Figure 10. AcTiV-D evaluation tool. The user can apply the evaluation procedure to the current frame "Evaluate CF" button or to all video frames "Evaluate All" button (**a**). The "Performance Value" button displays precision, recall and F-score values (**b**).

4.2. Recognition Protocols and Metrics

Table 5 depicts the recognition protocols.

- **Protocol 3** aims to evaluate the performance of OCR systems to recognize texts in HD frames.
- **Protocol 6** is similar to Protocol 3, differing only by the channel resolution. All SD (720×576) channels in our dataset can be targeted by this protocol which is split in four sub-protocols: three channel-dependent (Protocols 6.1, 6.2 and 6.3) and one channel-free (Protocol 6.4).
- **Protocol 6bis** is dedicated to the new added resolution (480×360) for the Tunisia Nat1 TV channel. The main idea of this protocol is to train a given system with SD (720×576) data i.e., Protocol 6.3 and test it with different data resolution and quality.
- **Protocol 9** is the generic version of Protocols 3 and 6 where text recognition is assessed without considering data quality.

Table 5. Recognition Evaluation Protocols. "Lns" and "Wds" respectively denote "Lines" and "Words".

Protocol	TV Channel	Training-Set			Test-Set			Closed Test-Set		
		#Lns	#Wds	#Chars	#Lns	#Wds	#Chars	#Lns	#Wds	#Chars
3	AlJazeeraHD	1909	8110	46,563	196	766	4343	262	1082	6283
6	France24	1906	5683	32,085	179	667	3835	191	734	4600
	Russia Today	2127	13,462	78,936	250	1483	8749	256	1598	9305
	TunisiaNat1	2001	9338	54,809	189	706	4087	221	954	5597
	All SD	6034	28,483	165,830	618	2856	16,671	668	3286	19,502
6bis	TunisiaNat1+	-	-	-	320	1487	8726	311	1148	6645
9	All	7943	36,593	212,393	814	3622	21,014	930	4368	25,785

Metrics: The performance measure for the recognition task is based on the Line Recognition Rate (*LRR*), Word Recognition Rate (*WRR*) at the line and words levels, respectively, and on the computation of insertion (*I*), deletion (*D*) and substitution (*S*) errors at the character level (*CRR*) that are defined as:

$$CRR = \frac{\#characters - I - S - D}{\#characters} \qquad (4)$$

$$WRR = \frac{\#words\_correctly\_recognized}{\#words} \tag{5}$$

$$LRR = \frac{\#lines\_correctly\_recognized}{\#lines} \tag{6}$$

Figure 11 shows an example explaining the impact on CRR and WRR metrics resulting from substitution and deletion errors.

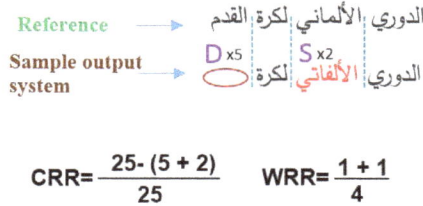

Figure 11. Example of CRR and WRR computation based on output errors.

It is worth noting that the proposed protocols help us understanding how generic is the system, i.e., if a system performs well for Protocols 7 and 9 (independently of the TV channel). For instance, in the AcTiVComp contest, we observed that some participating systems perform well in HD resolution only, some others are quite generic (i.e., good in both SD and HD resolutions). Other systems are incompatible with a specific resolution. Various examples of using these evaluation protocols will be presented in the next section.

5. Application of AcTiV Datasets

The proposed datasets have been used to build and evaluate two systems for Arabic video text detection and recognition. The text detector is based on a hybrid approach composed of CC-based heuristic phase and a machine learning verification procedure. The recognizer system consists of a Multi-Dimensional RNNs (MDRNNs) [49] coupled with a Connectionist Temporal Classification (CTC) layer [50].

5.1. LADI Detector

The LADI text detection system is based in our previous work [14,46], with new added enhancements considering the color consistency of near text regions. Our text detector represents a hybrid approach consisting of two stages: a CC-based heuristic algorithm and a machine learning classification. The main idea of this system is to combine two techniques: an adapted version of the SWT algorithm and a convolutional auto-encoder (CAE). As shown in Figure 12, the first stage starts with a preprocessing step to decrease noise and fine detail. It then computes the edge map and X&Y gradients from the processed frame using Canny and Sobel operators, respectively. After that, the SWT operator is performed as follow.

- Gradient direction d_p is calculated, at each edge pixel p, which is roughly perpendicular to the stroke orientation.
- A search ray $r = p + n * d_p$ $(n > 0)$ starting from an edge pixel p along the gradient direction d_p is shot until we find another edge pixel q. If these two edge pixels have nearly opposite gradient orientations, the ray is considered valid. All pixels inside this ray are labeled by the length $|p - q|$.

The next step is to group adjacent pixels in the resulting SWT image into CCs. This is done by applying a flood-fill algorithm based on consistency in stroke width and color. The CCs are then filtered using a set of simple heuristic rules concerning the CC size, position, aspect-ratio and color uniformity. The remaining CCs are iteratively merged into words and textlines based on a proposed textline formation method (see [46] for more details). The second stage uses CAE to automatically produce features, instead of hard-coding them. These features have been learned in an unsupervised way from the textline candidates obtained in the first stage. Then, to discriminate text objects from non-text ones, an SVM classifier with RBF kernel is trained on the patches extracted from the textline candidates by using the generated CAE features.

Note that the whole algorithm is performed twice (for each image) to handle both dark-on-light and light-on-dark texts, once along the gradient direction and once along the inverse direction. The results of two passes are combined to make final decisions.

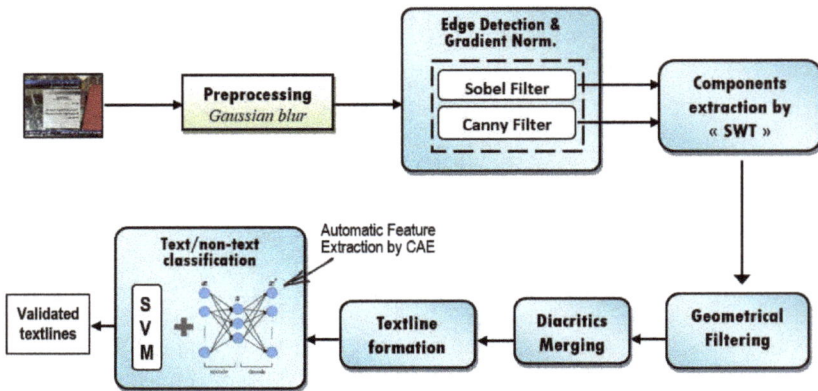

Figure 12. Pipeline of the text detection algorithm. Two passes are performed, one for each text polarity (Dark text on Light background or Light text on Dark background).

5.2. SID OCR

The SID OCR system [51] relies specifically on a Multi-Dimensional Long Short Term Memory (MDLSTM) with a CTC output layer. The proposed network is composed of three levels: an input layer, five hidden layers and an output layer. The hidden layers are MDLSTM that respectively have 2, 10, and 50 cells and separated by feedforward layers with 6 and 20 cells. In fact, we have created a hierarchical structure by repeatedly composing MDLSTM layers with feedforward layers. Firstly, the image is divided into small patches using a pixel window called the "input block", each of which is presented to the first MDLSTM layer as a feature vector of pixel intensities. These vectors are then scanned by four MDLSTM layers in different directions (i.e., up, down, left an right).

After that, the cells activation of the MDLSTM layers are sequentially fed to the first and second feed-forward layers through sub-sample windows, namely "hidden block". This can be seen as a subsampling step with trainable weights, in which the activation are summed and squashed by the hyperbolic tangent (tanh) function. This step aims to extremely reduce the number of weight connections between hidden layers.

The final level is the CTC output layer which labels the sequences of textlines. This layer has n cells, where n is the number of classes, in our case 165 (164 characters and one cell for the 'blank' output). The output activations are normalized at each time step with the softmax activation function. The use of such layer allows working on unsegmented input sequence, which is not the case for standard RNN objective functions. A separate network has been trained for each TV channel of the reference protocol. All input images have been scaled to common heights (70 pixels) and converted to gray-scale. The training is carried out with back-propagation through time (BPTT) algorithm and steepset optimizer has been used with a learning rate of 10^{-4} and with a momentum value of 0.9. We performed several experiments to find the optimal sizes of the MDLSTM layers, feedforward layers, input block and hidden block. Table 6 summarizes the best obtained values of the network parameters.

Note that the size of the input block is set to 1×4 for Protocols 6.1 and 3 (not 2×4), respectively. To fine-tune these parameters we just pick out a set of 2000 labeled images from AcTiV-R, in which 190 are used as a validation set.

Table 6. Best parameters for training the network.

Parameters	Values
MDLSTM Size	2, 10 and 50
Feed-forward Size	6 and 20
InputBlock Size	2×4
HiddenBlock Sizes	1×4 and 1×4
Learn rate	10^{-4}
Momentum	0.9

5.3. Experimental Results

Several experiments have been conducted using the AcTiV-D and AcTiV-R subsets. These experiments can be divided into two categories: The first one concerns the comparison of our systems with two recent methods. The second category aims at analyzing the effect of increasing the training data on the accuracy of the LADI text detector.

5.3.1. Comparison with Other Methods

As proof of concept of the proposed benchmark, we compare our systems with two recent methods. The first one was proposed by Gaddour et al. [52] to basically detect Arabic texts in natural scene images. The main steps involved are:

- Pixel-color clustering using k-means to form pairs of thresholds for each RGB channel.
- Creation of binary map for each pair of thresholds.
- Extraction of CCs.
- Preliminary filtering according to "area stability" criterion.
- Second filtering based on a set of statistical and geometric rules.
- Horizontal merging of the remaining components to form textlines.

The second method was put forward by Iwata et al. [53] to recognize artificial Arabic text in video frames. It operates as follows:

- Textline segmentation into words by thresholding gaps between CCs.
- Over-segmentation of characters into primitive segments.
- Character recognition using 64-dimensional feature vector of chain code histogram and the modified quadratic discriminant function.
- Word recognition by dynamic programming using total likelihood of characters as objective function.
- False word reduction by measuring the average of the character likelihoods in a word and comparing it to a predefined threshold.

The detection systems have been trained on the training-set1 of Table 4. The evaluation has been done on the test set for the detection and recognition tasks. Table 7 presents evaluation results of the detection protocols in terms of precision, recall and F-measure. The best results are marked in bold. The LADI system scores best for all protocols with an F-measure between 0.73 and 0.85 for AllSD protocol (p4.4) and AljazeeraHD protocol (p1) respectively. In contrast to the SysA that represents a fully heuristic-based method, the LADI system increased the F-measure by 11% for Protocol 1. For Protocols 4.1, 4.2, 4.3 and 4.4 (SD channels), the results are higher, with a gain of, respectively, 11%, 17%, 14% and 24%. This reflects the effectiveness of using a machine-learning solution to filter the results given by the SWT algorithm. The Gaddo system has strong fragmentation and miss detection tendency as depicted by its obtained numerical results. Table 8 presents evaluation results of the recognition protocols in terms of CRR, WRR and LRR metrics. The SID-OCR system has shown superiority in all protocols. The best accuracies are achieved on the TunisiaNat1 channel subset (p6.3) with 0.94 as a CRR and 0.62 as a LRR. The IWATA system performs well for all SD protocols especially for the CRR/WRR metrics. However its current version is incompatible with HD resolution. The result shows that our system has low recognition rate when facing different text patterns and resolutions, i.e., global Protocol 9. Based on our knowledge about the shapes of Arabic characters, we divide the causes of errors into two classes: character similarity and insufficient samples of punctuation, digits and symbols. Several measures can be taken to minimize the character error rate, for instance by integrating language models or dropout mechanism.

Table 7. Performance of text detection systems evaluated on the test set of AcTiV-D.

Protocol	System	Precision	Recall	Fmeasure
1	LADI [46]	**0.86**	**0.84**	**0.85**
	SysA [14]	0.77	0.76	0.76
	Gaddo [52]	0.52	0.49	0.51
4.1	LADI [46]	**0.74**	**0.76**	**0.75**
	SysA [14]	0.69	0.6	0.64
	Gaddo [52]	0.47	0.61	0.54
4.2	LADI [46]	**0.8**	**0.75**	**0.77**
	SysA [14]	0.66	0.55	0.6
	Gaddo [52]	0.41	0.5	0.45
4.3	LADI [46]	**0.85**	**0.82**	**0.83**
	SysA [14]	0.68	0.71	0.69
	Gaddo [52]	0.34	0.49	0.41
4.4	LADI [46]	**0.71**	**0.76**	**0.73**
	SysA [14]	0.5	0.49	0.49
	Gaddo [52]	-	-	-

Table 8. Performance of the recognition systems evaluated on the test set of AcTiV-R.

Protocol	System	CRR	WRR	LRR
3	SIDOCR [51]	0.90	0.71	0.51
	IWATA [53]	-	-	-
6.1	SIDOCR [51]	**0.89**	**0.70**	**0.51**
	IWATA [53]	0.88	0.67	0.46
6.2	SIDOCR [51]	**0.94**	0.68	**0.41**
	IWATA [53]	0.9	**0.68**	0.39
6.3	SIDOCR [51]	0.94	**0.81**	**0.62**
	IWATA [53]	**0.94**	0.77	0.56
6.4	SIDOCR [51]	**0.93**	0.73	**0.52**
	IWATA [53]	0.9	**0.73**	0.48
9	SIDOCR [51]	0.73	0.58	0.32
	IWATA [53]	-	-	-

5.3.2. Training with AcTiV 2.0

To examine the effect of increasing the number of training samples on the accuracy of our text detector, we conduct the same experiment of Protocol 6.1, in Table 7, using training-set2, which includes roughly the double of samples (600 frames) than training-set1 (see Table 4). We observed that the detection rates of our text detector have been increased as expected. Specifically, the recall increases by 2% and the precision increases by 5%. This can be explained by the increase in the number of samples provided to the CAE, which leads to more robust feature representations and subsequently better classification results.

6. Conclusions

In this paper, we have presented a new version of the AcTiV dataset for the development and evaluation of text detection and recognition systems targeting Arabic news video. This dataset is freely available to research institutions. We have provided details about the characteristics and statistics of the database. We have also reported about our ground-truthing software used to semi-automatically annotate the video clips and our open text detection evaluation tool. We have evaluated five text detection and recognition algorithms as proof-of-concept of the new dataset. Additionally, a set of evaluation protocols has been made to measure the systems performance under different situations. The experimental results have shown that there is still room for improvement in both detection and recognition of Arabic video text. We look forward to more researchers joining the challenging research topic of Arabic video texts detection and recognition.

Acknowledgments: The researchers would like to thank all the TV channels for providing us with data and multimedia files, especially the archive staff of ElWataniya1 (TunisiaNat1) TV.

Author Contributions: Oussama Zayene conceived the dataset, developed the tools and realized the experiments under the supervision and help of professors Najoua Essoukri Ben Amara, Jean Hennebert and Rolf Ingold. Sameh Masmoudi Touj contributed to the collect, design and annotation of the dataset, and verified the annotated data. All the co-authors have substantially revised the manuscript.

Conflicts of Interest: The authors declare no conflict of interest.

References

1. Lu, T.; Palaiahnakote, S.; Tan, C.L.; Liu, W. *Video Text Detection*; Springer Publishing Company, Incorporated: London, UK, 2014.
2. Ye, Q.; Doermann, D. Text detection and recognition in imagery: A survey. *IEEE Trans. Pattern Anal. Mach. Intell.* **2015**, *37*, 1480–1500.
3. Yin, X.C.; Zuo, Z.Y.; Tian, S.; Liu, C.L. Text Detection, Tracking and Recognition in Video: A Comprehensive Survey. *IEEE Trans. Image Process.* **2016**, *25*, 2752–2773.
4. Lienhart, R. Video OCR: A survey and practitioner's guide. In *Video Mining*; Springer: Boston, MA, USA, 2003; pp. 155–183.
5. Yang, H.; Quehl, B.; Sack, H. A framework for improved video text detection and recognition. *Multimed. Tools Appl.* **2014**, *69*, 217–245.
6. Poignant, J.; Bredin, H.; Le, V.B.; Besacier, L.; Barras, C.; Quénot, G. Unsupervised speaker identification using overlaid texts in TV broadcast. In Proceedings of the Interspeech 2012—Conference of the International Speech Communication Association, Portland, OR, USA, 9–13 September 2012; p. 4.
7. Märgner, V.; El Abed, H. *Guide to OCR for Arabic Scripts*; Springer: Berlin, Germany, 2012.
8. Touj, S.M.; Amara, N.E.B.; Amiri, H. Arabic Handwritten Words Recognition Based on a Planar Hidden Markov Model. *Int. Arab J. Inf. Technol.* **2005**, *2*, 318–325.
9. Lorigo, L.M.; Govindaraju, V. Offline Arabic handwriting recognition: A survey. *IEEE Trans. Pattern Anal. Mach. Intell.* **2006**, *28*, 712–724.
10. Chammas, E.; Mokbel, C.; Likforman-Sulem, L. Arabic handwritten document preprocessing and recognition. In Proceedings of the 2015 13th International Conference on Document Analysis and Recognition (ICDAR), Tunis, Tunisia, 23–26 August 2015; pp. 451–455.
11. Jamil, A.; Siddiqi, I.; Arif, F.; Raza, A. Edge-based features for localization of artificial Urdu text in video images. In Proceedings of the 2011 International Conference on Document Analysis and Recognition, Beijing, China, 18–21 September 2011; pp. 1120–1124.
12. Halima, M.B.; Karray, H.; Alimi, A.M. Arabic text recognition in video sequences. *arXiv* **2013**, arXiv:preprint/1308.3243

13. Yousfi, S.; Berrani, S.A.; Garcia, C. Arabic text detection in videos using neural and boosting-based approaches: Application to video indexing. In Proceedings of the 2014 IEEE International Conference on Image Processing (ICIP), Paris, France, 27–30 October 2014; pp. 3028–3032.
14. Zayene, O.; Hennebert, J.; Touj, S.M.; Ingold, R.; Amara, N.E.B. A dataset for Arabic text detection, tracking and recognition in news videos-AcTiV. In Proceedings of the 2015 13th International Conference on Document Analysis and Recognition (ICDAR), Tunis, Tunisia, 23–26 August 2015; pp. 996–1000.
15. Elagouni, K.; Garcia, C.; Mamalet, F.; Sébillot, P. Text recognition in videos using a recurrent connectionist approach. In Proceedings of the International Conference on Artificial Neural Networks, Lausanne, Switzerland, 11–14 September 2012; pp. 172–179.
16. Khare, V.; Shivakumara, P.; Raveendran, P. A new Histogram Oriented Moments descriptor for multi-oriented moving text detection in video. *Expert Syst. Appl.* **2015**, *42*, 7627–7640.
17. Lucas, S.M.; Panaretos, A.; Sosa, L.; Tang, A.; Wong, S.; Young, R.; Ashida, K.; Nagai, H.; Okamoto, M.; Yamamoto, H.; et al. ICDAR 2003 robust reading competitions: Entries, results, and future directions. *Int. J. Doc. Anal. Recognit. (IJDAR)* **2005**, *7*, 105–122.
18. Lucas, S.M. ICDAR 2005 text locating competition results. In Proceedings of the Eighth International Conference on Document Analysis and Recognition, Seoul, Korea, 31 August–1 September 2005; pp. 80–84.
19. Huang, W.; Lin, Z.; Yang, J.; Wang, J. Text localization in natural images using stroke feature transform and text covariance descriptors. In Proceedings of the IEEE International Conference on Computer Vision, Sydney, Australia, 1–8 December 2013; pp. 1241–1248.
20. Zhu, Y.; Yao, C.; Bai, X. Scene text detection and recognition: Recent advances and future trends. *Front. Comput. Sci.* **2016**, *10*, 19–36.
21. Jaderberg, M.; Simonyan, K.; Vedaldi, A.; Zisserman, A. Reading text in the wild with convolutional neural networks. *Int. J. Comput. Vis.* **2016**, *116*, 1–20.
22. Shahab, A.; Shafait, F.; Dengel, A. ICDAR 2011 robust reading competition challenge 2: Reading text in scene images. In Proceedings of the 2011 International Conference on Document Analysis and Recognition, Beijing, China, 18–21 September 2011; pp. 1491–1496.
23. Liao, M.; Shi, B.; Bai, X.; Wang, X.; Liu, W. TextBoxes: A Fast Text Detector with a Single Deep Neural Network. In Proceedings of the 2017 AAAI Conference on Artificial Intelligence, San Francisco, CA, USA, 4–9 February 2017; pp. 4161–4167.
24. Karatzas, D.; Shafait, F.; Uchida, S.; Iwamura, M.; i Bigorda, L.G.; Mestre, S.R.; Mas, J.; Mota, D.F.; Almazan, J.A.; de las Heras, L.P. ICDAR 2013 robust reading competition. In Proceedings of the 2013 12th International Conference on Document Analysis and Recognition, Washington, DC, USA, 25–28 August 2013; pp. 1484–1493.
25. Karatzas, D.; Gomez-Bigorda, L.; Nicolaou, A.; Ghosh, S.; Bagdanov, A.; Iwamura, M.; Matas, J.; Neumann, L.; Chandrasekhar, V.R.; Lu, S.; et al. ICDAR 2015 competition on robust reading. In Proceedings of the 2015 13th International Conference on Document Analysis and Recognition (ICDAR), Tunis, Tunisia, 23–26 August 2015; pp. 1156–1160.
26. Yao, C.; Bai, X.; Liu, W.; Ma, Y.; Tu, Z. Detecting texts of arbitrary orientations in natural images. In Proceedings of the 2012 IEEE Conference on Computer Vision and Pattern Recognition (CVPR), Providence, RI, USA, 16–21 June 2012; pp. 1083–1090.
27. Liu, Z.; Li, Y.; Qi, X.; Yang, Y.; Nian, M.; Zhang, H.; Xiamixiding, R. Method for unconstrained text detection in natural scene image. *IET Comput. Vis.* **2017**, *11*, 596–604.
28. Wang, K.; Belongie, S. Word spotting in the wild. In Proceedings of the European Conference on Computer Vision, Heraklion, Greece, 5–11 September 2010; pp. 591–604.
29. Shi, B.; Bai, X.; Yao, C. An end-to-end trainable neural network for image-based sequence recognition and its application to scene text recognition. *IEEE Trans. Pattern Anal. Mach. Intell.* **2017**, *39*, 2298–2304.
30. Mishra, A.; Alahari, K.; Jawahar, C. Unsupervised refinement of color and stroke features for text binarization. *Int. J. Doc. Anal. Recognit. (IJDAR)* **2017**, *20*, 105–121.
31. Lee, S.; Cho, M.S.; Jung, K.; Kim, J.H. Scene text extraction with edge constraint and text collinearity. In Proceedings of the 2010 20th International Conference on Pattern Recognition (ICPR), Istanbul, Turkey, 23–26 August 2010; pp. 3983–3986.
32. Zhu, Y.; Zhang, K. Text segmentation using superpixel clustering. *IET Image Process.* **2017**, *11*, 455–464.

33. Nagy, R.; Dicker, A.; Meyer-Wegener, K. NEOCR: A configurable dataset for natural image text recognition. In Proceedings of the International Workshop on Camera-Based Document Analysis and Recognition, Beijing, China, 22 September 2011; pp. 150–163.

34. Veit, A.; Matera, T.; Neumann, L.; Matas, J.; Belongie, S. Coco-text: Dataset and benchmark for text detection and recognition in natural images. *arXiv* **2016**, arXiv:preprint/1601.07140.

35. Gomez, R.; Shi, B.; Gomez, L.; Numann, L.; Veit, A.; Matas, J.; Belongie, S.; Karatzas, D. ICDAR2017 Robust Reading Challenge on COCO-Text. In Proceedings of the 2017 International Conference on Document Analysis and Recognition, Kyoto, Japan, 10–15 November 2017; pp. 1435–1443.

36. Ch'ng, C.K.; Chan, C.S. Total-Text: A Comprehensive Dataset for Scene Text Detection and Recognition. In Proceedings of the 2017 International Conference on Document Analysis and Recognition, Kyoto, Japan, 10–15 November 2017; pp. 935–942.

37. Pechwitz, M.; Maddouri, S.S.; Märgner, V.; Ellouze, N.; Amiri, H. IFN/ENIT-database of handwritten Arabic words. In Proceedings of the Colloque International Francophone sur l'Ecrit et le Document (CIFED), Hammamet, Tunisia, 21–23 October 2002; pp. 127–136.

38. Mahmoud, S.A.; Ahmad, I.; Al-Khatib, W.G.; Alshayeb, M.; Parvez, M.T.; Märgner, V.; Fink, G.A. KHATT: An open Arabic offline handwritten text database. *Pattern Recognit.* **2014**, *47*, 1096–1112.

39. Slimane, F.; Ingold, R.; Kanoun, S.; Alimi, A.M.; Hennebert, J. A new arabic printed text image database and evaluation protocols. In Proceedings of the 2009 International Conference on Document Analysis and Recognition, Barcelona, Spain, 26–29 July 2009; pp. 946–950.

40. Kherallah, M.; Tagougui, N.; Alimi, A.M.; El Abed, H.; Margner, V. Online Arabic handwriting recognition competition. In Proceedings of the 2011 International Conference on Document Analysis and Recognition, Beijing, China, 18–21 September 2011; pp. 1454–1458.

41. Halima, M.B.; Alimi, A.; Vila, A.F.; Karray, H. Nf-SAVO: Neuro-fuzzy system for arabic video OCR. *arXiv* **2012**, arXiv:preprint/1211.2150.

42. Moradi, M.; Mozaffari, S. Hybrid approach for Farsi/Arabic text detection and localisation in video frames. *IET Image Process.* **2013**, *7*, 154–164.

43. Yousfi, S.; Berrani, S.A.; Garcia, C. Deep learning and recurrent connectionist-based approaches for Arabic text recognition in videos. In Proceedings of the 2015 13th International Conference on Document Analysis and Recognition, Tunis, Tunisia, 23–26 August 2015; pp. 1026–1030.

44. Yousfi, S.; Berrani, S.A.; Garcia, C. ALIF: A dataset for Arabic embedded text recognition in TV broadcast. In Proceedings of the 2015 13th International Conference on Document Analysis and Recognition (ICDAR), Tunis, Tunisia, 23–26 August 2015; pp. 1221–1225.

45. Zayene, O.; Hajjej, N.; Touj, S.M.; Ben Mansour, S.; Hennebert, J.; Ingold, R.; Amara, N.E.B. ICPR2016 Contest on Arabic Text Detection and Recognition in Video Frames AcTiVComp. In Proceedings of the 23rd International Conference on Pattern Recognition (ICPR), Cancun, Mexico, 4–8 December 2016; pp. 187–191.

46. Zayene, O.; Seuret, M.; Touj, S.M.; Hennebert, J.; Ingold, R.; Amara, N.E.B. Text detection in arabic news video based on SWT operator and convolutional auto-encoders. In Proceedings of the 2016 12th IAPR Workshop on Document Analysis Systems (DAS), Santorini, Greece, 11–14 April 2016; pp. 13–18.

47. Zayene, O.; Touj, S.M.; Hennebert, J.; Ingold, R.; Amara, N.E.B. Semi-automatic news video annotation framework for Arabic text. In Proceedings of the 2014 4th International Conference on Image Processing Theory, Tools and Applications, Paris, France, 14–17 October 2014; pp. 1–6.

48. Zayene, O.; Touj, S.M.; Hennebert, J.; Ingold, R.; Amara, N.E.B. Data, protocol and algorithms for performance evaluation of text detection in Arabic news video. In Proceedings of the 2016 2nd International Conference on Advanced Technologies for Signal and Image Processing (ATSIP), Monastir, Tunisia, 21–23 March 2016; pp. 258–263.

49. Graves, A. Offline arabic handwriting recognition with multidimensional recurrent neural networks. In *Guide to OCR for Arabic Scripts*; Springer: Berlin, Germany, 2012; pp. 297–313.

50. Graves, A.; Fernández, S.; Gomez, F.; Schmidhuber, J. Connectionist temporal classification: Labelling unsegmented sequence data with recurrent neural networks. In Proceedings of the 23rd International Conference on Machine Learning, Pittsburgh, PA, USA, 25–29 June 2006; pp. 369–376.

51. Zayene, O.; Essefi, S.A.; Amara, N.E.B. Arabic Video Text Recognition Based on Multi-Dimensional Recurrent Neural Networks. In Proceedings of the International Conference on Computer Systems and Applications (AICCSA), Hammamet, Tunisia, 30 October–3 November 2017; pp. 725–729.

52. Gaddour, H.; Kanoun, S.; Vincent, N. A New Method for Arabic Text Detection in Natural Scene Image Based on the Color Homogeneity. In Proceedings of the International Conference on Image and Signal Processing, Trois-Rivières, QC, Canada, 30 May–1 June 2016; pp. 127–136.
53. Iwata, S.; Ohyama, W.; Wakabayashi, T.; Kimura, F. Recognition and transition frame detection of Arabic news captions for video retrieval. In Proceedings of the 2016 23rd International Conference on Pattern Recognition (ICPR), Cancun, Mexico, 4–8 December 2016; pp. 4005–4010.

MDPI

St. Alban-Anlage 66

4052 Basel

Switzerland

Tel. +41 61 683 77 34

Fax +41 61 302 89 18

www.mdpi.com

Journal of Imaging Editorial Office

E-mail: jimaging@mdpi.com

www.mdpi.com/journal/jimaging

www.ingramcontent.com/pod-product-compliance
Lightning Source LLC
Chambersburg PA
CBHW051844210326
41597CB00033B/5775